Alginates

Scrivener Publishing
100 Cummings Center, Suite 541J
Beverly, MA 01915-6106

Publishers at Scrivener
Martin Scrivener (martin@scrivenerpublishing.com)
Phillip Carmical (pcarmical@scrivenerpublishing.com)

Alginates

Applications in the Biomedical and Food Industries

Edited by

Shakeel Ahmed

Scrivener Publishing

WILEY

This edition first published 2019 by John Wiley & Sons, Inc., 111 River Street, Hoboken, NJ 07030, USA and Scrivener Publishing LLC, 100 Cummings Center, Suite 541J, Beverly, MA 01915, USA

For more information about Scrivener publications please visit www.scrivenerpublishing.com.

Wiley Global Headquarters
111 River Street, Hoboken, NJ 07030, USA

For details of our global editorial offices, customer services, and more information about Wiley products visit us at www.wiley.com.

Library of Congress Cataloging-in-Publication Data

ISBN 978-1-119-48791-3

Cover image: Pixabay.Com
Cover design by Russell Richardson

Set in size of 11pt and Minion Pro by Manila Typesetting Company, Makati, Philippines

This book
is
dedicated to:

The woman who raised me—Amma
The woman who guided me—Prof Saiqa Ikram
The women who loved me—Nasreen Akhter, Parveen Akhter, Shamim Akhter, and Naida Bhabhi
The men who supported me during ups and downs of my life—Mohd Shabbir, Abdul Hamid, Mohd Aslam, and Naseeb Bajraan
The friends who stood with me—Wahid ul Rehman, Faheem Rasool, Aamir Mushtaq, and Maqbool Bajjar

Contents

Preface

Alginates are linear biopolymers consisting of 1,4-linked β-D-mannuronic acid and 1,4 α-L-guluronic acid residues. These groups of naturally occurring polysaccharides, which are derived from brown algal cell walls and several bacterial strains, have found numerous applications in biomedical sciences and pharmaceutical and food industries. Although there are currently many books available with chapters referencing alginates, this is the first of its kind solely devoted to their properties, modification, and characterization, with particular emphasis on their applications in the biomedical and food industries.

The wide-ranging topics discussed in this book are as follows. Chapter 1 gives an overview of alginates, their structures, and properties, and a detailed account of the modification of alginates, various characterization techniques, and methods of processing is given in chapter 2. Chapter 3 covers the dynamic properties of alginates and their innovative application in various materials, namely, the nanomaterial or the polymer. Chapters 4 and 5 discuss the biomedical applications of alginates. The focus of chapter 6 is the wide use of alginates in pharmaceutical and biomedical industries and that of chapter 7 is the evolution of alginate materials in restorative dentistry. Chapter 8 discusses applications of different cross-linked alginate networks, their microspheres, and hydrogel in relation to drug encapsulation and delivery processes and includes a brief introduction of the chemistry and pharmaceutical chemistry of alginates.

In chapter 9, biomedical applications of alginates—particularly wound care application in the various forms of alginate-based wound dressings—are discussed. Chapter 10 discusses the present use and future potential of alginates as a tool in drug formulation and regenerative medicine. Chapters 11, 12, and 13 focus on food packaging, beverage industry, and comestible applications of alginates, respectively. The last chapter of the book discusses the current uses and future prospects of alginates in food packaging and biomedical applications.

I hope that this book will be helpful to research scholars and scientists working in the area of alginates. I hope that it will also be helpful to beginners and undergraduate and graduate students, as it gives a full description of alginate structural details, history, properties, processing, etc. I am very grateful to the contributors of this book for their valuable contributions and Scrivener-Wiley for its publication.

Shakeel Ahmed
December 2018
Jammu & Kashmir, India

Part 1

ALGINATES—INTRODUCTION, CHARACTERIZATION AND PROPERTIES

1

Alginates: General Introduction and Properties

Rutika Sehgal[1], Akshita Mehta[1] and Reena Gupta[1*]

[1]Department of Biotechnology, Himachal Pradesh University, Summerhill, Shimla, India

Abstract

Alginates (ALGs) are a group of naturally occurring anionic polysaccharides derived from brown seaweeds. They are linear biopolymers of 1,4-linked β-D-mannuronic acid (M) and 1,4 α-L-guluronic acid (G) residues that are arranged in homogenous (poly-G, poly-M) or heterogenous (MG) block-like patterns. The physiological and chemical characteristics of ALGs depend on this arrangement of residues. Alginates are primarily used as thermally stable cold-setting gelling agents, which are formed in presence of divalent cations. They are more efficient gelling agents than gelatin and can gel at far lower concentrations as compared to other agents. This ability to create a chemically set, irreversible gel has proved to be useful in many food applications. Among various ALGs, sodium ALG is most widely studied in the pharmaceutical and biomedical field. Its various properties favor its use for viscosity enhancement, encapsulation polymer, matrixing agent, stabilizer, bioadhesive, and film former in transdermal and transmucosal drug delivery. With well-established uses in dentistry, the ALGs also offer interesting possibilities in the field of medicine and cosmetics as a skin care ingredient. This chapter will include general introduction, understanding of structure and properties of ALGs, and different forms of ALGs used in industries.

Keywords: Alginates, biopolymer, polysaccharide, medicines, cosmetics

**Corresponding author*: reenagupta_2001@yahoo.com

Shakeel Ahmed (ed.) Alginates, (3–20) © 2019 Scrivener Publishing LLC

1.1 Introduction

Alginates (ALGs) are naturally occurring anionic polysaccharides that are present as a structural component in cell walls of brown algae, mainly from *Macrocystis pyrifera*, *Ascophyllum nodosum*, and *Laminaria hyperborea* and as a capsular polysaccharide in bacterial strains like *Azotobacter* and *Pseudomonas*. It is present in the cell wall of brown algae as the calcium, magnesium, and sodium salts; therefore, it is usually referred to as "alginic acid and its salts." Alginates are available commercially as sodium, potassium, or ammonium salts in filamentous, granular, or powdered forms. Their color ranges from white to yellowish-brown. The molecular weight of ALG generally ranges from 60,000 to 700,000 Da depending on the application [1]. The size (diameter) of ALG gel particles can be macro (greater than 1 mm), micro (from 0.2 to 1,000 mm), or nano (less than 0.2 mm). These gel particles have high water holding capacity to form a viscous gum and have adjustable chemical and mechanical properties that are dependent on the type of cross-linking agent used. As a natural ingredient, ALG gel particles are attractive for various biological applications because they are biocompatible, nontoxic, biodegradable, and relatively cheap [2, 3]. Alginate is also a significant component of the biofilms produced by the bacterium *Pseudomonas aeruginosa*, the major pathogen in cystic fibrosis, that confers it a high resistance to antibiotics and killing by macrophages.

1.2 History

Alginate was discovered in the late 19th century by a British Pharmacist, E.C.C. Stanford, who called it "algin," which was a viscous solution obtained initially from Laminariaceae. Since its discovery in 1883, it has become an important industrial product that is commercially obtained from coastal brown seaweeds. Later its extracts were termed as "alginic acid." Its commercial production started in 1929. It has been estimated that algal ALGs are produced nearly 38,000 tons worldwide annually, and their major part contributes to food and pharmaceutical industries because of their increased demand [4].

1.3 Structure

Alginates are linear biopolymers of 1,4-linked β-D-mannuronic acid (M) and 1,4 α-L-guluronic acid (G) residues (Figure 1.1) organized in homogenous (poly-G, poly-M) or heterogenous (MG) block patterns. The G and M block pattern and sequence may be different in commercial ALG depending on the source of seaweed used, harvesting season, and geographical location of the seaweed source [5]. The random sequence of M and G block chains (Figure 1.2) are composed of regions of alternating MG blocks whose monad, diad, and triad frequencies are determined. Rigid six-membered

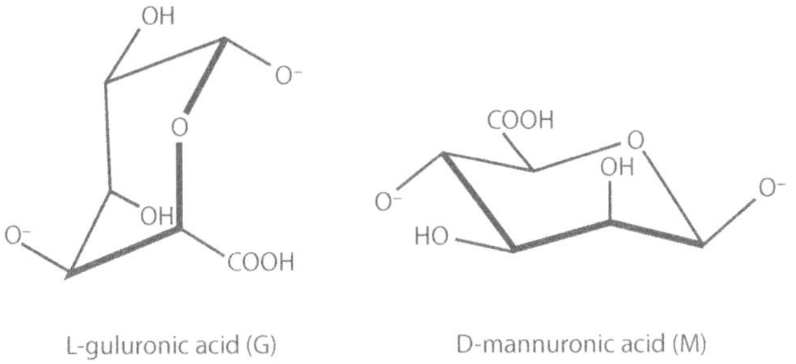

Figure 1.1 Structure of ALG monomers (L-guluronic acid and D-mannuronic acid).

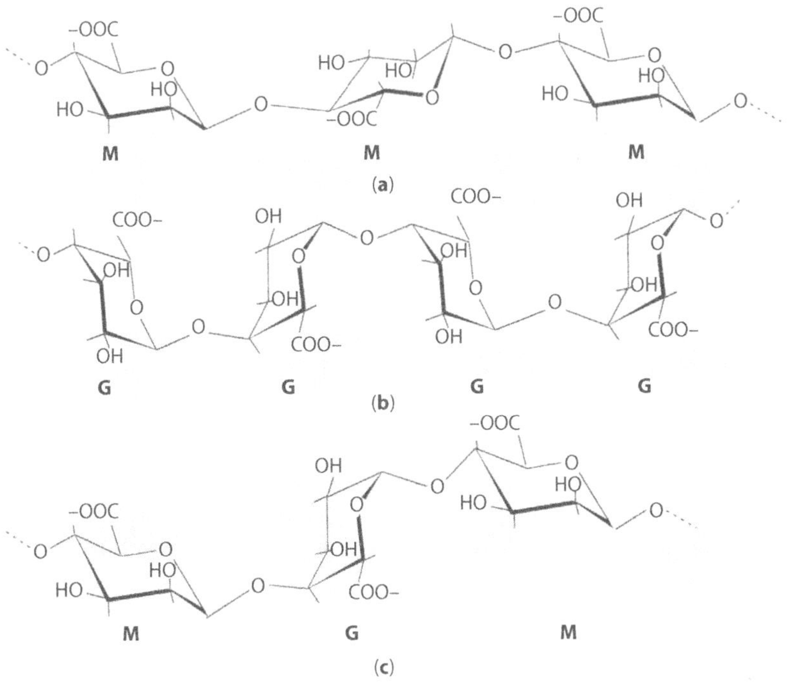

Figure 1.2 (a) Homopolymeric blocks of poly-β-1,4-D-mannuronic acid (MM blocks); (b) homopolymeric blocks of poly-α-1,4-L-guluronic acid (GG blocks); (c) heteropolymeric blocks of MG monomers in random pattern [6].

sugar rings and restricted rotation around the glycosidic linkage make ALG molecules stiff. The rigidity of the chains further is due to electrostatic repulsion between the charged groups on the polymer chain and on ALG composition. It increases in the order MG < MM < GG; therefore, G-rich ALGs generally form hard and brittle gels, while soft and elastic gels are produced by M-rich samples. Hence, the physicochemical properties and degree of polymerization of the ALG depend on the arrangement of these blocks [5].

1.4 Alginates and Their Properties

1.4.1 Gel Formation

Alginate can form gel independent of temperature as compared to other polysaccharides such as gelatin or agar. The ALG gels can either be ionic gels (formed by cationic cross-linking) or acidic gels (formed by acid precipitation).

1.4.1.1 Ionic Alginate Gels

The ability of ALG to form ionic gel in the presence of multivalent cations is mostly desired in food industries. The process of binding of ALG to divalent cation is very specific, and the affinity of ALG toward cations is in the order Mn < Zn, Ni, Co < Fe < Ca < Sr < Ba < Cd < Cu < Pb [7, 8], and it depends on the number of G blocks present in the structure [9]. The cooperative binding of G block and divalent cations results in gelation of ALGs. The use of highly toxic cations such as Pb, Cu, and Cd is limited for practical applications, but less toxic cations like Sr and Ba have been reported to be used in cell immobilization applications at limited concentrations [10]. Calcium being nontoxic is widely accepted to form ionic ALG gels. Calcium-ALG gel is the most commonly used ALG gel. Interactions between Ca ions and G residues result in gelation of ALG, which leads to chain–chain association and to the formation of junction zones. The two G chains bind on opposite sides with the addition of Ca ions to the ALG polymer, which results in a diamond-shaped structure with a hydrophilic cavity. The oxygen atoms from the carboxyl groups form multicoordination with the Ca ions in the hydrophilic cavity. This tightly bound complex forms a junction zone that is shaped like an "egg box" (Figure 1.3). In this egg box, a 3-D network is formed by the binding of each cation with four G residues [11]. In case of Ca ALG gels, there should be 8 to 20 adjacent G residues in order to form a stable junction [12].

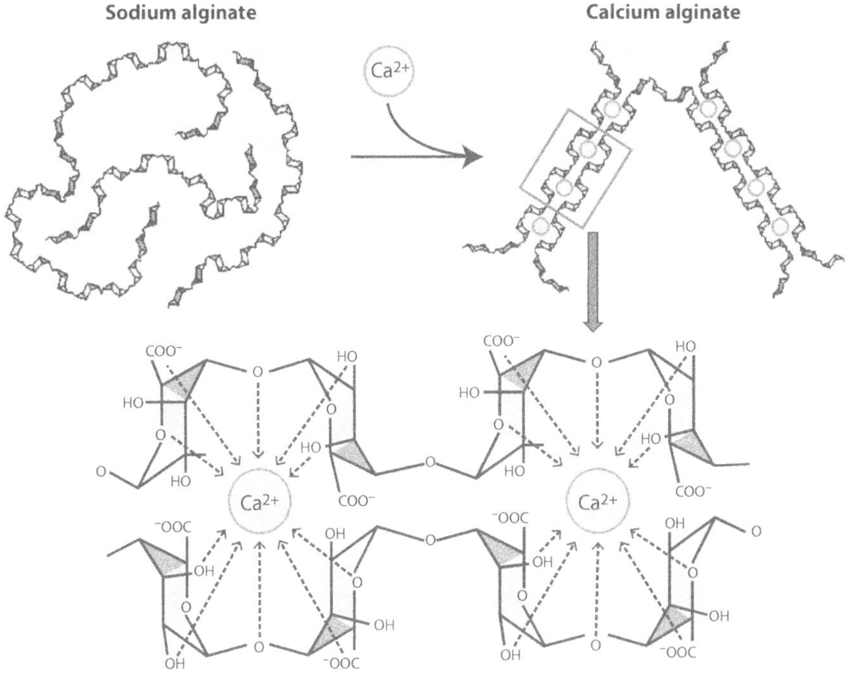

Figure 1.3 Egg-box structure formation during the ionic gelation of sodium ALG [17].

Although it is generally observed that most divalent cations form ALG gels by the "egg-box" formation, it is still not known if other divalent cations follow the same mechanism for gel formation [13–16]. Binding of Ca ion enhances with increasing content of G residues in the chains, while poly-M blocks and alternating MG blocks have lower affinity toward the ion. Generally, by raising the ALG G block content or molecular weight, more strong and brittle ALG gels may be achieved [4]. The affinity of ALG toward Ca ions increases with increasing content of the ion in the gel due to an autocooperative zipper mechanism. This first stage of dimerization is followed by a second stage of lateral association of the dimers at higher Ca^{2+} concentrations. Isolated and purified G blocks have been shown to act as gel modulators, forming higher-order junction zones composed of two or more chains.

Studies have shown previously that there could be different block sequence than G blocks to which cations can bind in ALG. For example, binding studies have recognized that Ca is able to bind to G and MG blocks, Ba can bind to G and M blocks, and Sr can bind to G blocks only [8, 12]. Trivalent cations such as Al^{3+} and Fe^{3+} can also be used to gel ALG.

In fact, they generally have an increased affinity of binding with ALGs as compared to divalent cations. They form a more compact gel network by binding in a 3-D structure due to their ability to bind with three carboxyl groups from different ALG biopolymers at the same time [18].

The ionic gels are widely used in various industries; like in the food industry, these are used in encapsulation of bioactives, in pharmaceuticals for making drugs, and in the biotechnology industry for cell immobilization.

1.4.1.2 Alginic Acid Gels

Alginic acid gels are formed when pH less than the dissociation constant (pKa) of the polymer is used for making the solution [12]. Alginate is negatively charged across a wide range of pH because M and G residues have pKa of 3.38 and 3.65, respectively [19, 20]. Alginate solution is affected in two ways by the rate of decrease in pH. A rapid decrease in pH leads to precipitation of alginic molecules into aggregates, while a low rate of decrease in pH leads to the formation of continuous alginic acid bulk gel [21]. The strength of the gel is correlated to the G block content in the polymer chain like in case of ionic gels [1], while they differ from ionic gels in that the hydrogen bonding in acid gels of ALG is known to stabilize them and M block residues have an important role in gelation. Although alginic acid gels have not got as much importance as compared to ionic gels due to their limited application [21], they are commonly used as antacid to relieve gastric reflux heartburn [22].

1.4.2 Molecular Weight

Alginate is a linear polymer whose viscosity is determined by molecular weight, rigidity, and extension of the chain of the polymer. Alginates may be prepared with a wide range of average molecular weights (50–100,000 residues), which depends on the application. Generally, the molecular weight of commercially available sodium ALGs ranges between 32,000 and 400,000 g/mol.

1.4.3 Solubility and Viscosity

Alginic acid is insoluble in water and organic solvents, whereas its monovalent salts and esters are water-soluble and form a stable, viscous solution [1–4]. Physical properties of ALG gels can be modified and further improved by increasing the molecular weight of ALG. However, it becomes highly viscous on increasing the molecular weight, which is often not desirable in further processing [23]. For example, there is a risk of damage due to high shear forces generated during mixing and injection of proteins or cells mixed with an ALG solution into the body [24]. The 1% *w/v* aqueous solution of sodium ALG has a dynamic viscosity of 20–400 mPa·s at 20°C. The solubility of ALGs is dependent on the solvent pH (a decrease in pH below pKa 3.38–3.65 may result in polymer precipitation), ionic strength, and the gelling ions used [2]. It also depends on the polymer structure, like ALG with more MG blocks (heterogeneous structure) is soluble at low pH as compared to poly-M or poly-G ALG molecules, which tend to precipitate under such conditions [18]. According to the Mark–Houwink relationship ($[\eta] = KMv\ a$), the parameters for sodium ALG in 0.1 M NaCl solution at 25°C are $K = 2 \times 10^{-3}$ and $a = 0.97$, where $[\eta]$ is the intrinsic viscosity (mL/g) and Mv is the viscosity-average molecular weight (g/mol). With decrease in pH, the viscosity of ALG solutions increases and reaches around pH 3–3.5, which is due to the protonation of carboxylate groups in the ALG backbone and formation of hydrogen bonds [23]. Therefore, manipulation of the molecular weight and its distribution can independently control the viscosity of the solution before gel formation and gelling stiffness after gel formation. By changing the combination of high and low molecular weight ALG polymers, the elasticity of gels can be increased significantly with the least increase in viscosity of the solution [24].

1.4.4 Ionic Cross-Linking

Alginate forms hydrogels by chelating divalent cations. Ionic cross-linking agents like divalent cations are combined with the aqueous solution of ALGs in order to make hydrogels [21]. The cations are taken in high concentration in a solution, and ALG microdroplets are dropped into the cationic solution to form heterogenous microcapsules structured in the shape of an egg box. This results in formation of a gel by cross-linking of ALG to divalent cations (Figure 1.4).

1.4.5 Chemical Properties

Polysaccharides get cleaved hydrolytically under acidic conditions. The mechanism of acid hydrolysis of the glycosidic bond involves three steps: (1) formation of conjugate acid due to protonation of the glycosidic oxygen, (2) formation of a nonreducing end group and a carbonium-oxonium ion due to the heterolysis of the conjugate acid, and (3) formation of a reducing end group due to the rapid addition of water to the carbonium-oxonium ion. Sodium ALG can be stored as a dry

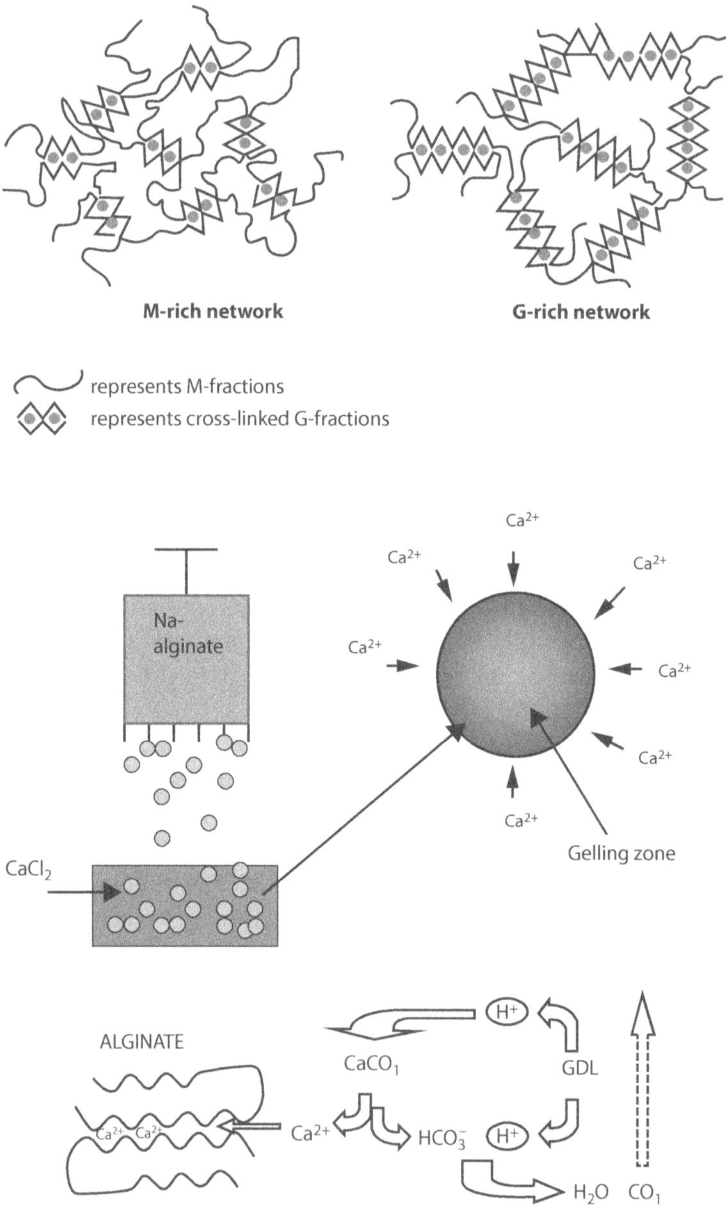

Figure 1.4 Gelation process of ALG [25].

powder for several months in a cool, dry place and away from sunlight. However, the shelf life can be increased for several years by storing it in a freezer. Sodium salt of ALG is more stable than its acidic form, which can degrade rapidly. The reason for this rapid degradation rate of alginic acid is thought to be intramolecular catalysis by the C-5 carboxyl groups [25].

1.5 Sources

Alginate is extracted from the brown seaweeds by methods that can convert the insoluble ALG (present in the seaweed cell walls as calcium and magnesium ALG) to a soluble form, usually sodium ALG during extraction [26]. Different seaweeds used for extraction of ALGs are *Laminaria digitata, Laminaria brasiliensis, Sargassum filipendula, L. hyperborean*, and *M. pyrifera*. Worldwide ALG is derived from various industrial sources like from *Macrocystis* from the USA, *Laminaria* and *Ascophyllum* from Northern Europe, *Durvillaea* in Australia and Chile, and *Sargassum* and *Turbinaria* in India, the Philippines, and other tropical countries.

1.6 Biosynthesis of Bacterial Alginate

P. aeruginosa has been studied first by Darzins and Chakrabarty (1984) by using complementation studies for the genes involved in the production of ALGs. Till now, at least 24 genes have been identified in *P. aeruginosa*, which are directly involved in production of ALG. Chitnis and Ohman (1993) proposed that all the structural genes involved in ALG biosynthesis are clustered in a single operon except algC. There are 12 genes in the cluster, namely algD, alg8, alg44, algK, algE, algG, algX, algL, algI, algJ, algF, and algA, which are located at approximately 3.96 Mb on the PAO1 genome map. The promoter is located upstream of algD, which tightly regulates the operon.

Pindar and Bucke, in 1975, proposed the first bacterial ALG biosynthesis pathway in *Azotobacter vinelandii*. They studied that ALG is first synthesized as a linear homopolymer of D-mannuronic acid residues. The process can be broken down into four stages: (1) precursor synthesis, (2) polymerization and cytoplasmic membrane transfer, (3) periplasmic transfer and modification, and (4) export through the outer membrane.

1.6.1 Precursor Synthesis

First the six-carbon substrate enters into the Entner–Douderoff pathway also known as the KDPG pathway, where pyruvate is formed, which is then channeled to the tricarboxylic acid (TCA) cycle, while oxaloacetate from the TCA cycle is converted to fructose-6-phosphate via gluconeogenesis. The fructose-6-phosphate is converted to mannose-6-phosphate by the phosphomannose isomerase (PMI) activity of the bifunctional protein AlgA (PMI-GMP). Mannose-6-phosphate is directly converted to its isomer form, mannose-1-phosphate, by AlgC (phosphomannomutase). The activated mannose-1-phosphate is converted to GDP-mannose with the hydrolysis of GTP by the GDP-mannose pyrophosphorylase (GMP) activity of AlgA (PMI-GMP). The GMP activity of this enzyme favors the reverse reaction, but AlgD (GDP-mannose-dehydrogenase) constantly converts GDP-mannose to GDP-mannuronic acid, and the reaction is shifted toward GDP-mannuronic acid and ALG production. This AlgD catalyzed reaction is essentially irreversible and provides the direct precursor for polymerization, GDP-mannuronic acid. The high intracellular levels of GDP-mannose indicate that this AlgD catalyzed step is a limiting step and/or is an important kinetic control point in ALG biosynthesis [27].

The two genes, *algA* and *algD*, are found on the ALG operon, whereas *algC* is located elsewhere in the genome at PA5322 [28]. AlgC plays an important role in general exopolysaccharide biosynthesis, i.e., not only ALG biosynthesis but it is also required for precursor synthesis of Psl, as well as LPS and rhamnolipids [29, 30]. The crystal structures of these two enzymes, AlgD and AlgC, have been determined [31, 32]. A common structural feature of enzymes involved in nucleotide binding, such as in the generation of activated sugars, is the presence of at least one β/α/β nucleotide binding domain. This domain is known as a Rossmann fold, which has a secondary structure consisting of alternating β-strands and α-helices arranged such that they form a central six-stranded parallel β-sheet linked to five surrounding α-helices. An example of many variations of the "classical" Rossmann fold or nucleotide binding domain is AlgD. This protein forms a dimer with each individual subunit containing one complete N-terminal nucleotide binding domain and a C-terminal nucleotide-like binding domain, which lacks the third β-strand and final α-helix of this motif [32]. In spacerhelix, the two nucleotide binding domains are separated by a long 33 residue α-helix.

Interestingly, the protein forms a domain-swapped dimer, whereby the N-terminal nucleotide binding domain of one subunit interacts with the C-terminal nucleotide binding domain of the second subunit. The interface of these two domains forms the active site; the location of which was verified by the structure of AlgD in complex with its substrate, NAD(H) and product GDP-mannuronate (ManUA). Two dimers of AlgD likely interact to form a tetrameric structure in the cell cytoplasm, creating the GDP-ManUA product, which is the irreversible step in ALG precursor formation.

The second enzyme, which is involved in the production of ALG precursor, is magnesium-dependent mutase, AlgC (PDB ID:3CO4). The structure of AlgC has been determined. It shows specificity for both phosphomannose and phosphoglucose substrates [31]. This protein contains four domains, which are approximately of equal size. The first three domains share a common topological core consisting of a four-stranded β-sheet sandwiched between 2 α-helices, while the fourth domain is a member of the TATA-box binding protein-like fold superfamily. This domain consists of a four-stranded antiparallel β-sheet, flanked by two α-helices and two short β-strands. All of the four domain residues help in the formation of a large active site cleft at the center of this "heart"-shaped molecule. The specificity for glucose vs mannose in this class of enzymes is thought to be determined by a conserved sequence motif GEMS(G/A) found in domain 3, which has been postulated to act as the sugar binding loop.

While the structure for AlgA has not been determined, it is predicted by structural modeling to have extensive similarity to other proteins with GMP activity, such as the *Thermotoga maritima* guanosine-diphospho-D-mannose pyrophosphorylase (PDB ID:2X65) and a putative mannose-1-phosphate guanyltransferase from *Thermus thermophilus* (PDB ID:2CU2). Both of these proteins contain Rossmann-like β/α/β nucleotide binding domains characteristic of proteins that generate or bind sugar-nucleotide precursors [33].

1.6.2 Polymerization and Cytoplasmic Membrane Transfer

In the periplasm, the activity of AlgI, AlgJ, and AlgF alters the obtained polymer through selective O-acetylation, and epimerization is carried out by AlgG (Figure 1.5) [34, 35]. At the polymer level, D-mannuronic acid residues can be converted to L-guluronic acid by AlgG, and acetylation can occur at the hydroxyl groups of either the C2 or C3 position; ALG can have somewhat random structure. This random structure distinguishes ALG from capsular polysaccharides of many of the *Escherichia coli* and from Psl, as these polymers are composed of regular repeating subunits.

It has been studied that another crucial role in the formation of polymerase complex is of Alg8 and Alg44. Some kind of periplasmic scaffold is formed to guide and protect the nascent ALG chain from degrading from lyase and is considered to be provided by AlgG, AlgK, and AlgX, along with the outer

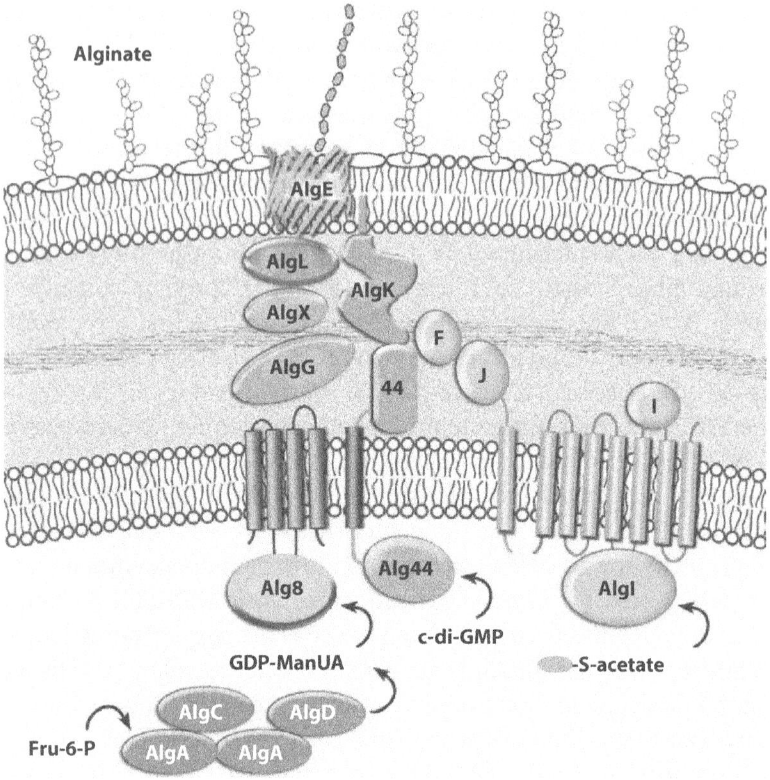

Figure 1.5 Structure of the ALG biosynthetic complex [33].

membrane protein AlgE. Alg8 is thought to be the bottleneck for ALG biosynthesis. Alg8 has a large cytoplasmic glycosyltransferase (GT) domain and four transmembrane (TM) domains [36, 37]. It contains two closely abutting β/α/β Rossmann-like nucleotide binding domains or a GT-A fold. The protein has been classified as a member of a family of inverting glycosyltransferases (GT-2 family) that include cellulose, chitin, and hyaluronan synthases [38].

Another cytoplasmic membrane protein needed for ALG production is Alg44. It has a single transmembrane domain located near the middle of the protein. The protein contains a cytoplasmic N-terminal PilZ domain [39], which plays an important role in binding the secondary messenger bis-(3-5)-cyclic dimeric guanosine monophosphate and suggests an additional regulatory role for Alg44. In the assembly of the multiprotein complex, the C-terminal periplasmic domain of Alg44 plays a role that, therefore, functions as a part of the periplasmic scaffold and provides a bridge between the cytoplasmic membrane proteins.

1.6.3 Periplasmic Transfer and Modification

Alginate is modified almost exclusively at the periplasm in bacteria, which suggests that ALG is synthesized as polymannuronate and modification occurs at the polymer level. A number of enzymes (AlgI/AlgJ/AlgF, the polymannuronan epimerase, AlgG, and AlgX) including the O-acetylation complex catalyze the modification of polymannuronic acid to the mature ALG polymer, in the periplasm [34, 35, 40, 41]. The function of AlgX is not clear, but it is likely to be associated with the multiprotein complex and polymer modification. Alkaline phosphatase fusion proteins were used to map membrane topology of AlgI, and it is found to contain seven transmembrane domains and two large cytoplasmic domains AlgF and AlgJ, which were both localized to the periplasm; AlgJ is anchored in the cytoplasmic membrane by an uncleaved signal peptide [42, 44].

The cellular location of AlgI/AlgJ/AlgF suggested a model for ALG O-acetylation, where AlgI transfers the acetyl group contained on a donor molecule (possibly an acyl carrier protein or Coenzyme A) across the membrane, and then the acetyl group is transferred to AlgJ or AlgF for O-acetylation of the mannuronate residues at the polymer level.

There are three classes of ALG modifying enzymes, which have been described as

1. Transacetylases
2. Mannuronan C 5-epimerases
3. Lyases

1.6.3.1 Transacetylases

Transacetylation occurs only at mannuronic acid residues at the O-2 and/or O-3 position. Acetylation of these residues prevents their epimerization to guluronic acid residues by AlgG. It also prevents the ALG chain degradation by AlgL. Therefore, the acetylation of ALG is indirectly responsible for controlling epimerization and length of the ALG polymer. The water binding capacity of ALG can be strongly enhanced by increasing the degree of acetylation, which may be particularly crucial for survival under dehydrating conditions [43]. Although the genes *algI*, *algJ*, and *algF* are required for the addition of O-acetyl groups to the ALG polymer, acetylation itself is not required for ALG biosynthesis [44].

1.6.3.2 *Mannuronan C 5-Epimerases*

Epimerization of the D-mannuronate to L-guluronate is catalyzed by AlgG at the polymer level [35]. This epimerization process modifies the structural properties of ALG, including its gelling ability and its ability to bind divalent ions such as calcium. Recently, six different C5-epimerase encoding genes have been identified in the genome of *L. digitata* [43].

1.6.3.3 *Lyases*

The ALG lyases, also known as ALG depolymerases or alginases, catalyze the β-elimination reaction, which leads to degradation of ALG. It functions as an editing enzyme in ALG-producing bacteria, controlling the length and the molecular weight of the polymer. These enzymes have different residue specificities and cellular localizations. Among *Pseudomonas* species, PA1167 is identified as one other protein to have ALG lyase activity. The epimerase AlgE7 and three others, AlyA1, AlyA2, and AlyA3, are the four lyases that have been identified among *Azotobacter* species.

1.6.5 Export through the Outer Membrane

Alginate is secreted through a putative porin known as AlgE (called AlgJ in *Azotobacter*). It forms an anion-selective pore through the outer membrane, and the pore is partially blocked by GDP-mannuronic acid. Homology modeling showed that the protein is a β-barrel consisting of 18 antiparallel β-strands with 8 periplasmic and 9 surface-associated loops. This protein is responsible for the secretion of intact ALG and can be detected in the outer membrane of mucoid, ALG-overproducing strains of *P. aeruginosa* but is absent in non-mucoid strains.

1.7 Conclusion

Alginates are naturally occurring anionic polysaccharides, which are present as a structural component in brown algae. They consist of linear biopolymers consisting of 1,4-linked β-D-mannuronic acid (M) and 1,4 α-L-guluronic acid (G) residues arranged in homogenous (poly-G, poly-M) or heterogenous (MG) block-like patterns. The pattern of residues determines the physicochemical properties of ALGs. Alginate has an excellent functionality as a thickening agent, gelling agent, emulsifier, stabilizer, texture improver, and many more. Due to these qualities, ALGs have various applications in different areas like in food industries (such as ice cream, jelly, lactic drinks, dressings, instant noodle, beer), textile printing industries, animal feed, pharmaceuticals (forming tablets, dentistry, wound dressing), and cosmetic industries. Moreover, an increased understanding of ALG composition and material properties will help meet medical and pharmaceutical specifications, thus providing enormous opportunity for the use of engineered bacteria for the production of ALGs.

Acknowledgment

Authors are highly grateful to the Department of Biotechnology for providing all necessary facilities and to CSIR for providing financial assistance to Ms. Rutika Sehgal. Financial assistance from DEST (Department of Environment, Science and Technology), Government of Himachal Pradesh, in the form of a research project is also thankfully acknowledged.

Conflict of Interests

The authors declare that there is no conflict of interests regarding the publication of this chapter.

References

1. Draget, K.I., Bræk, G.S., Smidsrød, O., Alginic acid gels: The effect of alginate chemical composition and molecular weight. *Carbohydr. Polym.*, 25, 31–38, 1994.
2. Orive, G., Carcaboso, A.M., Hernandez, R.M., Gascon, A.R., Pedraz, J.L., Biocompatibility evaluation of different alginates and alginate-based microcapsules. *Biomacromolecules*, 6, 927–931, 2005.
3. Andersen, T., Strand, B.L., Formo, K., Alsberg, E., Christensen, B.E., Alginates as biomaterials in tissue engineering, in: *Carbohydrate Chemistry*, A.P. Rauter, T. Lindhorst (Eds.), pp. 227–258, The Royal Society of Chemistry, London, 2011.
4. Helgerud, T., Gåserød, O., Fjæreide, T., Andersen, P.O., Larsen, C.K., Alginates, in: *Food Stabilizers, Thickeners and Gelling Agents*, A. Imeson (Ed.), pp. 50–72, Wiley-Blackwell, Oxford, UK, 2009.
5. Knudsen, N.R., Ale, M.T., Fatemeh, A., Meyer, A.S., Characterization of alginates from Ghanaian brown seaweeds: *Sargassum* spp. and *Padina* spp. *Food Hydrocoll.*, 71, 236–244, 2017.

6. Kupper, F.C., Kloareg, B., Guern, J., Potin, P., Oligoguluronates elicit an oxidative burst in the brown algal kelp *Laminaria digitata. Plant Physiol.*, 125, 278–291, 2001.
7. Smidsrød, O. and Skjak-Bræk, G., Alginate as immobilization matrix for cells. *Trends Biotechnol.*, 8, 71–78, 1990.
8. Nedovic, V., Wallaert, R., (Eds.), *Fundamentals of Cell Immoblilisation Biotechnology*, Kluwer Academic, Dordrecht, 2004.
9. Clare, K., Algin, in: *Industrial Gums: Polysaccharides and Their Derivatives*, R.L. Whister, J.N. BeMiller (Eds.), pp. 105–143, Academic Press, Toronto, 1993.
10. Donati, I. and Paoletti, S., Material properties of alginates, in: *Alginates: Biology and Applications*, B.H.A. Rehm (Ed.), Springer-Verlag, Berlin, 2009.
11. Haug, A., Affinity of some divalent metals to different types of alginates. *Acta Chemic. Scand.*, 15, 1794–1795, 1961.
12. Haug, A. and Smidsrod, O., Effect of divalent metals on properties of alginate solutions. II. Comparison of different metal ions. *Acta Chemic. Scand.*, 19, 341–351, 1965.
13. Jang, L.K., Lopez, S.L., Eastman, S.L., Pryfogle, P., Recovery of copper and cobalt by biopolymer gels. *Biotechnol. Bioeng.*, 37, 266–273, 1991.
14. Gombotz, W.R. and Wee, S.F., Protein release from alginate matrices. *Adv. Drug Deliv. Rev.*, 31, 267–285, 1998.
15. Yang, C.H., Wang, M.X., Haider, H., Yang, J.H., Sun, J.Y., Chen, Y.M., Zhou, J., Suo, Z., Strengthening alginate/polyacrylamide hydrogels using various multivalent cations. *ACS Appl. Mat. Interf.*, 5, 10418–10422, 2013.
16. Szekalska, M., PuciBowska, A., SzymaNska, E., Ciosek, P., Winnicka, K., Alginate: Current use and future perspectives in pharmaceutical and biomedical applications. *Int. J. Polym. Sci.*, 2016, 1–17, 2016.
17. Kashima, K. and Masanao, I., Advanced membrane material from marine biological polymer and sensitive molecular-size recognition for promising separation technology, in: *Advancing Desalination*, R.Y. Ning (Ed.), pp. 3–36, InTech, 2012.
18. Draget, K.I., Alginates, in: *Handbook of Hydrocolloids*, G.O. Phillips, P.A. Williams (Eds.), pp. 807–828, Woodhead Publishing, Cambridge, 2009.
19. Draget, K.I., Skjak-Braek, G., Stokke, B.T., Similarities and differences between alginic acid gels and ionically cross-linked alginate gels. *Food Hydrocoll.*, 20, 170–175, 2006.
20. Smith, A.M. and Miri, T., Alginates in foods, in: *Practical Food Rheology*, I.T. Norton, F. Spyropoulos, P. Cox (Eds.), pp. 113–132, Wiley-Blackwell, Oxford, 2011.
21. Lee, K.Y. and Mooney, D.J., Alginate: Properties and biomedical applications. *Prog. Polym. Sci.*, 37, 106–126, 2012.
22. Hay, I.D., Rehman, Z.U., Ghafoor, A., Rehm, B.H.A., Bacterial biosynthesis of alginates. *J. Chem. Technol. Biotechnol.*, 85, 752–759, 2010.
23. Leroux, M.A., Guilak, F., Setton, L.A., Compressive and shear properties of alginate gels: Effects of sodium ions and alginate concentration. *J. Biomed. Mater. Res.*, 47, 46–53, 1999.

24. Kong, H.J., Lee, K.Y., Mooney, D.J., Decoupling the dependence of rheological/mechanical properties of hydrogels from solids concentration. *Polymer*, 43, 6239–6246, 2002.
25. Juarez, G.A.P., Spasojevic, M., Faas, M.M., de Vos, P., Immunological and technical considerations in application of alginate-based microencapsulation systems. *Front. Bioeng. Biotechnol.*, 2, 1–15, 2014.
26. Goldberg, J.B., Hatano, K., Pier, G.B., Synthesis of lipopolysaccharide O side chains by *Pseudomonas aeruginosa* PAO1 requires the enzyme phosphomannomutase. *J. Bacteriol.*, 175, 1605–1611, 1993.
27. Olvera, C., Goldberg, J.B., Sanchez, R., Soberon-Chavez, G., The *Pseudomonas aeruginosa* algC gene product participates in rhamnolipid biosynthesis. *FEMS. Microbiol. Lett.*, 179, 85–90, 1999.
28. Regni, C., Tipton, P.A., Beamer, L.J., Crystal structure of PMM/PGM: An enzyme in the biosynthetic pathway of *P. aeruginosa* virulence factors. *Structure*, 10, 269–279, 2002.
29. Snook, C.F., Tipton, P.A., Beamer, L.J., Crystal structure of GDP-mannose dehydrogenase: A key enzyme of alginate biosynthesis in *P. aeruginosa*. *Biochemistry*, 42, 4658–4668, 2003.
30. Franklin, M.J., Nivens, D.E., Weadge, J.T., Howell, P.L., Biosynthesis of the *Pseudomonas aeruginosa* extracellular polysaccharides, alginate, Pel, and Psl. *Front. Microbiol.*, 2, 167, 2011.
31. Franklin, M.J. and Ohman, D.E., Identification of algF in the alginate biosynthetic gene cluster of *Pseudomonas aeruginosa* which is required for alginate acetylation. *J. Bacteriol.*, 175, 5057–5065, 1993.
32. Franklin, M.J., Chitnis, C.E., Gacesa, P., Sonesson, A., White, D.C., Ohman, D.E., *Pseudomonas aeruginosa* AlgG is a polymer level alginate C5-mannuronan epimerase. *J. Bacteriol.*, 176, 1821–1830, 1994.
33. Remminghorst, U. and Rehm, B.H.A., Alg44, a unique protein required for alginate biosynthesis in *Pseudomonas aeruginosa*. *FEBS Lett.*, 580, 3883–3888, 2006.
34. Oglesby, L.L., Jain, S., Ohman, D.E., Membrane topology and roles of *Pseudomonas aeruginosa* Alg8 and Alg44 in alginate polymerization. *Microbiology*, 154, 1605–1615, 2008.
35. Cantarel, B.L., Coutinho, P.M., Rancurel, C., Bernard, T., Lombard, V., Henrissat, B., The carbohydrate-active enzymes database (CAZy): An expert resource for glycogenomics. *Nucleic Acids Res.*, 37, 233–238, 2009.
36. Merighi, M., Lee, V.T., Hyodo, M., Hayakawa, Y., Lory, S., The second messenger bis-(3-5)-cyclic-GMP and its PilZ domain-containing receptor Alg44 are required for alginate biosynthesis in *Pseudomonas aeruginosa*. *Mol. Microbiol.*, 65, 876–895, 2007.
37. Chitnis, C.E. and Ohman, D.E., Cloning of *Pseudomonas aeruginosa* algG, which controls alginate structure. *J. Bacteriol.*, 172, 2894–2900, 1990.
38. Weadge, J.T., Yip, P.P., Robinson, H., Arnett, K., Tipton, P.A., Howell, P.L., Expression, purification, crystallization and preliminary X-ray analysis of

Pseudomonas aeruginosa AlgX. *Acta Crystallogr. Sect. F Struct. Biol. Cryst. Commun.*, 66, 588–591, 2010.
39. Franklin, M.J. and Ohman, D.E., Mutant analysis and cellular localization of the AlgI, AlgJ, and AlgF proteins required for O acetylation of alginate in *Pseudomonas aeruginosa*. *J. Bacteriol.*, 184, 3000–3007, 2002.
40. Nivens, D.E., Ohman, D.E., Williams, J., Franklin, M.J., Role of alginate and its O acetylation in formation of *Pseudomonas aeruginosa* microcolonies and biofilms. *J. Bacteriol.*, 183, 1047–1057, 2001.
41. Franklin, M.J. and Ohman, D.E., Identification of algI and algJ in the *Pseudomonas aeruginosa* alginate biosynthetic gene cluster which are required for alginate O acetylation. *J. Bacteriol.*, 178, 2186–2195, 1996.
42. Gacesa, P., Alginates. *Carbohydr. Polym.*, 8, 161–182, 1988.
43. Nyvall, P., Corre, E., Boisset, C., Barbeyron, T., Rousvoal, S., Scornet, D., Kloareg, B., Boyen, C., Characterization of mannuronan C-5-epimerase genes from the brown alga *Laminaria digitata*. *Plant Physiol.*, 133, 726–735, 2003.
44. Remminghorst, U. and Rehm, B.H.A., Bacterial alginates: From biosynthesis to application. *Biotechnol. Lett.*, 28, 1701–1712, 2006.

2

Alginates Production, Characterization and Modification

Pintu Pandit[1]*, T. N. Gayatri[1] and Baburaj Regubalan[2]

[1]Department of Fibres and Textile Processing Technology
[2]Department of Food Engineering and Technology,
Institute of Chemical Technology, N. P. Marg, Matunga, Mumbai, India

Abstract

Alginate, naturally occurring in brown seaweeds, is an anionic polysaccharide that has been deeply researched, and its properties of biocompatibility with human tissue, bone, and teeth, and mild gelation along with minimal toxicity while being relatively cheap, have found application in many biomedical science and engineering products and processes. Globally, seaweeds provide the major raw material source for economically producing alginate and related polysaccharides and industrial products. The physicochemical properties of alginates vary as they are composed of various proportions of β-d-mannuronic acid (M) and α-l-guluronic acid (G) residues in their polymeric chains, measured by the G/M ratio, and the specific distribution of molecular weight of the different polymers in the product. The sequence and nature of extraction and precipitation procedures of alginates from seaweed species not only affect the purity and amount of yield but also control the chemical constitution and rheological properties of the product alginate. This chapter will provide a detailed overview of alginate's production, characterization, and modification, which could indicate new starting points for future studies.

Keywords: Alginates, production, characterization, modification

**Corresponding author*: pintupanditict@gmail.com

Shakeel Ahmed (ed.) Alginates, (21–44) © 2019 Scrivener Publishing LLC

2.1 Introduction

Alginate is an anionic polysaccharide, naturally occurring in brown seaweeds, and has been extensively studied and used in many biomedical applications due to its biocompatibility, minimal toxicity, and mild gelation achieved through addition of divalent cations such as Ca^{2+}, along with its availability at a relatively cheap price [1]. Alginate hydrogels can be prepared by different cross-linking techniques and agents; depending on the nature of cross-linking and density of the network, drugs ranging in size from small molecules to large macromolecular proteins can be released in a controlled manner from the alginate gel, which encapsulates them or carries them. Their structure has features similar to the extracellular matrices of biological structural tissues promoting their extensive application in wound healing medical devices, cell transplantation and in the delivery of bioactive drugs and proteins. Alginate wound dressings favor wound healing by maintaining a physiologically moist microenvironment, while inhibiting bacterial infection at the wound site.

Alginate gels find important and numerous applications in the pharmaceutical field as they may be orally ingested or given as injections into the body, without causing much discomfort to the patient. In tissue engineering, cell and organ transplants and implants made from alginate gels provide a substitute for regeneration for patients with damaged, nonfunctioning organs and tissues [2]. Hydrogels function by carrying a payload of regenerative cells and drugs to the site of damage, while also acting as a substrate for the growth of new tissue, whose structure and function can be guided by the mechanical flexibility, stiffness, and pore size of the alginate gel [3].

Alginate has been utilized in biomedical applications such as wound healing, drug delivery, and *in vitro* cell culture and has unexplored potential as a biomaterial for many tissue engineering problems. The suitability and fit of alginate for these applications are because of its biocompatibility, gelation under mild conditions, and the ease of synthetic modifications required to make alginate derivatives that have the required properties. Chemically modified alginate has found a crucial use as a carrier for dental follicle cells and suitable growth factors to initiate and promote periodontal regeneration while sustaining osteogenesis [4]. Similar to other hydrogels from agarose, polyvinyl alcohol, and acrylates, alginate gels have a limited mechanical stiffness and degrade gradually on exposure to physiological fluids, and the more general physical properties of absorption, swelling, and ion and small molecule binding and release could be modulated with the structure and compositional variation. The encapsulation of cells by covalent cross-linking reactions can cause toxic harm to the cells encapsulated, but through a suitable choice of cell-compatible chemical reagents (e.g., initiator), and complete removal of unreacted reagents and by-products, cross-linked alginate can be fitted such applications. Looking to the future, the alginate-based materials used in medicine are likely to develop exponentially.

The clinical application of alginate gels in wound healing involves their passive role as resorbable matrices. The use and design of alginate gels as reservoirs of drugs and progenitor cells for growth, which release on external mechanical signals or tuned magnetic fields, is feasible [5, 6]. The ability of alginates to interact with cells is pivotal to tissue engineering applications, since without modification with protein signaling motifs, alginates are not recognized for mammalian cell adhesion. The nature of adhesion ligands and their spatial arrangement in gels are key parameters that can regulate and select the growth of cell phenotype in regenerated tissue, eventually determining its resultant function. Genetic engineering techniques to modify bacterial synthesis of alginates could forge new pathways to the design and creation of alginate polymers with tailor-made properties. Various polypeptides and proteins that improve structural properties and engender novel functions in alginate gels have been prepared and examined for biomedical applications [7, 8]. The ability to design and synthesize new classes of alginates with control of specific physical and chemical characteristics, unlike the limited range of properties available from natural alginate sources, tuned to a particular application could herald a revolution in the use of these materials. In this chapter, general properties, production, and modification of alginate have been discussed.

Biopolymer demand continues to grow annually by 3–4% because of their enhanced application in different areas, but production needs to keep pace with it. The annual growth rate over 2007–2011 in volume terms of seaweed produced has been 9%. Emerging producers in China, Eastern Europe, Brazil, etc. have driven most of this growth. Prices have tended to remain stable, even though demand has outstripped production because the markets do not support increased prices where substitution or extension with cheaper materials is possible. The higher seaweed material and process costs of energy and chemicals reflect the production of alginic acid on a large scale. The islands of the Philippines and Indonesia are the major producers of cultivated *Laminaria digitata*, *Laminaria hyperborea*, and *Laminaria saccharina* because their oceanic geography assists the growth of alginate-producing seaweed throughout the year. In comparison, in India, the available alginate produce much less and the viscosity of the produced alginate does not readily fit the requirements of the textile industry. Therefore, there is a great need to discover bacterial resources for alginate production with higher viscosity, which may be exploited with little capital cost. Production of alginate depends on the demand from various industries, and the viscosity of the alginate production can be enhanced, for example, by using recombinant strains of *Azotobacter*, obtained by genetic manipulation [9].

2.2 Alginate: Production

Seaweeds are the major source for producing value-added polysaccharides and industrial by-products globally. Alginates are one of the chief products extracted from seaweeds, mainly brown algae. These polysaccharides constitute the structural composition of the cell walls and the intercellular matrix in seaweeds.

The ratio M/G varies with the source of alginates, but its physical properties, for similar M/G ratio, are almost the same even when extracted from different biosources. Alginates are most often used in food processing as a stabilizer, viscosity agent, and gelling agent, with annual industrial requirements of alginates reaching ~30,000 metric tons [10]. Alginates form 40% of the dry matter of the commercially harvested species of seaweeds, such as *Laminaria* spp. and *Macrocystis* spp. [11]. Recently, this polysaccharide has been used for wide range of pharmaceutical and biological products, as in wound dressings [12]. Prior to 1975, commercial alginate production was based on seaweeds, where they were treated with alkali solution followed by filtration. The alginates were precipitated with chloride salts of sodium/calcium cations, so that the alginate salts can be converted to alginic acid, on acidifying with dilute HCl. On purification, the obtained alginic acid gave water-soluble sodium alginates [13]. During the late 1980s, two genera of bacteria, the nonpathogenic *Pseudomonas* and *Azotobacter*, were identified to be major producers of alginate and alginic acids. In nature, microbes produce alginates through different metabolic processes with various material functions. *Pseudomonas aeruginosa* produces alginates that constitute the thick highly structured biofilm, characteristic of the species [9], whereas *Azotobacter* produces rigid alginate, essential for the formation of water-conserving cysts [14] resistant to dessication and stress.

2.2.1 Screening of Alginate-Producing Microbes

Alginates are hydrocolloids derived from seaweed that interact with water to form colloid systems, like a gel or solubilized particles. Alginates are extracted in disparate ways depending on the application, but the most generally applied procedure [9] involves extracting the alginate as sodium alginate. The insoluble calcium and magnesium alginates present in the brown seaweed cell walls are extracted by maceration and converted to soluble sodium alginates, eventually obtained as alginic acid or calcium alginate. The consecutive addition of acid, alcohol, and sodium carbonate affects the conversion. The extraction procedures applicable to alginate extraction encounter problems such as insolubility from seaweed residuals interfering with the ease of separation. Filtration of the solution of dissolving alginate as sodium alginate requires large volumes of water, as the increased viscosity of the solution makes the separation onerous. The fine particulate nature of the seaweed residuals can block the filter; so filter aids are required to ease the process and make it cost effective. Also, the chemicals utilized for extraction influence the physicochemical properties of alginates. There exists a need to establish extraction and processing by alternate and milder methods, so as to overcome the problems faced in the traditional extraction procedures along with the detrimental effects on the quality and quantity of alginate yield. Enzymatic extraction techniques of alginate from seaweed using enzymes such as alginate lyase, laminarinase, which could degrade the seaweed cell wall to release free alginate, have been studied, but not standardized to routine extractions. The main hurdles facing alginate producers are the varying available areas for alginate farming and production due to increasing non-alginate uses for the same types of seaweed, increasing government proscriptions on the harvesting of natural seaweeds, and decreasing ease of access to large natural seaweed resources. Natural seaweed releases alginate alone into the surrounding sea water, but in the marine environment, it is converted to sodium salt of alginates. Hence, potassium alginate also is present in the extract from cells of marine seaweed. Calcium alginate is obtained from sodium alginate, where sodium is substituted with calcium.

2.2.2 Production of Alginate by Bacteria

Alginate-producing microbes can be screened by the cetylpyridinium chloride (CPC) method and the plate assay method. In the CPC method, enrichment culture technique is employed to screen for microorganisms capable of producing alginate lyase enzyme through their ability to grow on alginate-containing solid media plates; a clearance zone is formed after flooding the plates with agents such as 10% (w/v) CPC, which can form complexes with alginate, absent in the presence of digested alginate.

In the plate assay method, alginate-containing agar plates are flooded with Gram's iodine in place of CPC. Gram's iodine forms a bluish black complex with alginate but not with hydrolyzed alginate, giving rise to sharp, distinct zones around the alginate lyase producing microbial colonies within 2–3 min. Gram's iodine method was found to be better than the CPC method for the clarity of visualization and ease of measurement of zone size. The alginate-lyase-activity region obtained using Gram's iodine method was larger and sharper than the CPC method.

2.2.3 Production of Alginate by *Pseudomonas*

In the presence of *in vivo* conditions, nonpathogenic *Pseudomonas* creates an alginate layer, which prevents the permeation of heavy metal ions into the cell and protects against heavy metal toxicity; this type of alginate secretion could be increased in the presence of NaCl and ethanol in fluorescent *Pseudomonas* species. It is thus possible to deduce that osmolarity and dehydration experienced by the algal cell may be metabolic cues to kick-start production of alginate polysaccharide [15].

2.2.4 Production of Alginate by *Azotobacter* spp.

Azotobacter vinelandii continues to produce significant amounts of acetylated alginate steadily, despite variations of the *in vitro* and *in vivo* conditions. Alginate, an excess metabolite by-product, has a barrier role preventing permeation of toxic heavy metal ions, and conserves cell conditions internally when subjected to environmental and chemical degradative stress. *A. vinelandii* growing diazotrophically in the presence of varying partial pressures of available oxygen, in a fermenter under high shear stress, still forms alginate capsules [16]. When *A. vinelandii* is subjected to higher dissolved oxygen concentration (DO), more compact alginate capsules are formed at higher density; the reason deduced is the increased need for a diffusion barrier to oxygen to seclude the oxygen-sensitive enzyme, nitrogenase, by surrounding the cell with a denser, more impermeable alginate layer.

2.2.5 Influence of Medium Components

2.2.5.1 Effect of Nutrients on Bacterial Alginate Production

Numerous trials have been attempted with changes in the fermentation medium of *Azotobacter* to achieve optimum alginate production. While maintaining the nitrogen level in the culture medium at a fixed level, both favorable and detrimental effects were studied in detail. For *Azotobacter* spp., the accumulated nitrogen may retard the production of alginate [14]. In contrast, peptone used in the medium will alter the alginate production favorably by decreasing the associated formation of poly-β-hydroxybutyric acid, an alternate storage polymer, to 30% of the dry cell mass, suggesting a more specific role for nitrogenous nutrients [17].

2.2.5.2 *Effect of Phosphate on Bacterial Alginate Production*

The study of the effect of phosphate on alginate production by *A. vinelandii* has also given divergent results. When a phosphate-rich medium (7.5 g K_2HPO_4) was used in culture media, the phosphate was reported to merely function as a buffer agent in the medium [17]. It has also been observed that a medium with excess phosphate (200–400 g/L) leads to maximum growth rate and greater biomass production; alternately alginate yield from biomass is highest at low phosphate levels (100 mg/L). The calculated RQ value (respiratory quotient) was optimally ~0.8 at the low end of phosphate concentrations. The kinetics of alginate production are affected by culture conditions, in addition to growth rate.

2.2.5.3 *Effect of Dissolved Oxygen on Bacterial Alginate Production*

Dissolved oxygen has a key function in the alginate production by *Azotobacter*, in the presence of nitrogen- and phosphate-rich medium. The kinetics of biomass growth and alginate production by *A. vinelandii* from glucose in a nitrogen- and phosphate-rich medium were studied based on the use of a laboratory fermenter at pH 7 and 35°C. Batch fermentations in presence of uncontrolled DO and at controlled levels of 1%, 2%, 5%, and 10% DO were carried out. At higher DO, growth was speedy, but the maximum biomass concentration was reduced. At 10% DO, the highest controlled level, production of alginate was nil. Alginate production was rapid at 5% and 2% DO, but increased alginate concentrations and yields were obtained in the absence of DO control. The DO level of 3–5% is found to be ideal for significant alginate production by *Azotobacter*; at this DO level, reducing the activity of nitrogenase aids the microbe to increase alginate production. At a fixed growth rate, the pO_2 (partial pressure of oxygen) at 2–5% air saturation gave the peak algal production rate [14].

2.2.5.4 *Effect of Agitation in the Medium for the Production of Alginate*

The effect on alginate production by *Azotobacter* spp. by varying shaking rates (revolutions per minute [rpm]) was significant in the fermenter studies at controlled pH conditions. At a shaking rate of 500–600 rpm, the maximum level of alginate production was 2 g/L [18]. At a shaking rate of 170 rpm, and in the absence of pH control, a mutant strain of A. vinelandii NCIB 9068 produces 6.22 g/L alginate [9]. On increasing the rate of agitation from 200 to 400 rpm, the growth rate of bacteria was almost doubled from 0.165 to 0.239 h^{-1}. However, alginate production was found to peak at 400 rpm with the highest yield of 4.51 g/L obtained close to the termination of fermentation. Increased agitation or shaking speed along with variation of other determinants like aeration and phosphate concentration was often employed by different researchers to achieve optimal production of alginate.

2.2.6 Commercial Production of Alginate

The extraction of alginate from algal material is carried out commercially to produce alginate. Initially, acidification of insoluble counterion of alginate is effected by extracting the milled algal tissue with 0.1–0.2 M mineral acid. Alginic acid thus brought into solution is neutralized with an alkali such as sodium carbonate or sodium hydroxide to form water-soluble sodium alginate. Different separation techniques may be employed to remove algal materials, other than alginate, such as sifting, flotation, centrifugation, and filtration. After this elimination of impurities, sodium alginate is precipitated by addition of alcohol, calcium chloride, or mineral acid, reconverted to the sodium form as required, and finally dried and milled. About 35 years ago, a similar route for microbial production of alginate was suggested [19]. But this requires that the biological pathway to alginate formation be clarified, if maximal microbial alginate production is to be achieved commercially. In this regard, numerous questions are still in search of answers [14–18].

2.3 Characterization of Physicochemical Properties of Alginate

Alginates are linear unbranched polysaccharides containing different numbers of β-d-mannuronic acid (M) and α-l-guluronic acid (G) residues. The monomer composition G/M ratio and the molecular weight of alginates are known to have effects on their properties. The chemical composition and sequence of the two types of residues are affected by the biological source and the growth and seasonal conditions of the marine plants. The seaweed species, harvesting region, season, sea current, water temperature, and processing method are all factors that affect important structural features of the alginates.

Slightly more than 200 different grades of sodium alginate are commercially available from manufacturers [20]. The molecular weight average of commercial alginate varies between 33 and 400 kDa. With this abundance of choice, it is essential to be aware of how differences in alginates can affect the performance parameters of pharmaceutical compositions. The variability of alginates available with respect to molecular weight, composition, and sequence of M-block and G-block in their copolymer chain changes the viscosity property, features of sol/gel transition, and capacity to uptake and absorb water and swelling response [21].

2.3.1 Composition of Alginate Polymer Chains

Structural linkage composition of alginates is determined to be made of two different uronic acid residues existing as blocks of homopolymeric sequences of either D-mannuronic acid residues (M-blocks) or L-guluronic acid residues (G-blocks), interspersed almost alternately by long sequences of heteropolymeric material (MG-blocks) [22]. The three types of blocks have been evaluated through decomposition by partial hydrolysis with hydrochloric acid (HCl). The material solubilized corresponds to the MG-block. The resistant part on fractionation at pH 2.9 yields a soluble fraction derived from the M-block and an insoluble fraction arising from the G-block [23].

From the standpoint of commercial use in pharmaceutical and food processing industries, the ability of alginates to form viscous solutions in aqueous media is crucial, and hence alginate portions are often classified on the basis of their intrinsic viscosity. The rate and degree of gelling and the nature of chemical derivatization feasible are determined by the composition of the alginic acid. The index called the M/G ratio, which is the ratio of constituent monomers in the alginic acid, is of much importance in deciding the industrial application. The viscosity of its solutions is directly affected by the alginate molecular weight, the number of M or G residues, and the solution strength. The intrinsic viscosity of alginates is influenced by the conformation (sequence of M and G residues) and ionic strength of the solution. The viscosity increases as the stiffness of the constituent chain blocks increases in the order MG<MM<GG, on the basis of intramolecular steric hindrance. With alginate being a polyelectrolyte, the electrostatic repulsion present between the charged groups on the polymer chain increases with chain extension, and the intrinsic viscosity follows with an increase [24]. Physical properties of gels can be enhanced by increasing the molecular weight of the alginate, but such high MW alginates become increasingly viscous, which is not favorable during processing, such as in destructive high shear generated during mixing and injection. So tuning formulations with a favorable mix of high and low MW alginates to manipulate average MW and its polydispersity becomes essential.

Another property of much industrial use is the ability of alginates to form cross-linked gels in the presence of divalent calcium ions. The formation of gels is explained on the basis of autocooperatively formed junctions between chain regions enriched in GG sequences. Such polyguluronate regions are also able to chelate the metal ions by the inherent spatial arrangement of the ring and hydroxyl oxygen atoms and thus develop stronger interactions. Polyguluronate intrinsically forms the junction zones so that they are terminated by zones of zero interaction, i.e., polymannuronate sequences and to a lesser degree the random blocks [24, 25]. The chemical structure and polyelectrolyte nature are the reasons for their ion binding, ion exchange, and gelation properties.

Using atomistic molecular dynamic simulation, a relation between chain stiffness and G content of alginate is observed at high alginate concentrations in Na^+ solutions, explained by chain association. For Ca^{2+}- alginate, chain stiffness seems to increase with G content. Stiffer homopolymer GGGG and MMMM chains associate laterally, while the more flexible heteropolymer chains, exemplified by the poly-G2M2 sequence, associate in coil-like configuration at high concentrations. In the presence of calcium ion, mannuronic acid residues act to moderate elasticity by favoring chain association. As alginate concentrations increase, poly-M chains transfer to lateral association of dimers and higher order clusters, displaying an absence of zipper mechanism. Poly-G chain association initially is created through zipping first of dimers followed by higher order clusters as solutions increase in concentration. The heteropolymeric compositions show the formation of stiff zones in the sequence regions richer in G, linked by flexible junctions. Further small-angle X-ray scattering studies on alginate gels indicate that lateral association of G-blocks occurs to a degree greater than mere dimerization, with increasing $[Ca^{2+}]$ and G-content of the alginate [26, 27].

The release rate from gel beads made from sodium alginate material with similar G/M ratios but different viscosities was identical in acidic conditions but completely different at neutral pH [28]. The release of furosemide from formulations composed of varying viscosity grades of sodium alginate revealed that increasing viscosity lowered the release rate [29]. Regarding the dissolution of theophylline from low and high viscosity sodium alginate matrices, a faster dissolution rate of theophylline was measured from the low viscosity matrix in water, whereas the trend was opposite in acidic conditions [30].

The strength of the gel formed from alginate in the presence of divalent metal ions is much affected by the G/M ratio, as the G residues "polyguluronate" exhibit a greater attraction to chelate the ions than the M residues [30]. Alginates rich in L-guluronate form strong but brittle gels, but those rich in ᴅ-mannuronate are weaker but more flexible [31, 32]. Rapid swelling and dissolution of gel beads made from alginates with a low G/M ratio occurred at neutral pH; in contrast, beads made from a high G/M ratio alginate did not disintegrate under identical conditions [33–34].

2.3.2 XRD, FTIR, and NMR Spectroscopy for Alginate Structure Analysis

The structure of alginates has been investigated using X-ray diffraction and infrared spectroscopy. Alginic acid fibers give sharp diffraction peaks, which indicate a high crystallinity and a more perfect ordering. Comparatively, calcium alginate fibers give broad diffraction peaks, pointing to defective crystals and low crystallinity [32]. According to Atkins's assignment, the Miller indexes for the two main diffraction peaks of alginate acid (110) and (002) are at 2θ of ~13.4° and ~22.9°, which correspond to *d*-spacing of 0.69 and 0.39 nm, respectively, while the shoulder at the 2θ of about 10.6° corresponds to the (001) position [33].

In FTIR spectra, -C-H str of mannuronic acid functional group is assigned at wavenumber 884 cm^{-1} and the uronic acid at wavenumber 939 cm^{-1}, while the asymmetric stretching of carboxylate -O-C=O str is at 1603 cm^{-1} and C–OH deformation vibration with contribution of O–C–O symmetric stretching vibration of carboxylate group is present at 1410 cm^{-1} (occurrence of carboxyl and carbonyl groups, typical to alginate); -OH functional group has peak at wavenumber 3200–3400 cm^{-1}, and $-CH_2$ stretching bands are present at wavenumber 2928 cm^{-1} [22]. FTIR spectroscopy has also been used as a tool to determine chemical composition of the M/G ratio in the alginates [35–36]. Separation of M-block and G-block from sodium alginate as highly purified polymannuronic acid and polyguluronic acid was possible through acid hydrolysis using oxalic acid, followed by isolation using the difference of solubility with pH value. In the FTIR spectra, the ratio of specific absorbance combinations (A_{1030}/A_{1080}) and (A_{1010}/A_{1025}) in the spectra of Ca and Mn salts of alginic acid (presence of cations enhances the specific peaks) indicates the M/G ratio of the sample.

Historically, the determination of monomer content began with total acid hydrolysis of the glycosidic bonds followed by separation techniques such as paper chromatography, thin layer chromatography, anion-exchange liquid chromatography, and gas–liquid chromatography [37]. Nowadays, the most common method used for the structural analysis of sodium alginates is ^{1}H and ^{13}C solution-state nuclear magnetic resonance (NMR) spectroscopy. From solution-state NMR data, the G/M ratio can be measured and in-depth sequence information (fractions of GG, MM, GGM, MGM, etc.) can be gained [38].

The chemical shifts in ^{13}CNMR spectra assigned to distinctive peaks of mannuronate (MM) and guluronate (GG) in each block are as follows (in ppm): $\delta_{MM\text{-}1} = 98.47$, $\delta_{MM\text{-}2} = 68.78$, $\delta_{MM\text{-}3} = 70.06$, $\delta_{MM\text{-}4} = 76.05$, $\delta_{MM\text{-}5} = 74.40$, $\delta_{MM\text{-}6} = 173.99$, $\delta_{GG\text{-}1} = 99.71$, $\delta_{GG\text{-}2} = 63.55$, $\delta_{GG\text{-}3} = 67.71$, $\delta_{GG\text{-}4} = 78.99$, $\delta_{GG\text{-}5} = 68.58$, $\delta_{GG\text{-}6} = 174.21$ [39].

In proton NMR spectra, the characteristic peak of the guluronic acid anomeric proton (G-1) occurs at 5.06 ppm (peak I); guluronic acid H-5 (G-5) at 4.4 ppm (peak III); and mannuronic acid anomeric proton (M-1) and the C-5 of alternating blocks (GM-5) overlapped at 4.7 ppm (peak II) [40, 41]. The individual guluronic acid (F_G) and the doublet (F_{GG}) fractions can be calculated from the area under the peak (I to III) using the following equation.

$$F_G = A_I/(A_{II} + A_{III});\ F_{GG} = A_{III}/(A_{II} + A_{III}) \tag{2.1}$$

The fraction M can be determined by Equation (2.2),

$$F_M = 1 - F_G \text{ and the M/G ratio is given by } M/G = (1 - F_G)/F_G \tag{2.2}$$

The disadvantage of solution-state NMR of alginates is that it needs partial acid hydrolysis, which results in tedious sample preparation and the chance for some insoluble material to be lost to quantitation [40].

2.3.3 Rheology and Mechanical Characterization of Alginate Gels and Solutions

Since alginates have a gelling behavior, the physics of deformation and flow of matter is of interest. The rheological behavior of alginates in aqueous dispersions was determined at various alginate concentrations and temperatures in the ranges 0.12.5–1.5% w/v and 278.16–308.16 K, respectively. The shear diagrams displayed typical pseudo plastic behavior. The effect of temperature on the consistency index (K) at constant alginate concentration was explained by a better-fitting power-law model than the Arrhenius relationship. Estimation of the thickening capability of any alginate solely based upon its average molecular weight was possible. The apparent viscosity of pseudo plastic alginate solutions depends on concentration, shear rate, and temperature. A 0.5% medium viscosity Na-alginate solution behaved in a Newtonian manner at low shear rates 1–100 sec^{-1}, but changed to pseudo plastic behavior only at high shear rates greater than 1000 sec^{-1} [42].

Mechanical properties, as measured by uniaxial compression and stress relaxation tests, for a range of high-M or high-G alginate gels, prepared from the alginate concentration in the range 0.75 ± 1.5% w/w, could be empirically modelled [43]. The alginates were fundamentally characterized by their percentage composition of G- and M-residues using NMR spectroscopy and by their viscosity. The specimen sizes (cross-section diameter of around 23 mm and height of 25 mm) were subjected to testing on Instron Universal testing machine with a 100 N load cell. Results showed that the yield stress (σ_{ER}) at failure increased along with increase in the effective alginate content, in high-G alginates, in concentration, and Mn, the average molecular weight values. The corresponding rupture strain was seen to be independent of alginate concentration and almost constant $\varepsilon_{ER} = 0.536$ for high-M alginates and $\varepsilon_{ER} = 0.503$ for high-G alginates. Square cylindrical specimens of alginate gels at 1.0% and 1.5% w/w nominal concentrations were loaded at a constant deformation rate of 2 mm sec^{-1} to assess the effect of stress relaxation on the compressive stress ± strain relationship. After a strain of 25%—about 50% of the average strain at failure for the gels tested—was generated, the cross-head was stopped and the stress relaxation was monitored for as much as 30 min (far longer than the time required [0.05 min] to deform the specimens). M-rich alginates formed gels with a greater residual stress (A_0), practically constant and equal to about 34% of the initial relaxation stress G (0), while G-rich ones formed gels with a smaller residual stress of the order of ca. 9% of G (0). High-G alginate gels were stronger and brittle owing to the higher G (0) and lower A_0 values.

With respect to mechanical properties, variation in the compression modulus of alginate gels and their shear moduli is due to the alginate composition, cross-linked nature, gelling conditions, and storage environment. At lower temperatures, Ca^{2+} cross-linking of alginate proceeds slowly, forming more ordered networks, with improved mechanical properties [44].

2.4 Modification of Alginates

2.4.1 Chemical Modification

Alginates have both hydroxyl and carboxyl functional groups, which are the sites for modification, and the presence of these polar moieties makes alginates hydrophilic in nature. Alginates can undergo chemoselective or stereoselective reaction due to the chemical reactivity of the functional group. It is essential to know completely the exact chemical reaction that happens in the modification process, as otherwise controlled modification is difficult to achieve. The sensitivity of alginates with acidic, basic, and reductive environments is also important while performing the reaction. Unfavorable conditions could ultimately lead to lowering the molecular weight and poor functional performance of the alginates. The chemical modification can be classified into two categories, which include modification in hydroxyl group and modification in carbonyl group.

2.4.2 Oxidation

The oxidation reaction creates aldehyde groups on the alginate. The reactivity of aldehydes is disparate from the alcohols and carboxyl groups, which would help in the synthesis of novel alginate derivatives. Generally, the oxidation is performed in aqueous media using periodate as a reagent. The oxidation reaction at C-2 and C-3 positions of the uronic units of sodium alginate with sodium periodate is shown in Figure 2.1. The number of aldehyde groups introduced can be reduced by limiting the concentration of the oxidant. Oxidation of alginate could possibly decrease the stiffness of the polymer [13]. Tyramine can attach to oxidized alginate by reductive amination that improves the immobilization of peroxidase enzyme onto Ca-alginate beads [45].

2.4.3 Sulfation

Sulfated alginates can be used as anticoagulants due to their structural similarities to heparin. The chemical sulfation of polysaccharide reaction should be under control; otherwise, excess sulfation can occur, which leads to various side effects. In order to avoid oversulfation, quaternary amine groups are attached to the alginates. The decreasing anticoagulant activity is in direct proportion to the number of quaternary amine groups attached onto the alginates.

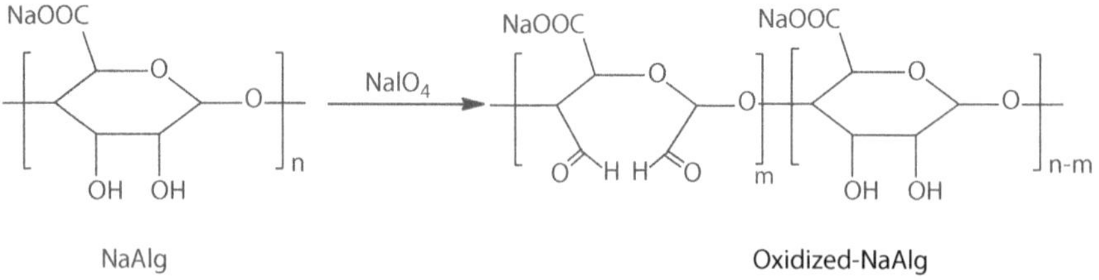

Figure 2.1 Oxidation of sodium alginate.

2.4.4 Phosphorylation

Generally, phosphorylation reaction in alginate occurs in a dipolar aprotic solvent with urea/phosphoric acid as a reagent system [46]. Decrease in molar mass and lowering of molecular weight happen during the course of reaction due to the phosphoric acid. The degree of substitution of the products is up to 0.26 by this method. The blending of phosphorylated alginate and native alginate gives physical gels that resist the leaching of calcium ions.

2.4.5 Graft Copolymerization

The physicochemical properties of the alginate were changed by grafting with synthetic polymers. Poly(acrylonitrile), poly(methyl acrylate), or poly(methyl methacrylate) was used with ceric ammonium nitrate as an initiator for the grafting of alginate. Polyacrylamide, when grafted onto alginate backbone, showed high molecular weight and functioned as a better flocculant in the sedimentation of coal, kaolin, and iron ore slurries (refer to Figure 2.2).

2.4.6 Esterification

Esterification is a facile method of attaching alkyl groups to molecules. Generally, esterification of alginates occurs with several alcohols in the presence of the catalyst. The addition of alkyl groups to the alginate increases the hydrophobic nature [47, 48]. The esterification of alginate with propylene oxide gives propene glycol esters of alginates, which showed commercially other alginates derivatives. Figure 2.3 depicts the esterification of alginate.

Butyl ester of alginate, a novel material, has the ability to encapsulate both hydrophilic and hydrophobic molecules. It is obtained from

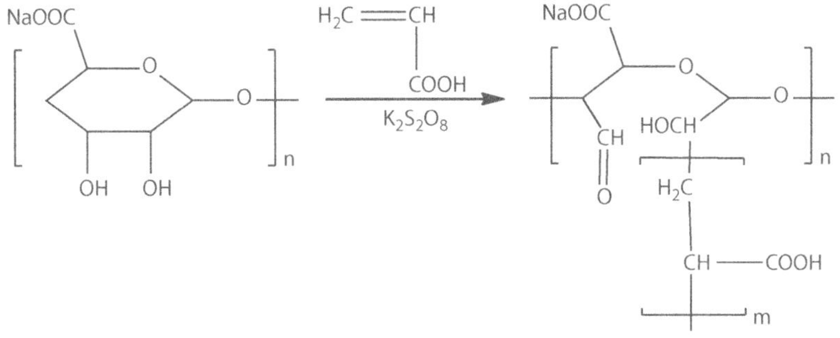

Figure 2.2 Copolymerization of sodium acrylate with sodium alginate.

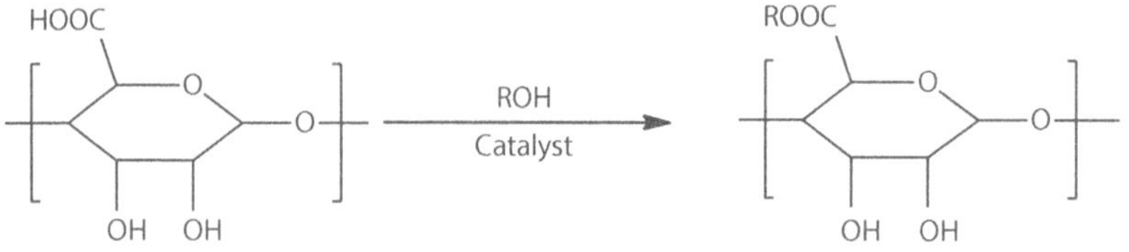

Figure 2.3 Esterification of alginate.

esterification of sodium alginate with butanol, catalyzed by the presence of sulfuric acid. It showed nontoxic properties similar to its native alginate [49].

2.4.7 Carbodiimide Coupling

It is one of the popular methods to form amide groups by connecting the carboxyl group of alginates with amine functional substituents. The reaction is carried out in an aqueous system, and this reaction does not need harsh environments like high acid or base. It has been possible to achieve the reversal of predominant hydrophobic interactions between the alkyl chains by mixing β-cyclodextrin with hydrophobically modified alginates [50]. β-Cyclodextrin is highly hydrophobic in its interior and external entities also showed similar hydrophobicity so the small size compounds can be encapsulated within the β-cyclodextrin core. Carbodiimide coupling can achieve the grafting of polyglycolic acid (PGLA) onto the alginate backbone. It has been suggested that graft of PGLA alginate is feasible by direct condensation of alginate and glycolic acid by carbodiimide coupling reagent [51].

2.4.8 Covalent Cross-Linking

Figure 2.4 depicts the cross-linking of alginate using a covalent cross-linker, epichlorohydrin. The covalently cross-linked alginate was obtained by reacting calcium alginate gels with epichlorohydrin in the presence of base catalyst. The calcium ions may be exchanged with sodium ions without

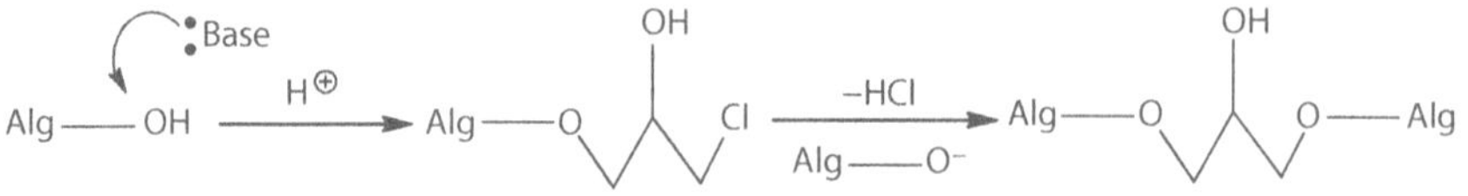

Figure 2.4 Alginate cross-linking using epichlorohydrin.

gel dissolution. The carboxylate group of the alginate is preserved in this method; these modified alginates can be used as ion exchange resin. The network of the gels is insoluble at high ionic strength and when exposed to polar solvents. Even the stability of these gels is intact at the wide range of pH of 1–13 and temperature up to 100°C. The water absorption capacity of the modified materials is larger than the native alginates [52]. When the covalent cross-linked alginate is additionally exposed to multivalent ion solutions, the gels' strength is higher.

The C-2 and C-3 positions of sodium alginate are cross-linked with another alginate using the bifunctional aldehyde such as glutaraldehyde in the presence of HCl to form a cross-linked membrane as shown in Figure 2.5. The glutaraldehyde cross-linked alginates can be used as a superabsorbent material for sanitary napkins and diapers [53]. The sodium alginate cross-linked membranes are also useful for the separation of isomers [54]. The efficiency of the separation depends on the degree of cross-linking. The covalently cross-linked alginate gels can be utilized to encapsulate active compounds in the pharmaceutical field.

There exist many potential modification pathways that are as yet unexplored, although extensive work on synthesis of alginate derivatives continues, and meanwhile, applications of synthetically modified alginate derivatives in various fields remain open to trial [55].

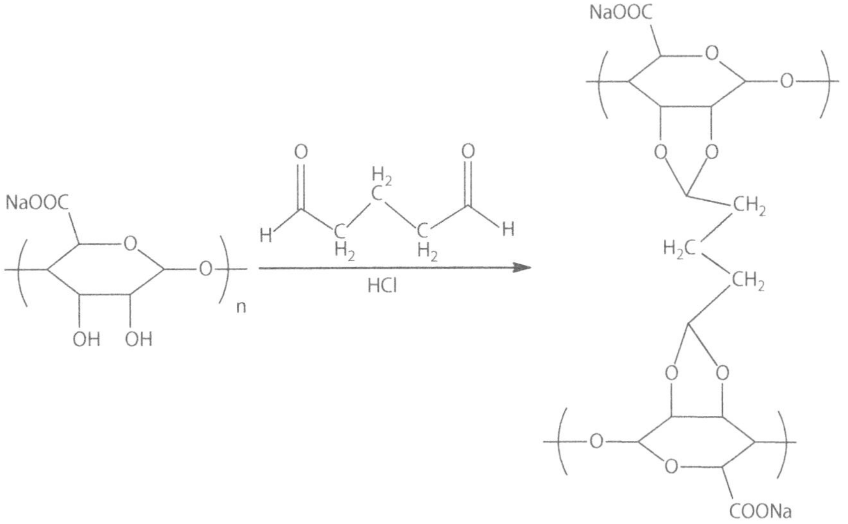

Figure 2.5 Alginate cross-linking using glutaraldehyde.

2.5 Future Perspectives

Alginate has shown significant utility and promise as a biomaterial for innumerable biomedical applications, specifically related to wound healing, drug delivery, *in vitro* cell culture, and tissue engineering. The exact suitability and match of the properties of alginate for these applications depend on its biocompatibility with negligible toxicity to mammalian tissue, gelation under mild reaction conditions, and the ease of effecting modifications to prepare alginate derivatives with requisite tuned properties. Alginate has a history of safe clinical use in wound healing dressing material and as a pharmaceutical ingredient, and has been implanted in different organ sites, including islet transplantation for treatment of type 1 diabetes [55] and chondrocyte transplantation for treatment of urinary incontinence and vesicoureteral reflux with minimal immunotoxic repercussions [56]. A chemically modified alginate has also been widely used as a carrier for progenitor cells and associated growth factors to heal periodontal damage by replacement. Gels suitable for each application may be generated by a judicious choice of available cross-linking strategies, in conjunction with molecules with favorable chemical structures, molecular weights, and apt cross-linking functionality. Many of the cross-linking and synthetic modification strategies are also employed in cell encapsulation strategies. It is known that covalent cross-linking reactions can be toxic to the cells encapsulated, and a careful use of cell-compatible chemical reagents (e.g., initiator) and complete removal of unreacted reagents and by-products need to be ensured in such cell delivery applications. Looking forward, the use of alginate-based materials in medicine is likely to expand. Alginate gels are much in use clinically in wound-healing sponges, where they act passively absorbing fluid and serving as a basis structure for new cell growth. Next-generation dressings could have greater active function. Single or multiple bioactive drugs and proteins that accelerate wound healing can be embedded into alginate dressings, as these gels support controlled release in the tissue microenvironment sustaining local concentrations of biological factors, such as proteins, for the required time duration. For the purposes of wound healing, and drug delivery, the ability to possess fine control over the delivery of number and dosage of drugs, such as constant or pulsed release in response to the external physiological conditions, is much favored. Active control over delivery can enhance the safety and efficacy of drugs, and can provide new therapeutic solutions. Modification of alginate scaffolds with RGD (arginylglyclaspartic acid) peptide motifs has been widely employed to function as cell adhesion ligand, multiple ligands, and/or a combination of ligands, and soluble factors which may be needed to promote layered growth of cells in replacement tissues and organs. Expanding our present understanding of the core properties of alginate, and developing innovative types of cell and tissue-interactive alginate gels, would cultivate future growth in biomedical science and engineering.

2.6 Conclusions

Recent developments in the control of alginate biosynthesis in microbes, and the comparative simplicity of bacteria modification, may allow production of alginate with tailor-made characteristics, suitable to specific application in biomedical procedures and products. X-ray diffraction and infrared spectroscopy have been used to delineate the structure of alginates, and more advanced analytical methods are being standardized. Characterization of properties such as cross-linked nature, which affect solubility at different pH, and storage stability of dried powder forms has become essential to achieve pharmaceutical, biomedical, and food processing applications, with the present commercial availability of widely varying grades of alginic acid and sodium, calcium salts of alginic acid from different sources and extraction procedures. A significant portion of research has targeted the chemical modification of alginate, at the free hydroxyl and carboxyl groups occurring along the backbone, to favorably change the viscosity, gelling, water absorption, and mechanical strength relative to the native alginates. However, only a few chemically modified derivatives of alginate such as modified esters, amides, and specifically propylene glycol alginate have come into industrial use, and many more are expected to mainstream into commercial use in the near future in biomedical, health, and cosmetic products.

References

1. Gombotz, W.R. and Wee, S., Protein release from alginate matrices. *Adv. Drug Deliv. Rev.*, 31, 267–285, 1998.
2. Vacanti, J.P. and Langer, R., Tissue engineering: The design and fabrication of living replacement devices for surgical reconstruction and transplantation. *Lancet*, 354, S32–S34, 1999.
3. Lee, K.Y. and Mooney, D.J., Hydrogels for tissue engineering. *Chem. Rev.*, 101, 1869–1880, 2001.

4. Sculean, A., Auschill, T.M., Donos, N. *et al.*, Effect of an enamel matrix protein derivative (Emdogain) on *ex vivo* dental plaque vitality. *J. Clin. Periodontol.*, 28, 1074–1078, 2001.
5. Lee, K.Y., Peters, M.C., Anderson, K.W. *et al.*, Controlled growth factor release from synthetic extracellular matrices. *Nature*, 408, 998, 2000.
6. Zhao, X., Kim, J., Cezar, C.A. *et al.*, Active scaffolds for on-demand drug and cell delivery. *Proc. Natl. Acad. Sci.*, 108, 67–72, 2011.
7. Petka, W.A., Harden, J.L., McGrath, K.P. *et al.*, Reversible hydrogels from self-assembling artificial proteins. *Science (80-)*, 281, 389–392, 1998.
8. MacKay, J.A., Chen, M., McDaniel, J.R. *et al.*, Self-assembling chimeric polypeptide–doxorubicin conjugate nanoparticles that abolish tumours after a single injection. *Nat. Mater.*, 8, 993, 2009.
9. Nagarajan, A., Shanmugam, A., Zackaria, A., Mini review on alginate: Scope and future prospectives. *J. Algal. Biomass Utln.*, 7, 45–55, 2016.
10. Rhein-Knudsen, N., Ale, M.T., Meyer, A.S., Seaweed hydrocolloid production: An update on enzyme assisted extraction and modification technologies. *Marine Drugs*, 13, 6, 3340–3359, 2015.
11. Mchugh, D.J., *A Guide to the Seaweed Industry*, FAO Fisheries Technical Paper 441; Food and Agriculture Organization of the United Nations, Rome, Italy, 2003.
12. Thomas, S., Alginate dressings in surgery and wound management—Part 1. *J. Wound Care*, 9, 56–60, 2000.
13. Gomez, C.G., Rinaudo, M., Villar, M.A., Oxidation of sodium alginate and characterization of the oxidized derivatives. *Carbohydr. Polym.*, 67, 296–304, 2007.
14. Sabra, W. and Zeng, A.P., Microbial production of alginates: Physiology and process aspects, in: *Alginates: Biology and Applications*. pp. 153–173, Springer, 2009.
15. Kidambi, S.P., Sundin, G.W., Palmer, D.A. *et al.*, Copper as a signal for alginate synthesis in *Pseudomonas syringae* pv. *syringae*. *Appl. Environ. Microbiol.*, 61, 2172–2179, 1995.
16. Sabra, W.A., Zeng, A.P., Lünsdorf, H. *et al.*, Function and variation of alginate production in *Azotobacter vinelandii* under nitrogen fixation conditions. *Appl. Environ. Microbiol.*, 66, 4037–4044, 2000.
17. Brivonese, A.C. and Sutherland, I.W., Polymer production by a mucoid stain of *Azotobacter vinelandii* in batch culture. *Appld. Microbiol. Biotechnol.*, 30, 97–102, 1989.
18. Chen, W.-P., Chen, J.-Y., Chang, S.-C. *et al.*, Bacterial alginate produced by a mutant of *Azotobacter vinelandii*.. *Appl. Environ. Microbiol.*, 49, 543–546, 1985.
19. Young, M., Microbial polysaccharide, in: *Comprehensive Biotechnology*, vol. 3, pp. 1005–1044, Pergamon Press, Toronto, 1983.
20. Tønnesen, H.H. and Karlsen, J., Alginate in drug delivery systems. *Drug Dev. Ind. Pharm.*, 28, 621–630, 2002.

21. Szekalska, M., Puciłowska, A., Szymańska, E. *et al.*, Alginate: Current use and future perspectives in pharmaceutical and biomedical applications. *Int. J. Polym. Sci.*, 2016.
22. Larsen, B., Salem, D.M.S.A., Sallam, M.A.E. *et al.*, Characterization of the alginates from algae harvested at the Egyptian Red Sea coast. *Carbohydr. Res.*, 338, 2325–2336, 2003.
23. Haug, A., Larsen, B., Smidsrød, O., Uronic acid sequence in alginate from different sources. *Carbohydr. Res.*, 32, 217–225, 1974.
24. Smidsrod, O., Haug, A., Lian, B., Properties of poly (1, 4-hexuronates) in gel state. 1. Evaluation of a method for determination of stiffness. *Acta Chem. Scand.*, 26, 71, 1972.
25. Gacesa, P., Alginates. *Carbohydr. Polym.*, 8, 161–182, 1988.
26. Hecht, H. and Srebnik, S., Structural characterization of sodium alginate and calcium alginate. *Biomacromolecules*, 17, 2160–2167, 2016.
27. Stokke, B.T., Draget, K.I., Smidsrød, O. *et al.*, Small-angle x-ray scattering and rheological characterization of alginate gels. 1. Ca– alginate gels. *Macromolecules*, 33, 1853–1863, 2000.
28. Imai, T., Kawasaki, C., Nishiyama, T. *et al.*, Comparison of the pharmaceutical properties of sustained-release gel beads prepared by alginate having different molecular size with commercial sustained-release tablet. *Pharmazie*, 55, 218–222, 2000.
29. Efentakis, M. and Koutlis, A., Release of furosemide from multiple-unit and single-unit preparations containing different viscosity grades of sodium alginate. *Pharm. Dev. Technol.*, 6, 91–98, 2001.
30. Efentakis, M. and Buckton, G., The effect of erosion and swelling on the dissolution of theophylline from low and high viscosity sodium alginate matrices. *Pharm. Dev. Technol.*, 7, 69–77, 2002.
31. Penman, A. and Sanderson, G.R., A method for the determination of uronic acid sequence in alginates. *Carbohydr. Res.*, 25, 273–282, 1972.
32. Rees, D.A., Shapely polysaccharides. The eighth Colworth medal lecture. *Biochem. J.*, 126, 257, 1972.
33. Liew, C.V., Chan, L.W., Ching, A.L. *et al.*, Evaluation of sodium alginate as drug release modifier in matrix tablets. *Int. J. Pharm.*, 309, 25–37, 2006.
34. Mackie, W., Perez, S., Rizzo, R. *et al.*, Aspects of the conformation of polyguluronate in the solid state and in solution. *Int. J. Biol. Macromol.*, 5, 329–341, 1983.
35. Atkins, E.D.T., Nieduszynski, I.A., Mackie, W. *et al.*, Structural components of alginic acid. I. The crystalline structure of poly-β-D-mannuronic acid. Results of X-ray diffraction and polarized infrared studies. *Biopolymers*, 12, 1865–1878, 1973.
36. Sakugawa, K., Ikeda, A., Takemura, A. *et al.*, Simplified method for estimation of composition of alginates by FTIR. *J. Appl. Polym. Sci.*, 93, 1372–1377, 2004.
37. Usov, A.I., Alginic acids and alginates: Analytical methods used for their estimation and characterisation of composition and primary structure. *Russ. Chem. Rev.*, 68, 957–966, 1999.

38. Sperger, D.M., Fu, S., Block, L.H. *et al.*, Analysis of composition, molecular weight, and water content variations in sodium alginate using solid-state NMR spectroscopy. *J. Pharm. Sci.*, 100, 3441–3452, 2011.
39. Grasdalen, H., Larsen, B., Smisrod, O., 13C-NMR studies of monomeric composition and sequence in alginate. *Carbohydr. Res.*, 89, 179–191, 1981.
40. Grasdalen, H., High-field, 1H-nmr spectroscopy of alginate: Sequential structure and linkage conformations. *Carbohydr. Res.*, 118, 255–260, 1983.
41. Torres, M.R., Sousa, A.P.A., Silva Filho, E.A.T. *et al.*, Extraction and physicochemical characterization of *Sargassum vulgare* alginate from Brazil. *Carbohydr. Res.*, 342, 2067–2074, 2007.
42. Salomonsen, T., Jensen, H.M., Stenbæk, D. *et al.*, Chemometric prediction of alginate monomer composition: A comparative spectroscopic study using IR, Raman, NIR and NMR. *Carbohydr. Polym.*, 72, 730–739, 2008.
43. Mancini, M., Moresi, M., Sappino, F., Rheological behaviour of aqueous dispersions of algal sodium alginates. *J. Food Eng.*, 28, 283–295, 1996.
44. Augst, A.D., Kong, H.J., Mooney, D.J., Alginate hydrogels as biomaterials. *Macromol. Biosci.*, 6, 623–633, 2006.
45. Prodanovic, O., Spasojevic, D., Prokopijevic, M. *et al.*, Tyramine modified alginates via periodate oxidation for peroxidase induced hydrogel formation and immobilization. *React. Funct. Polym.*, 93, 77–83, 2015.
46. Coleman, R.J., Lawrie, G., Lambert, L.K. *et al.*, Phosphorylation of alginate: Synthesis, characterization, and evaluation of *in vitro* mineralization capacity. *Biomacromolecules*, 12, 889–897, 2011.
47. Leonard, M., De Boisseson, M.R., Hubert, P. *et al.*, Hydrophobically modified alginate hydrogels as protein carriers with specific controlled release properties. *J. Control Release*, 98, 395–405, 2004.
48. Pelletier, S., Hubert, P., Payan, E. *et al.*, Amphiphilic derivatives of sodium alginate and hyaluronate for cartilage repair: Rheological properties. *J. Biomed. Mater. Res. Part A*, 54, 102–108, 2001.
49. Broderick, E., Lyons, H., Pembroke, T. *et al.*, The characterisation of a novel, covalently modified, amphiphilic alginate derivative, which retains gelling and non-toxic properties. *J. Colloid Interface Sci.*, 298, 154–161, 2006.
50. Burckbuchler, V., Kjøniksen, A.-L., Galant, C. *et al.*, Rheological and structural characterization of the interactions between cyclodextrin compounds and hydrophobically modified alginate. *Biomacromolecules*, 7, 1871–1878, 2006.
51. Chen, M.-J. and Shi, Q.-S., Transforming sugarcane bagasse into bioplastics via homogeneous modification with phthalic anhydride in ionic liquid. *ACS Sustain. Chem. Eng.*, 3, 2510–2515, 2015.
52. Moe, S.T., Skjaak-Braek, G., Elgsaeter, A. *et al.*, Swelling of covalently crosslinked alginate gels: Influence of ionic solutes and nonpolar solvents. *Macromolecules*, 26, 3589–3597, 1993.

53. Kim, Y., Yoon, K., Ko, S., Preparation and properties of alginate superabsorbent filament fibers crosslinked with glutaraldehyde. *J. Appl. Polym. Sci.*, 78, 1797–1804, 2000.
54. Yang, J.-S., Xie, Y.-J., He, W., Research progress on chemical modification of alginate: A review. *Carbohydr. Polym.*, 84, 33–39, 2011.
55. Soon-Shiong, P., Heintz, R.E., Merideth, N. *et al.*, Insulin independence in a type 1 diabetic patient after encapsulated islet transplantation. *Lancet*, 343, 950–951, 1994.
56. Diamond, D.A. and Caldamone, A.A., Endoscopic treatment of vesicoureteric reflux in children using autologous chondrocytes—Preliminary results. *Pediatrics*, 102, 868, 1998.

3

Alginate: Recent Progress and Technological Prospects

Tanvir Arfin* **and Kamini Sonawane**

Environmental Materials Division, CSIR-NEERI, Nehru Marg, Nagpur, Maharashtra, India

Abstract

Alginate, a form of biopolymer, is used for industrial application due to its properties such as biocompatibility, nontoxicity, and biodegradability. It is called alginic acid. It can absorb water very quickly. Therefore, it is employed as an essential ingredient in manufacturing industries such as paper industry, textile industry, and food industry as a source of drinks, cosmetic products, ice cream, etc. It forms a hydrogel in the presence of millimolar concentration of calcium or even in the divalent cation such as barium or strontium. The properties of alginate are affected by its monomeric composition and molecular weight. Hence, necessary approaches are applied to yield alginates of different molecular weights, and reproducible physicochemical features are manipulating the culture state of fermentation. This chapter relates to the dynamic properties of alginate and also the innovative aspect of applying the alginate with various materials, such as nanomaterials or polymers.

Keywords: Alginate, alginate salts, calcium alginate, dietary fiber, cancer, cardiovascular, Parkinson's disease, marine pollution

3.1 Introduction

Polymers play a significant role to survive better in our daily lives. They are mainly made up of macromolecules comprising a large number of repeated monomers. They can be natural or synthetic in nature [1–4]. In the current era of science, polymer and its derivatives are considered necessary ingredients for good livelihood by replacing different artificial sources [5, 6]. There are various applications of polymers generally in different

**Corresponding author*: tanvirarfin@gmail.com

Shakeel Ahmed (ed.) Alginates, (45–58) © 2019 Scrivener Publishing LLC

fields such as domestic application, in aerospace, etc. Alginates are simply anionic polysaccharides in linear form. It is also called as alginic acid. But sometimes all derivatives of alginic acid are also called alginates. They are renewable polymers with two monomers.

3.2 Structure

Monomers are structures composed of six members where five are carbon atoms and one is oxygen. These monomers are also called pyranoses. These are composed of β(1-4) d-mannuronic acids and its C-5 epimer α-l-guluronic acid (G) as shown in Figure 3.1. This structure is arranged in a block form

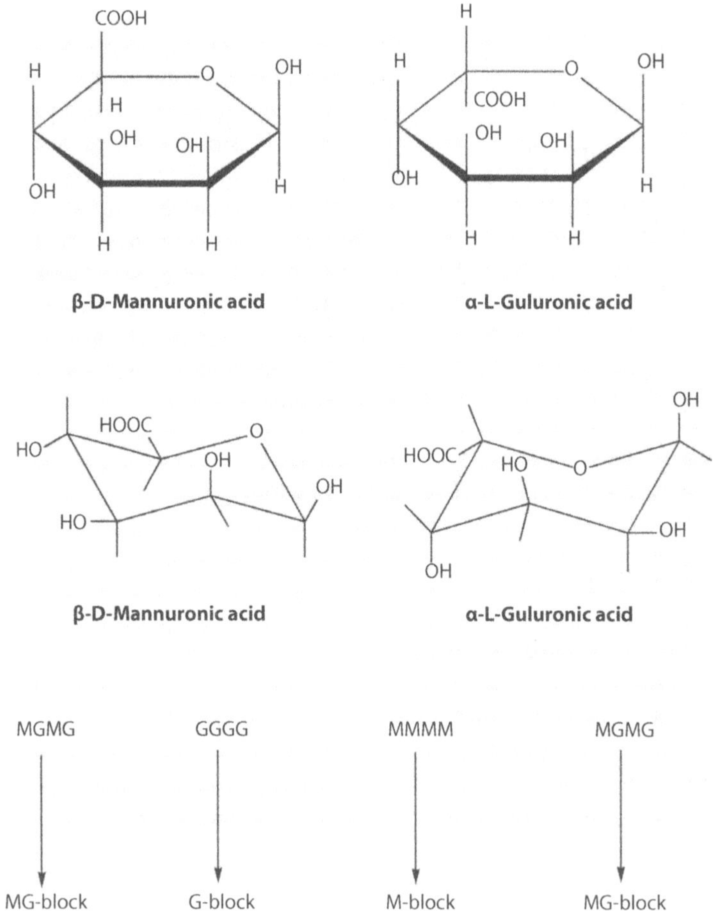

Figure 3.1 Alginate monomers in block distribution.

as (1-4) heterogeneous or homogenous pattern in which blocks relate to "G" and "M." This structure is present in the tetrahydropyran ring form. The sources of isolation are organs and tissues; thus, they differ in sequence as well as composition [7].

3.3 Sources

These are extracted from the plant's materials like starch and cellulose. But the production of alginate is also done by using algal source. Nowadays, the production is preferred from fermentation process through microbes for increasing their properties and applications and to provide a more advanced form of alginate. The bacterial genera and seaweed are required to produce this polysaccharide. Mostly the seawoods become the source when they are large brown algae. Specifically *Pseudomonas aeruginosa* and *Azotobacter vinelandii* are used as extracellular polymers [8]. The difference between the bacterial and algal alginates is that the bacterial alginate is acetylated. These bacterial alginates are not used mostly, but they can be the best alternative. Prolonged presence of seaweed alginates produces marine pollution; hence, use of bacterial alginates is the best approach. Bacterial alginate is available in the form of powder and paste. It can be mixed with water for the reaction. It can also be considered as a renewable resource and can be derived from brown algae [9]. It gives flexibility and strength to algal tissues. Alginate extraction is done by drying the raw material and then treating it with diluted mineral acid. Purification is done and it is converted into a water-soluble form [10]. The process of extraction was done in 1881 by Stanford. The product made was crude containing nitrogen in high levels. The procedure is shown in Figure 3.2.

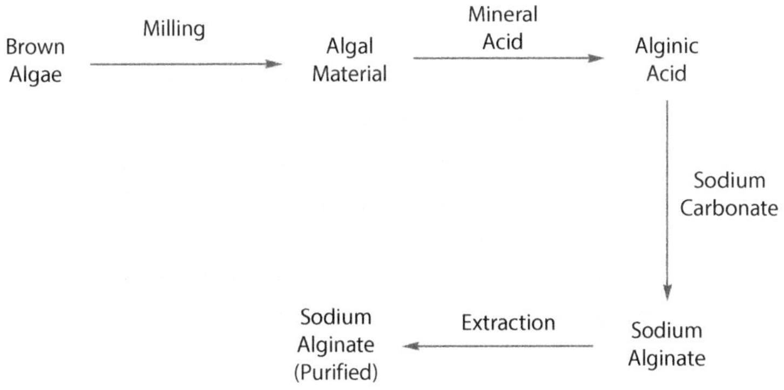

Figure 3.2 Procedure of alginate.

Alginates are just the salts of alginic acid. They are available in different forms such as ammonium alginate, sodium alginate, propylene glycol alginate, and calcium alginate. These are also available in different compositions and molecular grades; their patterns are M and G blocks. Due to such special features, it clarifies water solubility, sol–gel properties, and biological and physiochemical properties. The molecular form influences greatly as it ranges from 33,000 to 400,000 g/mol.

The preparation of alginate composite microparticles or polyglutamic acid microparticles can be done by coating or dropping the alginate mixture into a $CaCl_2$ solution. This is performed rapidly, but the disadvantage is that the size of the microparticle is not controllable. This uncontrolled and rapid formation is caused by Ca^{2+} ions. For this, the encapsulation method is the most suitable to prepare the alginate composite microparticles.

3.4 Characteristics of Alginate Salts

- *Presence of water.* Calcium salts and sodium salts generally have the same decomposition temperature of about 458 K and weight loss. Water content varies with salt. This means that there are two water molecules in the calcium and sodium salts of alginates, but in other salts, the amount decreases.
- *Crystallinity.* Compared to sodium salts, other salts do not have greater influence. There is no more change on the addition of other salts. To know this property, characterization is done by X-ray diffraction technique.
- *Measurement of mechanical strength.* This can be studied during storage or before wound dressing. Mechanical tensile stress involves Young's modulus, fracture stress, and fracture strain.

3.5 Properties

Alginate is well known for its properties like nontoxicity, biodegradability, cross-linking capability, pH sensitivity, and versatile chemical nature.

- Processes like preparation, purification, product formation, swelling, and degradation require pH at a specific range, which means pH should be maintained. pH variation is responsible for the change in swelling and polymer chain expansion. pH also affects the ionized and unionized form of carboxylic group.

- It is biocompatible and consists of hemostatic property, and therefore, mostly it can be applied for hemostatic wound dressing [11]. This is compatible by stimulating the clotting factor and shortening the time of blood coagulation. This wound dressing can be removed completely with the help of weak base or saline solution, but it will cause some pain. If the fibers remain with the wound, they will biodegrade, so complete removal of dressing is not necessary. It is the reason why the United States Food and Drug Administration has declared alginate as a grass or safe material.
- Mucoadhesiveness is the property of a material that adheres the mucous layer with the help of some forces like hydrogen bonds, electrostatic forces, hydrophobic interactions, and ligand-receptor actions. [12]. This mucoadhesion property depends on the chemical structure and nature of a material, so that interactions will be better with the mucous layer. This property is important in the sense of drug delivery treatment as drug absorption is essential.
- This alginate powder is not stable. This means in warm temperature or moisture conditions, it will create a problem while fast reaction is expected. It shows good details of the surface. The reaction becomes faster by the use of alginate.
- The best property is that this is not a toxic one and also does not cause any irritation. It shows hydrophilic nature, and this will make it more suitable to be used in the existence of blood and saliva. The properties of alginates get affected by composition and molecular weight [13].
- It can be directly converted into a solid or semisolid form because of their sol–gel property. The content of guluronic acid is responsible for the formation of a gel. This is the reason why alginate is used as a thickener, a suspension agent, or a stabilizer in emulsion process and also to increase viscosity especially in pharmaceutical and food industries.
- Viscosity and gel-forming capacity of alginate are most important. After this, the attention is toward the swelling property. With the increase in the cross-link density of alginate, gel swelling increases, based on the study by Moe *et al.*, [14]. Due to its swelling property, it can be used in a wider range. The point of concern is that high concentration of G increases the porosity but decreases the shrinkage. On the opposite side, high M concentration increases shrinkage and

decreases porosity. Thus, the concentration of guluronic acid should be maintained greater than 60% and not very high. The properties have much influence on its presence, use, quality, and potential applicability.

3.6 Applications

Alginate is a multifunctional substance. Large numbers of opportunities are there for alginate in medical and industrial fields. There is a large range of synthetic materials available for the treatment and protection of a bioactive agent. Among those, alginate serves as the naturally available material. Alginate composites or blends can be made into different forms as beads, microparticles, films, matrix, sponge, fibers, nanoparticles, etc. where different types are applicable in wide fields. These can be used in different fields such as pharmaceutical, food, paper making, adhesive, biomedicines, plastics, etc. as they possess versatile and specific physiochemical and biological properties. The alginate is an irreversible and elastic hydrocolloid impression substance. There are several reasons why alginate is preferable, including its solubility in water at normal room temperature, for gel formation does not need heating and cooling cycles. Thus, it serves as the best material for impression in dental practices, as it is simple in impression, low cost, and essential. It is applicable not only in dental practices but also in other clinical fields. Encapsulation is the process for which alginate serves as the best carrier. It acts as a matrix in case of bioartificial organs. The properties of alginate include promotion of cell proliferation, and biocompatibility makes it easy to use in different fields such as tissue engineering, wound dressing, and delivery of drugs. Biodegradability of alginate increases its application in cosmetic and food industries. The use of nonbiodegradable polymers, which are synthetic, can be replaced by producing a superabsorbent using such natural polysaccharides [15]. There are many ways available to prepare composites using alginate.

For the case of dental impression, it requires an alginate-like material. This material settles faster, is easy to use, and is economical. It can be operated easily by maintaining the temperature of liquid like water; the setting time can be managed.

In food industries, the use of alginic acid and their derivatives can be summarized as follows:

- Alginic acid: as a thickener, emulsifier, formulation aid, and stabilizer.

- Sodium alginate: in addition to above application, a texturizer, surface active agent, firming agent, etc.
- Ammonium alginate: a humectant, thickener, and stabilizer.
- Propylene glycol alginate: a flavoring adjuvant and surfactant.
- Calcium alginate: a stabilizer and thickener [16].

In pharmaceutical industries, the applications of alginate and their derivatives are as follows:

- Alginic acid: as a suspension and viscosity increasing agent, binder and disintegrates the modifying agent.
- Sodium alginate: a capsule and tablet disintegrating agent, sustained releasing agent, tablet binder, thickener, and in capsule formation, it can be used as a diluent.
- Ammonium alginate: an emulsifier, film former, color diluent, and humectant.
- Propylene glycol alginate: viscosity and suspension increasing agent.
- Calcium alginate: a tablet disintegrating agent.

Another way to increase the physiochemical properties is to form alginate hydrogels. Cell cross-linking, ionic cross-linking, free radical polymerization, phase transition, and covalent cross-linking are methods of cross-linking for alginate [17]. Wide applications have been studied for alginate in different fields including pharmaceutical, drug delivery, *in situ* gel formation, medical purposes, and controlled release [18]. For drug delivery, the alginate microparticles are used, which can be made by emulsification, droplet method, or atomization process. This drug delivery includes the encapsulation of growth factors, enzymes, peptides, proteins, antibiotics, antilabile drugs, etc. [19]. There may be some problems like erosion and swelling, but they can be managed by cross-linking. It has great importance in the medical field as it can be used in the treatment of esophagus and heartburn, meaning it is used for symptomatic treatment. Alginate and its derivatives can be used for human body treatment, as it can float on gastric juice because it has the carbonate group, which is converted into carbon dioxide [20]. These alginate products remain in the stomach for several hours to give relief [21].

Thrombin generation and platelet activation are achieved by the use of calcium alginate. Alginates also have an advantageous point as they are able to capture by binding heavy metals and toxins present in the intestine, thereby protecting the cells from carcinogenic agents. On the other hand, alginate oligosaccharides show anti-inflammatory and anti-oxidant properties.

Cell immobilization can be carried out for different purposes including medical and pharmaceutical. In case of monoclonal antibody production, hybridoma cells can be immobilized. In bioreactors, living or dead cells can be immobilized for the production of a specific product. To implant artificial organs, animal cells can be immobilized. In case of plants also, seeds can be immobilized for plant propagation. It means in order to resolve technological problems, encapsulation is the biotechnological tool. And alginate fulfills all the requirements required for the immobilization technique.

The simple way of using alginates is as a dietary fiber. Great effects have been observed including the following: that it can lower the absorption rate of the small intestine and that it reduces the insulin rate. The risk of diabetes and cardiovascular disease decreases with the increase in the concentration of alginate in the diet. The chances of problems occurring due to potential damaging bacteria are reduced even when the colonic mucosae are exposed to such agents. It will also reduce the glucose absorption rate. Hypertension also gets relief and cholesterol is under control by taking a high sodium alginate diet. The alginate-containing diet of pregnant women, children, and infants gives potential benefits related to their health. It absorbs chemical and food mutagens and lowers colon exposure to harmful agents. As cross-linking is a property of alginate, it is applied greatly to overcome the problems occurring during encapsulation. All the applications have been discussed along with their properties related to immobilization from many years. But there are obstacles coated as alginate gels after immobilization where the metabolic activities and cell viability are not discussed in brief.

Cancer treatment is possible with anticancer drug delivery using alginate materials. As it possesses ligand receptor interaction, it will be helpful for cancer patients. One of the stressful topics of today is the increasing number of breast cancer patients, which may be caused by the environment or food habitats. For this, alginate nanoparticles serve their application. By using a simple gelation method, the EXE drug is loaded with the alginate nanoparticles. Then *in vitro* release of the drug is a way to treat cancer. Characterization techniques were used to study and analyze the encapsulated nanoparticles, which are done by XRD, FTIR, UV-spectrophotometer, and SEM. This approach is a novel one toward cancer treatment.

3.7 Future Perspectives

There will be the option of making three-dimensional cell culture systems. The synthetic and natural polymers, their derivatives, and composites can be used for making a 3-D culture system. The properties of alginate to make this type of system involve mechanical strength, the ability of gel formation, and also the way of interaction to react with the cell. Alginate poses the property of giving support to the immobilized cell, which makes it compatible and processed at room temperature. Because of this, the material is the best option for tissue and cell culture and therefore for 3-D system also. Proliferation, differentiation, and cell attachment can be favored by this 3-D culture system because it possesses alginate microporous scaffolds in dried form and is ionically gelled. This will help to convert scaffolds into hydrogels.

The alginates can be used in the microparticle forms for therapeutic purposes. The common method used in tissue engineering is that living cells are immobilized. Immobilization should be done under mild conditions and when cells are compatible. Encapsulation is the best approach made in the treatment of neurodegenerative tissues and diabetic diseases using alginate. This system is very selective and protects the encapsulated cells. In case of Parkinson's disease, the NT cell is encapsulated to recover and regenerate damaged tissues. It can also be used in the case of animals, where brain implantation can be done into the impaired site. These NT cells have the ability to produce nerve growth factors and cerebrospinal fluid. Also other serious diseases like Alzheimer and neuronal diseases can be cured or improved up to some extent by this method of treatment.

The biological activity of alginate oligosaccharides is another approach. These alginate oligosaccharides also act as postbiotic and prebiotic. This is because the material can promote the *Bifidobacterium* species growth, and at the same time, it reduces or inhibits the colonization of *Salmonella enteritidis* present in the intestine [22].

Immobilization is the technique used for large amounts nowadays. In pharmaceutical industries, food industries, agricultural production, environmental protection, and drug development, immobilization is the best approach. In case of the coating technique involving immobilization of coating, alginate gel still has some problems. Thus, future interest should be toward the modification and development of coating materials for alginate capsules, so that they will remain more viable and effective. For this, the studies should be done on encapsulation, metabolic activities, metabolism, genomics, transcription, and modern technologies. The way should be toward the development of covalent and ionic interactions and cross-link. Also attention should be given for developing the functionalized and modified alginate material and coating. Still there should be improvement done in the chemical stability and mechanical strength of alginates and their products.

3.8 Advantages

The following are the advantages of alginates:

- Biocompatibility
- Biodegradability
- Naturally available
- Large application in different areas

3.9 Disadvantages

The following are the disadvantages of alginates:

- They have bad dimensional stability. They serve as a bad material or messy material to work with.
- They have low chemical stability. This is because most of the time, alginates are cross-linked with calcium ions. And the solution to that is to increase the content of G so that stability will also increase.
- Instead of using original forms, their composites give better results.

3.10 Conclusion

Studies in the field of polymers are necessary since they are innovative materials possessing a number of uses owing to their extraordinary properties. Since polymer creation today, scientific research has gathered lots of interest due to their benefits towards the environment [23–25]. Besides the general properties and functions of alginates, there are various approaches available in drug delivery, pharmaceutical, and biomedical fields by use of encapsulation and alginate microparticles. Initially this was used for wound dressing, but now it is widely used in the medical field.

Properties of alginates such as gel formation, biocompatibility, and biodegradability made them special and opened the door to a vast range of fields. Different forms of alginate composites are helpful in the medical field to save and protect the life of patients, like microparticles, which are best for inflammatory bowel disease (IBD) patients. Dental practices have been done by making impressions with good results. This summarizes the advanced form of alginates in the biotechnological field by immobilizing or encapsulating process. The coating technique of alginate material provides information about the interaction of coating materials and alginates. Also in the pharmaceutical field, the capsule formation of alginates and observation of the effects of the coating technique on the cells and metabolic activities have been discussed. This chapter describes the activity of alginates in different fields. The modifications are shown in alginates, their derivatives, their composites, and physiochemical properties. The development of alginate nanoparticle for encapsulation of the drug EXE is for effective cancer treatment, which has been discussed. This was characterized by FTIR, XRD, and SEM techniques. This technique will serve as a good carrier for drugs.

Acknowledgments

The authors acknowledge the Knowledge Resource Centre, CSIR-NEERI, (CSIR-NEERI/KRC/2018/MAY/EMD/1) for their support.

References

1. Arfin, T., Mohammad, F., Yusof, N.A., Applications of polystyrene and its role as base in industrial chemistry, in: *Polystyrene: Synthesis, Characteristics and Applications*, C. Lynwood (Ed.), pp. 269–280, Nova Publishers, New York, USA, 2014.
2. Athar, S. and Arfin, R., Commercial and prospective applications of gelatin, in: *Natural Polymers: Derivatives, Blends and Composite*, vol. 11, S. Ikram and S. Ahmed (Eds.), pp. 199–216, Nova Publishers, New York, USA, 2017.
3. Arfin, T. and Tarannum, A., Polymer materials: From the past to the future, in: *Polymeric Materials: Advances and Sustainable Development*, S. Ahmed, Annu, S. Ikram (Eds.), pp. 35–52, Nova Publishers, New York, USA, 2017.
4. Arfin, T., Applications of bentonite and its role as a base of engineering chemistry, in: *Bentonite: Characteristics, Uses and Implications for the Environment*, A.K. Mishra (Ed.), pp. 127–136, Nova Publisher, New York, USA, 2015.

5. Arfin, T. and Mohammad, F., Dendrimer and its role for the advancement of nanotechnology and bioengineering, in: *Advances in Materials Science Research*, vol. 21, M.C. Wythers (Ed.), pp. 157–174, Nova Publishers, New York, USA, 2015.
6. Arfin, T. and Mohammad, F., Chemistry and structural aspects of chitosan towards biomedical applications, in: *Natural Polymers: Derivatives, Blends and Composites*, vol. 1, S. Ikram and S. Ahmed (Eds.), pp. 265–280, Nova Publishers, New York, USA, 2016.
7. Skjak-Braek, G., Smidsrød, O., Larsen, B., Tailoring of alginates by enzymatic modification *in vitro*. *Int. J. Biol. Macromol.*, 8, 330, 1986.
8. Rehm, B.H., Bacterial polymers: Biosynthesis, modifications and applications. *Nat. Rev. Microbiol.*, 8, 578, 2010.
9. Szekalska, M., Puciłowska, A., Szymańska, E., Ciosek, P., Winnicka, K., Alginate: Current use and future perspectives in pharmaceutical and biomedical applications. *Int. J. Polym. Sci.*, 2016, 7697031, 2016.
10. Andersen, T., Melvik, J.E., Alsberg, E., Christensen, B.E., Ionically gelled alginate foams: Physical properties controlled by operational and macromolecular parameters. *Biomacromolecules*, 13, 3703, 2012.
11. Andersen, T., Melvik, J.E., Gåserød, O., Alsberg, E., Christensen, B.E., Ionically gelled alginate foams: Physical properties controlled by type, amount and source of gelling ions. *Carbohydr. Polym.*, 99, 249, 2014.
12. Ponchel, G. and Irache, J.M., Specific and non-specific bioadhesive particulate systems for oral delivery to the gastrointestinal tract. *Adv. Drug Deliv. Rev.*, 34, 191–219.
13. George, M. and Abraham, T.E., Polyionic hydrocolloids for the intestinal delivery of protein drugs: Alginate and chitosan—A review. *J. Control. Release*, 114, 1, 2006.
14. Moe, A-, Skjaak-Braek, G., Elgsaeter, A., Smidsroed, O., Swelling of covalently cross-linked alginate gels: Influence of ionic solutes and nonpolar solvents. *Macromolecules*, 26, 3589, 1993.
15. Rashidzadeh, A. and Olad, A., Slow-released NPK fertilizer encapsulated by NaAlg-g-poly (AA-co-AAm)/MMT superabsorbent nanocomposite. *Carbohydr. Polym.*, 114, 269, 2014.
16. Ching, S.H., Bansal, N., Bhandari, B., Alginate gel particles—A review of production techniques and physical properties. *Crit. Rev. Food Sci. Nutr.*, 57, 1133, 2017.
17. Sun, J. and Tan, H., Alginate-based biomaterials for regenerative medicine applications. *Materials*, 6, 1285, 2013.
18. Ching, A.L., Liew, C.V., Chan, L.W., Heng, P.W.S., Modifying matrix micro-environmental pH to achieve sustained drug release from highly laminating alginate matrices. *Eur. J. Pharm. Sci.*, 33, 361, 2008.
19. Hampson, F.C., Farndale, A., Strugala, V., Sykes, J., Jolliffe, I.G., Dettmar, P.W., Alginate rafts and their characterisation. *Int. J. Pharm.*, 294, 137, 2005.

20. Falkeborg, M., Cheong, L.Z., Gianfico, C., Sztukiel, K.M., Kristensen, K., Glasius, M., Xu, X., Guo, Z., Alginate oligosaccharides: Enzymatic preparation and antioxidant property evaluation. *Food Chem.*, 164, 185, 2014.
21. Hu, X., Jiang, X., Hwang, H., Liu, S., Guan, H., Antitumour activities of alginate-derived oligosaccharides and their sulphated substitution derivatives. *Eur. J. Phycol.*, 39, 67, 2004.
22. Chaluvadi, S., Hotchkiss, A.T., Jr., Call, J.E., Luchansky, J.B., Phillips, J.G., Liu, L.S., Yam, K.L., Protection of probiotic bacteria in a synbiotic matrix following aerobic storage at 4 C. *Benef. Microbes*, 3, 175, 2012.
23. Mogarkar, P.R. and Arfin, T., Chemical and structural importance of starch based derivatives and its applications, in: *Natural Polymers: Derivatives, Blends and Composites*, vol. 11, S. Ikram and S. Ahmed (Eds.), pp. 73–88, Nova Publishers, New York, USA, 2017.
24. Shoeb, A. and Arfin, T., Commercial and prospective applications of gelatin, in: *Natural Polymers: Derivatives, Blends and Composites*, vol. 11, S. Ikram and S. Ahmed (Eds.), pp. 199–216, Nova Publishers, New York, USA, 2017.
25. Arfin, T., Chitosan and its derivatives: Overlook of commercial application in diverse field, in: *Chitosan: Derivatives, Composites and Applications*, S. Ahmed and S. Ikram (Eds.), pp. 115–150, Scrivener Publishing LLC, 2017.

4

Alginate Hydrogel and Aerogel

Ajith James Jose*, Kavya Mohan and Alice Vavachan

Postgraduate and Research Department of Chemistry, St. Berchmans College, Changanassery, Kerala, India

Abstract

Alginate, a natural block polysaccharide with carboxylic functions, has increasingly drawn attention as an attractive compound in the world of research. Its excellent biocompatibility, biodegradation, and nontoxicity enable its application in a variety of fields like pharmaceutical and biomedical. Hydrogel microspheres of alginate play a significant role in wound dressing. Alginate-incorporated hydrogels play a well-defined role in bioengineering research to stimulate brain tissue enabling regeneration and also in bone reconstruction. Preservation of the dispersion of the gel by supercritical drying facilitates the formation of alginate aerogel. Unlike regular superabsorbents, alginate aerogel absorbs more saline compared to distilled water. This unique property opens a window for the development of nonsteroidal anti-inflammatory drugs. The biomedical applications of hydrogel and aerogel forms of the alginate are discussed in this chapter.

Keywords: Alginate, hydrogel, aerogel, biomedical applications

4.1 Introduction

Biomaterials traditionally remain inert and do not interact with the host. But recently, biomaterials began to be replaced with synthetic materials such as polymers, ceramics, and metal alloys, because of their improved performance, compared to natural materials. Recent studies are going on the evaluation, treatment, augmentation, replacement, and regeneration of function of the body. Naturally derived materials are regaining their attention because of their inherent biocompatibility properties [1]. Alginates are now trending. They are naturally occurring anionic polymers, obtained from brown seaweed

**Corresponding author*: ajithjamesjose@gmail.com

Shakeel Ahmed (ed.) Alginates, (59–78) © 2019 Scrivener Publishing LLC

and used for many medical applications because they possess high biocompatibility, low toxicity, relatively low cost, and mild conditions for gelation [2].

A hydrogel is a macromolecular polymer gel. It consists of a network of cross-linked polymer chains. Hydrogels are synthesized from certain hydrophilic monomers, either by chain or step growth along with a functional cross-linker to promote network formation. Hydrogels have greater affinity for water. They readily absorb water through H-bonding, since its network-like structure possesses voids. Due to structural properties and electrostatic attraction forces, new bond formation occurs through covalent bonding or hydrogen bonding. Their properties make them more applicable in the field of tissue engineering. Aerogels are synthetic porous materials, which are extremely thin, and they are derived from a gel where the liquid part of the gel has been replaced using gas. This replacement leads to the formation of a solid that has extremely low density and low thermal conductivity. It is also known in some other names such as frozen smoke, solid smoke, solid air, solid cloud, blue smoke, etc. By International Union of Pure and Applied Chemistry (IUPAC) definition, aerogel is "gel comprised of a macro porous solid in which the dispersed phase is a gas." Aerogels are solid, dry, and rigid materials. It is very strong structurally. The density and average size of pores in aerogel can be controlled during the manufacturing process.

In this chapter, general properties, current applications, and biomedical applications of alginate hydrogel and aerogel will be discussed.

4.2 Alginate Hydrogel

Alginic acid or algin, also called alginate, is an anionic polysaccharide distributed widely in the cell walls of brown algae, which binds with water to form vision gums. It is also a significant component of biofilms produced by a major pathogen in cystic fibrosis named *Pseudomonas aeruginosa*, showing high resistance to antibiotics and killing by macrophages. Generally, color ranges from white to yellowish brown. It is solid in filamentous granular or powdered form. Alginate is a linear copolymer containing homopolymeric blocks of (1,4) linked β-D mannuronate (M) and its C-5 epimer α-L-guluronate (G) residues covalently linked together in various sequences or blocks [3].

Alginates absorb water readily and hence are useful as an additive p_x dehydrated product such as slimming aids and in the manufacturing of paper and textiles. They are also incorporated in waterproofing and fireproofing fabrics, food industry, gelling agents, etc. Alginates found immense application in various pharmaceutical preparations. Alginates along with bicarbonate give the medicine Gaviscon, which inhibits reflux. The sodium derivative of alginate is used as an impression-making material in prosthetics, life casting, and dentistry. It is also used in reactive dye printing and as a thickener for reactive dyes used in textiles. Skin wound dressing products include materials of calcium alginate. They are widely used because they are easily removable. Alginate can be used in hydrogels consisting of microparticles or otherwise bulk gels combined with nerve growth factor. It found application in bioengineering research of brain tissue regenerations. Alginate composites are found to have suitable property for regeneration. Research on bone recreations is recently active. Alginates have better porosity, proliferation, and mechanical strength. Alginate hydrogels are found to be popular in applications in biomedical science and engineering because of their favorable properties, including biocompatibility and ease of gelation. Alginate hydrogels are important in case of wound healing. They are used in delivery system and tissue engineering applications, as these retain structural similarity to extracellular matrices in tissues.

Alginate hydrogels are prepared by various cross-linking methods. Wound dressings using alginate hydrogel can maintain a physiologically moist environment, which considerably reduces the bacterial infections at the wound site and facilitates fast wound healing. Alginate hydrogel can be used for the controlled release of drug molecules based on cross-linker types and cross-linking methods. In addition, alginate gels can be administrated or injected in a least invasive manner. Alginate gels also pave the way to cell transplantation in tissue engineering. Tissue engineering is related to manmade tissue and organ replacement in patients who suffer from organ failure and organ loss. Hydrogels are nowadays used to deliver cells to the decreed sites in a controlled manner and provide sufficient space for cell formation and tissue formation, and control the function and structure of the engineered tissue. Increasing the size of the alginate, we can also improve the properties of resultant gels. However, an alginate solution formed from a large polymer of high molecular size becomes highly viscous, which turns out to be undesirable for processing. Controlling the molecular weight and its distribution can independently control the pre gel solution viscosity and post gelling stiffness. The elastic modulus of the hydrogel can be easily increased. On the other hand, the viscosity changes are gradual by using combinations of high and low molecular weight alginate polymers.

4.2.1 Preparation of Alginate Hydrogels

Alginate is a naturally derived anionic polysaccharide and alginate hydrogels are prepared by various cross-linking methods.

4.2.1.1 *Ionic Cross-Linking*

The most common method for preparing alginate hydrogel is to combine a solution of alginate with a solution of conic cross-linking agents, such as divalent cation, e.g., Ca^{2+}. Alginate chains are believed to caution guluronate blocks, which can easily bind with divalent cations. The guluronate block of one alginate then combines with that of the other alginate chain and form a function, what is termed as cross-linking and resulting in gel structure [4]. Calcium chloride, calcium sulfate, and calcium carbonate can be used as cross-linking agents [5]. Gelation control, uniformity, strength structure and mechanical integrity of gels, and gelation temperature strongly affect the mechanical properties of gels [6]. At lower temperature, the reactivity of cross-linkers is reduced and cross-linking becomes slower. One of the important drawbacks of conically cross-linked alginate gel is the limited long-term stability in physiological conditions, because the divalent gel is released to the surrounding media due to some exchange reactions, and this makes the gel soluble; sometimes this feature may be beneficial and sometimes it may be negative.

4.2.1.2 *Covalent Cross-Linking*

Covalent cross-linking is studied to a great extent, and it finds application in tissue engineering. Covalent cross-linking in alginate can lead to elastic deformation of cells without much stress [7]. But covalent linking agents are generally toxic so the unreacted chemicals need to be removed from the gel. The covalently cross-linked alginate carrying polyethylene glycol (PEG) was first investigated. By the use of PEG, the loss of the hydrophilic character of hydrogel is compensated [8]. Later, alginate hydrogels containing poly(aldehyde guluronate) (PAG), multifunctional cross-linker such as adipic acid dihydrazide (AAD), etc. are introduced [9].

Latest cross-linking agents are photocross-linking agents that carry methacrylate. Alginate modified with α-phenoxy cinnamylideneace tyl chloride, etc. are important among them [10].

4.2.1.3 *Thermal Gelation*

Thermosensitive hydrogels are widely studied because of their applications in the field of drug delivery. Due to their adjustable swelling properties, there is modulation in drug release from gels [11]. One important example are poly(N-isopropylacrylamide) hydrogels. The transition temperature can be varied by copolymerization with hydrophilic monomers (e.g., acrylic acid, acrylamide). Semi-interpenetrating structures were prepared by copolymerization of N-isopropylacrylamide with polyethylene glycol macromer in the presence of sodium alginate in UV light [12]. The swelling ratio can be altered by changing the ratio of sodium alginate at a constant temperature. The mechanical strength shows improvement when combined with semi-IPN (interpenetrating polymer network) structures. Graft copolymerization of N-isopropylacrylamide (NIPAAm) in alginate after reaction with ceric ions also provided useful means for preparing temperature-sensitive alginate gels (Lee and Mooney, unpublished data).

4.2.1.4 Cell Cross-Linking

When the alginate gel is modified with ligands that show cell adhesion, the ability of cells to bind polymers can lead to a long distance without the help of any chemical-binding agents. Cell added to restriction site associated (RAD) modified alginate solution can generate cross-linked network structures via specific receptor–ligand interaction without using any other cross-linking in such modified alginate [13]. But cells added to a nonmodified alginate form non-uniform dispersion. Even if the cross-linking in the structures is broken, the structures are recovered within a few minutes due to weak and reversible ligand–receptor interaction in the system. Also it was reported that cells can provide additional mechanical integrity to Arginylglycylaspartic acid (RGD), alginate gels. RGD alginate gels are ionically cross-linked with calcium ion through a binding interaction between cells and adhesion ligands that are coupled to the alginate chain [14].

4.2.2 Biomedical Applications

4.2.2.1 Pharmaceutical Applications

Conventionally, alginates in pharmaceutical fields serve gel-forming and stabilizing agents, and they play a significant role in the controlled release of drugs to the desired site. Most common forms of dosage in pharmaceutical application is in the oral dosage form. But the use of hydrogel as a tissue-centered drug delivery system is the currently growing field. Here we briefly discuss the recent progress in drug delivery using alginate or its derivatives.

4.2.2.1.1 Delivery of Small Chemical Drugs

Alginate gels can be used for the controlled delivery of drugs having low molecular weight and are found to be highly useful when primary and secondary bond between the drug and the alginate can be used to regulate the kinetics of drug release. Alginate gels are porous, where pore size ranges in nanodimensions leading to fast diffusion of small molecules through the pores of gels [15]. For example, the release of flurbiprofen from an ionically cross-linked, partially oxidized alginate hydrogel is completed in 1.5 hours. However, the alginate formed into bead due to the combination of ionic and covalent bonds in the presence of calcium ions and adipic acid leading to a prolonged release because of increased cross-links and reduced swellings. The controlled and localized delivery of antineoplastic agents has been achieved by partially oxidized aerogels [16].

Multiple drug delivery is possible using alginate gels for simultaneous and sequential delivery, as the chemical structure of drug and the mode of interaction between the drug and the gel will affect the release kinetics. For example, methotrexate (non-interactive with alginate) was readily released by diffusion, while doxorubicin covalently attached to the alginate was released through chemical hydrolysis of cross-linkers. Mitoxantrone ionically attached to the alginate got released through the dissociation of gels [17].

Amphiphilic derivative of hydrogel can control the release of hydrophobic drugs. For example, alginate with poly(ε-caprolactone) (PCL) has been cross-linked with calcium ion and facilitates the controlled delivery of theophylline, a model drug with minimum water solubility. The length of PCL chain controls the swelling nature of the gel beads, and the PCL slows the rate of release of theophylline. The drug release will be completed within 2 hours for alginate-g-PCL/Ca^{2+} beads and within 1 hour for alginate/Ca^{2+} beads [18]. The incorporation of carbon nanotube (CNT) in alginates fulfills the sustained delivery of theophylline. The addition of CNT increases the mechanical strength and stability of gels, without affecting the structure and morphology of the microspheres, and no toxicity is observed, indicating the possibility of drug delivery to the intestine and colon [19]. Alginate in combination with chitosan is used in drug delivery as in combination, they form ionic complexes. Chitosan is a derivative of chitin, the second most abundant natural polymer with monomer (1,4) linked β-D glucosamine with pK of 6.5. Commercially available products have a composition of 80% β-D glucosamine and 20% N-acetyl-β-D glucosamine. Chitosan is a cationic polymer that is found to have extensive application in food industry, cosmetics, and biomedical and pharmaceutical fields due to its biocompatible nature and various other properties [20]. Multiparticulate systems of triamcinolone, containing alginate and chitosan, were prepared by a complex coacervation/ionotropic gelation method for colonic drug delivery. Compared to a stimulated gastric environment, a higher swelling degree and faster release of drug were observed from a simulated enteric environment [21]. Albendazole (ABZ) loaded magnetic alginate chitosan beads were also prepared for passive targeting in the gastrointestinal tract using physical capture mechanisms. The beads showed significant pH-dependent swelling behaviors and a continuous release of ABZ [22]. Alginate particles in combination with chitosan containing all trans retinoic acid (ATRA) have shown to enhance dermal localization and achieved a sustained release of ATRA into the skin; chitosan-treated alginate beads entrapped with metronidazole by an ionotropic gelation method were effective in the eradication of *Helicobacter pylori* when orally administrated into mice [23, 24]. Alginates are used to generate the matrix in which depots releasing small drugs can be incorporated. For the effective treatment of *H. pylori* infection, amoxicillin-loaded chitosan/poly(γ-glutamic acid) nanoparticles have been incorporated into alginate/Ca^{2+} hydrogels. Alginate gel outer layer surfaces protected amoxicillin loaded nanoparticles in the gastric environment and facilitated amoxicillin interactions with intercellular spaces, which is the center site for the infection of *H. pylori* [25].

4.2.2.1.2 Protein Delivery

Protein drug market is a rapidly growing area, and various protein drugs are now available due to the development of recombinant DNA technology. Alginates are excellent in protein drug delivery due to the fact that drugs can be easily incorporated in pores within the alginate without denaturation, and the gels can protect the proteins from degradation until their release. The protein release rate from alginate gels can be controlled using various methods. In general, the protein from alginate is rapid due to the hydrophilic nature and inherent porosity of the gels. Heparin binding growth factors such as vascular endothelial growth factor (VEGF) or basic fibroblast growth factor exhibit reversible binding property to the alginate, which results in a sustained and localized release of protein [26, 27]. The releasing rate in this case can be changed by changing the rate of degradation of the gels (e.g., using partially oxidized alginate), in order to make the drug release partially dependent on the degradation reaction [27]. The release of angiogenic molecules from alginate gels can be controlled by various methods, which are further developed. This is most important for heparin, which does not bind.

Alginate microspheres, which are ionically cross-linked, show high pI proteins such as lysozyme and chymotrypsin effective encapsulation; these problems physically cross-link with sodium alginate and allow more sustained release [28]. Amino group terminated F poly (C2-dimethylamino) ethyl methacrylate also reaches with oxidized alginate without using a catalyst, and the beaks were prepared by pouring the aqueous solution of the alginate derivative into the aqueous solution of $CaCl_2$ solution to form particles that can be released orally [29].

Alginate is used as a building block in the synthesis of a tetra-functional–actual linked polymer network that stimulates responsive gel with adjustable pore sizes. Acid labile proteins such as insulin can be protected from denaturation in the gastric environment using gels (pH 1.20), while they help in releasing the protein at near nonzero-order kinetics in neutral pH [30].

By enhancing the protein–hydrogel interactions through various cross-linking or encapsulation techniques, we can even incorporate proteins that show lower encapsulation efficiency and exhibit fast release from the alginate. Insulin-loaded alginate microspheres can be prepared by blending alginate with anionic polymers such as cellulose acetate phthalate polyphosphates, dextrin sulfate, etc. followed by coating with chitosan. Chitosan coating protects insulin at gastric pH and maintains its sustained release at intestinal pH [31]. Alginate microspheres can also be coated with *Bombyx mori* silk fibroin through layer-by-layer deposition techniques, which leads to the formation of mechanically stable shells as well as a diffusion barrier to the encapsulated proteins [32]. A combination of different microspheres that serves a deportation of protein and alginate hydrogels can also prove sustained protein release. Alginate hydrogels containing microspheres were prepared by encapsulation of a suspension of poly(d,l-lactide-co-glycolide) (PLGA) before ionic cross-linking. A homogeneous dispersion of PLGA microsphere was observed by SEM. The mixing of PLGA microsphere with alginate hydrogels can be studied using SEM. The release of a model protein called bovine serum from this combination is controlled by the mixing ratio between the PLGA microsphere and alginate hydrogels, which were independent of the total BSA content and the size of PLGA microspheres used [32]. Similarly, the release of TAT-HSP 27 (heat shock protein 27 fused with transcriptional activators) is regulated by changing the mixing ratios between gels and microspheres [33]. Alginate gel releasing proteins are explored in tissue engineering and regeneration.

4.2.2.1.3 Wound Dressing

It is necessary to treat painful and acute wounds. Alginate-based wound dressing offers many advantages. It is difficult to keep a wound dry using traditional wound dressing methods, but alginates allow evaporation of wounds and thereby prevent the entry of pathogens into wounds [34]. They also provide a moist wound environment and facilitate wound healing [35]. Alginate dressings are produced by the ionic cross-linking between alginate solution and calcium ion to form a gel that on further processing forms freeze-dried porous sheets and fibrous but unwoven dressings. Alginate dressings can provide a moist environment because in their dry form, they absorb wound fluid, and the gel can supply sufficient water to dry wounds; also it minimizes the bacterial infection at the wound site. Alginate can also facilitate granular tissue formation, rapid epithelialization, and thereby rapid healing. General alginate dressings include Algicell™, Algisite M™, Comfeel plus™, Sorbsan™, Kaltostat™, and Tegagen™. Similarly, partially ionized alginate gels can accelerate wound healing by the sustained release of human keratinocyte proliferation; di-butyryl cyclic adenosite monophosphate leads to complete re-epithelialization. Alginate gels releasing stromal cell derived factors and those containing silver can also accelerate wound healing. The incorporation of silver in the gel can improve antioxidant capacity, antimicrobial activity, and binding affinity to certain enzymes, whereas incorporation of zinc enhances immune modulatory and antimicrobial effect. The combination of alginate with chitin/chitosan and fucoidan gels is reported to provide moist healing in rats [36].

4.2.2.1.4 Cell Culture

Alginate gels are very widely used for mammalian cell culture. The gel can be adapted to serve as 2D and 3D culture systems. Highly specific and quantitative modes of cell adhesion can be incorporated in alginate gels because of their low protein adsorption and lack of mammalian cell receptors. Also the biocompatibility and ease of introduction of alginate are supporting factors.

RGD-containing alginate gels are frequently used in *in vivo* cell culture because they can control the phenotype of interacting myoblasts, chondrocytes, osteoblasts, ovarian follicle, and bone marrow stromal cells. The number of cells adheres to the gel, and its growth rate is dependent on the bulk RGD density in the gels. The presence of RGD in alginate gels enhances cell adhesion and differentiation, chondrogenic gene expression, etc. Similarly, alginate gels have been formed in a microfluidic device, and it provides a useful means for integrating 3D-culture microenvironments into microfluidic systems [37]. Among different types of RGDs, cyclic RGD peptides are more resistant to proteolysis and show higher binding affinity and are more selective than linear RGD [38]. Such alginate derivative having cyclic RGD may enhance tissue regeneration by reducing the need for an exogenous solubility factor.

Recent studies on using alginate gels as a 3D cell culture substrate, which gives insight to cancer biology, are growing. Integrin encapsulation in RGD alginate gels can give an idea about how cancer cells signal to recruit blood vessels, and this can lead to the development of new anti-angiogenic cancer therapies [39].

An important limitation of most 3D cell culture systems is the difficulty in analyzing and quantifying cell matrix interactions. However, this problem can be now solved by various fluorescence resonance energy transfer (more commonly referred to by the acronym FRET) techniques, which enable an ability to quantitatively probe the relation between cell adhesion and decision making. These types of analysis now help one to predict cell behavior, particularly in 3D culture, and to design proper 3D cell culture substrates for a number of applications.

4.2.3 Tissue Regeneration with Protein and Cell Delivery

For the past few decades, alginate gels are used as vehicles for the delivery of proteins or cell population which can direct the regeneration or engineering of various tissues and organs in the body. There exist some limits to the size of a regenerative agent that can be released from an alginate gel because the pore size is ≈5 nm. Most of the proteins can be delivered even without degradation, whereas larger molecules such as DNA can be released via degradation. Further studies are ongoing in this area.

4.2.3.1 Blood Vessels

Blood vessels are critical for the transport of oxygen and nutrients to all parts of tissues, removal of metabolic waste, and trafficking of stem and progenitor cells, which are important for the growth of organs in an embryo and wound repair in adults [40]. New blood vessel formation is crucial in tissue engineering, and it provides treatment to patients suffering from coronary and arterial diseases. Blood vessel formation is exploiting the sustained and localized realizing of heparin from alginate gels. Differential binding of various growth factors to alginate gels leads to sequential delivery of factors in the early and late stages of angiogenesis, which supports the maturation of new vessels and better functionality. Those growth factors that promote initiation of angiogenesis and maturation of capillaries are VEGF and platelet-derived growth factor (PDGF), respectively [41]. When host cells that are capable of responding to delivered proteins are lacking or dysfunctional, transplantation of cells is more attractive for the formation of new blood vessels. The delivery of VEGF from alginate–PLG carrier in concert with epithelial transplantation significantly increases the number of blood vessels formed by transplanted endothelial cells. Recent studies proved that alginate gel delivery can promote outward migration and dispersion throughout ischemic tissues [42].

4.2.3.2 Bones

Due to poor healing effects, bone injuries are treated only limitedly. But alginate gels have the capacity to regenerate bone by the delivery of osteo-inductive factors, bone forming cells, or a combination of both. They can be introduced into the body in a minimally invasive manner, and they have the ability to fill irregularly shaped defects; there is a case of chemical modification with adhesion ligands (e.g., RGD) and controlled release of tissue induction factors (e.g., BMP, TGF-β). But alginate gels do not have sufficient mechanical properties to allow load bearings in the initial stage of regeneration, and they are not inherently degradable in physiological conditions. The use of RGD alginate gels can lead to complete regeneration of critical sized femoral defects in rodents. Alginate gels that contain BMP (bone morphogenetic proteins) can result in tissue regeneration; they can also be used to transplant cell populations that directly participate in bone formation [43]. There is wide application for the use of alginate gels for the transplantation of stem cells. Alginates have also been combined with some inorganic materials to enhance bone tissue formation. Alginate gel beads containing cells presented a potential for bone tissue engineering in moderate stress-bearing conditions [44]. Alginate gels containing collagen type 1 and β-tricalcium phosphate exhibit adhesion and proliferation of human bone marrow stem cells than pure alginate gels [45].

4.2.3.3 Cartilage

The repair of degraded and damaged cartilage is still a major problem faced in the orthopedics field. But this challenge can be solved to a great extent by the use of alginate gels. Alginate gels have a capacity to transplant chondrogenic cells to restore damaged cartilage in animal models. Before, alginate solution containing chondrocytes is mixed with calcium sulfate and produced preshaped cartilage. But now, shape memory alginate gels are developed with which we can engineer the cartilage with the desired shape and size *in vivo*. In short, alginate gels, which have a macroporous structure, are introduced into mice with a catheter. Later, the gel is rehydrated *in vivo* to attain its original shape, which allows cartilage formation in mice with desired geometry [46]. The use of stem cells in alginate gel can help in regulating the differentiation. It is hypothesized that chondrogenesis of stem cells is related to the morphology of encapsulated cells. Human mesenchymal stem cells (MSCS), rabbit bone marrow stromal cells, human adipose derived stem cells, etc. along with alginate gels can promote cartilage regeneration. RGD-containing peptide along with alginate can enhance adhesive interaction with chondrocytes and can control the cell phenotype. These types of gels can be injected *in vivo* in a minimally invasive manner and are effective in cartilage engineering *in vivo* [47].

4.2.3.4 Muscle, Nerve, Pancreas, and Liver

Alginate gels are also studied for their ability to participate in the regeneration of various tissues and organs including skeletal muscle, nerve, pancreas, and liver. These studies include cell transplantation, growth factor delivery, or a combination of both [48, 49]. It is possible to regulate angiogenesis and myogenesis by the combined delivery of VEGF and insulin growth factor-1 (IGF-1) from alginate gels. Alginate gels have also been studied for the repair of central and peripheral nervous systems. Alginate gels covalently linked with ethylene diamine help in maintaining a 50-mm gap in cat static nerves, and it promoted the outgrowth of regenerating axons and astrocyte reaction in the spinal cords of young rats [50]. Alginate gels can also be used in the repair of peripheral nerve gaps [51]. They may be useful for cell-based neural therapies, sequence promoted adhesion of NB2a neuroblastoma cells, and neurite outgrowth from the cells.

Alginate gels encapsulated with hepatocytes may enhance the repair of failing liver. The hydrophilic nature of alginate gels allows the efficient seeding of hepatocytes into gels [52, 53]. Alginate gels encapsulated with pancreatic islet allographs are used to cure type 1 diabetes. This method helps to avoid the use of immunosuppressive drugs. These islets-encapsulated alginate beads are generally coated with poly amino acids (poly-L-lysine) to decrease the outer pore size [54]. However, alginate beads have mechanical and chemical instability, which decreases their survival period within the body. Research is ongoing on increasing their survival time by using various coating materials and various microfabrication methods [55].

4.3 Alginate Aerogel

Aerogels are a class of product materials that can be prepared using a variety of precursors ranging from inorganic synthetic to biopolymers. A research or polysaccharide-based aerogels brought an opportunity to expand the concept of medication. Polysaccharides are biopolymers. Biocompatible and biodegradable properties of polysaccharides enable their application in the world of pharmaceuticals and other biomedical applications; unlike hydrogel, alginate-based aerogels are well known for their open-porous structure, high surface area, and pore volume. Alginate hydrogels discussed above are more saline compared to distilled water and thus show a deviation from regular superabsorbent behavior. These special characteristics brought them into the field of aeronautics, energy conservation, biomedicine, drug delivery systems, photovoltaics, nanotechnology, acoustics, etc.

4.3.1 Properties of Alginate Aerogels

4.3.1.1 Bulk Density and Pore Volume

Bulk density and BJH (Brunauer-Joyner-Halenda) pore volume were determined from linear dimensions and desorption branches of N_2 isotherms, respectively. Values of specific volume lie between 4.17 and 16.7 $cm^3 g^{-1}$ indicating only 40–62% of the overall. Porosity can be attributed to the volume of mesopores. Thus, micro- and macroporosity may be present. Detailed studies show that only meso- and macropores are present in alginate aerogels. La–alginate aerogel has the highest BJM pore volume of 4.10 ± 0.78 $cm^3 g^{-1}$. Two factors affecting pore volume have been considered. First, pore volume increases when solution becomes more and more diluted since the volume of empty space possessed by the aerogel increases consequently. Second, stiffness of the backbone decreases with decreasing alginate concentration. The latter factor is probably applicable to gels derived from 0.25 wt% alginate.

4.3.1.2 Specific Surface Area

The potential adsorption capacity of aerogels depends not only on the pore volume but also on the specific surface area. Typical values for alginate aerogels were observed as 151.857 $cm^2 g^{-1}$. The highest surface area reported was 608 ± 60 $m^2 g^{-1}$. Both the surface area and pore volume were found to increase with increasing alginate concentration within the range of 1.5 to 3 wt%.

4.3.1.3 Compressibility

Compressibility is one of the interesting properties of alginate aerogels produced by 102 induced gelation. The sample retains almost the same surface area, while the BJH pore volume increases when compressed under a pressure of 10 kN m^{-2} to 20% of the original thickness. Thus, it indicates that compression affects the macroporous rather than the mesoporous structure. It was found that compressed aerogels become flexible and could be many times reversibly bent by an angle of 180°. Alginate aerogels are produced by 102 induced gelation because as viscoplastic material aerogels, biopolymer aerogels can be converted into flexible materials by compression retaining the porous characteristics of the original aerogels at the meso level.

4.3.1.4 Thermal Conductivity and Absorption

Low thermal conductivity is an attractive property. Thermal conductivity measurements were conducted by the hot-wire method. This method revealed excellent thermal insulation properties, i.e., thermal conductivity was measured to be in the range of (18–22) ± 2 Mw M^{-1} K^{-1}. A unique feature of alginate aerogels is that they absorb more saline compared to distilled water, in contrast to the behavior of regular superabsorbents. For example, a gram of alginate aerogel absorbs as much as 120 g of saline (0.9 wt% aq NaCl solution) solution compared to 20 g of distilled water.

4.3.2 Preparative Methods

Aerogel production consists of multiple steps that include gelation, soluent exchange, and supercritical drying (Sc-drying). Hydrogel produced through gelation acts as a precursor for the production of aerogel. Here we discuss the preparation of alginate aerogel under the assistance of CO_2 as an alternative catalyst.

Materials required are sodium alginate (Alg), calcium carbonate, ethanol 99.9%, and carbon dioxide with 99.9% purity. Deionized water was used throughout the study. Alginate solution was prepared by sodium alginate powder with an appropriate amount of water overnight and stored at 218 K. Calcium carbonate was suspended in water by vigorous mixing for about 5 min. While mixing, a certain part of the suspension was bled off and immediately transferred into the alginate solution to reach the target Alg–$CaCo_3$ ratio. The suspension obtained was again mixed for 2 min until they became homogenous. All suspensions thus prepared were transferred into Petri dishes and placed into a 25-L autoclave for subsequent gelation. The autoclave was pressurized with gaseous carbon dioxide up to 5 MPa at 298 K. The pressure was maintained for 12 hours and then slowly released. Hydrogels formed were either transparent or translucent. The hydrogels were left in the air until the formation of bubbler ceased, then transferred into an ethanol–water mixture to perform solvent exchange and thus producing alcogels.

Alcogels were packed into filtered paper parcels, placed on the preheated autoclave (313 K), and covered by ethanol to prevent premature solvent evaporation from the gel. Supercritical drying was performed using the same autoclave as for gelation. The autoclave was sealed and filled with CO_2 by a membrane pump. The system was depressurized in 1 hour followed by cooling down to room temperature resulting in the formulation of alginate aerogels [56].

4.4 Future Perspectives

Alginates as a biomaterial for clinical use have been established, although they are not free from issues. Alginate-based 3D cell culture systems are important in cellular physiological investigation, drug delivery evaluation, and tissue engineering. The 3D culture system; macroporous structures have the ability to reflect native extracellular metrics and natural physiological environment.

Cell-based microparticles are undergoing advanced research in the progress of cell transplantation therapy of diabetes and neurodegenerative disease treatment using alginate. Alginate oligosaccharides of low molecular polymer fragment formed by enzymatic depolymerization or high temperature acid hydrolysis is used as a drug vehicle.

References

1. Lee, K.Y. and Mooney, D.J., Alginate: Properties and Biomedical applications. *Prog. Polym. Sci.*, 37, 1, 106–126, 2012.
2. Ratner, B.D. and Bryant, S.J., Biomaterials: Where we have been and where we are going? *Ann. Rev. Biomed. Eng.*, 6, 41–75, 2004.
3. Haug, A., Fractionation of alginic acid. *Acta Chem. Scand.*, 13, 601–603, 1959.
4. Grant, G.T., Morris, E.R., Rees, D.A., Smith, P.J.C., Thom, D., Biological interactions between polysaccharides and divalent cations—Egg-box model. *FEBS Lett.*, 32, 195–198, 1973.
5. Crow, B.B. and Nelson, K.D., Release of bovine serum albumin from a hydrogel-cored biodegradable polymer fiber. *Biopolymers*, 81, 419–427, 2006.
6. Augst, A.D., Kong, H.J., Mooney, D.J., Alginate hydrogels as biomaterials. *Macromol. Biosci.*, 6, 623–633, 2006.

7. Zhao, X.H., Huebsch, N., Mooney, D.J., Suo, Z.G., Stress-relaxation behavior in gels with ionic and covalent cross-links. *J. Appl. Phys.*, 107, 063509/1–5, 2010.
8. Lee, K.Y., Rowley, J.A., Eiselt, P., Moy, E.M., Bouhadir, K.H., Mooney, D.J., Controlling mechanical and swelling properties of alginate hydrogels independently by cross-linker type and cross-linking density. *Macromolecules*, 33, 4291–4294, 2000.
9. Lee, K.Y., Bouhadir, K.H., Mooney, D.J., Controlled degradation of hydrogels using multifunctional cross-linking molecules. *Biomaterials*, 25, 2461–2466, 2004.
10. Tanaka, H. and Sato, Y., Photosensitivity of polyvinyl esters of substituted cinnamylidene acetic acids. *J. Polym. Sci.*, 10, 3279–3287, 1972.
11. Roy, D., Cambre, J.N., Sumerlin, B.S., Future perspective and recent advances in stimuli responsive materials. *Progr. Polym. Sci.*, 35, 278–301, 2010.
12. Zhao, S.P., Cao, M.J., Li, H., Li, L.Y., Du, W.L., Synthesis and characterization of thermosensitive semi-IPN hydrogels based on poly(ethylene glycol)-copoly-(epsilon-caprolactone) macromer, N isopropylacrylamide, and sodium alginate. *Carbohydr. Res.*, 345, 425–431, 2010.
13. Lee, K.Y., Kong, H.J., Larson, R.G., Mooney, D.J., Hydrogel formation via cell cross-linking. *Adv. Mater.*, 15, 1828–1832, 2003.
14. Drury, J.L., Boontheekul, T., Mooney, D.J., Cellular cross-linking of peptide modified hydrogels. *J. Biomech. Eng.—Trans. ASME*, 127, 220–228, 2005.
15. Boontheekul, T., Kong, H.J., Mooney, D.J., Controlling alginate gel degradation utilizing partial oxidation and bimodal molecular weight distribution. *Biomaterials*, 26, 2455–2465, 2005.
16. Maiti, S., Singha, K., Ray, S., Dey, P., Sa, B., Adipic acid dihydrazide treated partially oxidized alginate beads for sustained oral delivery of flurbiprofen. *Pharm. Develop. Technol.*, 14, 461–470, 2009.
17. Bouhadir, K.H., Alsberg, E., Mooney, D.J., Hydrogels for combination delivery of antineoplastic agents. *Biomaterials*, 22, 2625–2633, 2001.
18. Colinet, I., Dulong, V., Mocanu, G., Picton, L., Le Cerf, D., New amphiphilic and pH-sensitive hydrogel for controlled release of a model poorly water-soluble drug. *Eur. J. Pharm. Biopharm.*, 73, 345–350, 2009.
19. Zhang, X.L., Hui, Z.Y., Wan, D.X., Huang, H.T., Huang, J., Yuan, H., Yu, J.H., Alginate microsphere filled with carbon nanotube as drug carrier. *Int. J. Biol. Macromol.*, 47, 389–395, 2010.
20. Rinaudo, M., Chitin and chitosan: Properties and applications. *Progr. Polym. Sci.*, 31, 603–632, 2006.
21. Lucinda-Silva, R.M., Salgado, H.R.N., Evangelista, R.C., Alginate chitosan systems: *In vitro* controlled release of triamcinolone and *in vivo* gastrointestinal transit. *Carbohydr. Polym.*, 81, 260–268, 2010.
22. Wang, F.Q., Li, P., Zhang, J.P., Wang, A.Q., Wei, Q., A novel pH-sensitive magnetic alginate 74 chitosan beads for albendazole delivery. *Drug Dev. Ind. Pharm.*, 36, 867–877, 2010.

23. Lira, A.A.M., Rossetti, F.C., Nanclares, D.M.A., Neto, A.F., Bentley, M.V.L.B., Marchetti, J.M., Preparation and characterization of chitosan treated alginate microparticles incorporating all-trans retinoic acid. *J. Microencapsul.*, 26, 243–250, 2009.
24. Ishak, R.A.H., Awad, G.A.S., Mortada, N.D., Nour, S.A.K., Preparation, *in vitro* and *in vivo* evaluation of stomach-specific metronidazole-loaded alginate beads as local anti-*Helicobacter pylori* therapy. *J. Control Release*, 119, 207–214, 2007.
25. Chang, C.H., Lin, Y.H., Yeh, C.I., Chen, Y.C., Chiou, S.F., Hsu, Y.M., Chen, Y.S., Wang, C.C., Nanoparticles incorporated in pH-sensitive hydrogels as amoxicillin delivery for eradication of *Helicobacter pylori*. *Biomacromolecules*, 11, 133–142, 2010.
26. Lee, K.Y., Peters, M.C., Mooney, D.J., Comparison of vascular endothelial growth factor and basic fibroblast growth factor on angiogenesis in SCID mice. *J. Control Release*, 87, 49–56, 2003.
27. Silva, E.A. and Mooney, D.J., Effects of VEGF temporal and spatial presentation on angiogenesis. *Biomaterials*, 31, 1235–1241, 2010.
28. Wells, L.A. and Sheardown, H., Extended release of high pI proteins from alginate microspheres via a novel encapsulation technique. *Eur. J. Pharma. Biopharm.*, 65, 329–335, 2007.
29. Gao, C.M., Liu, M.Z., Chen, S.L., Ji, S.P., Chen, J., Preparation of oxidized sodium alginate-graft-poly((2-dimethylamino) ethyl methacrylate) gel beads and *in vitro* controlled release behavior of BSA. *Int. J. Pharm.*, 371, 124, 2009.
30. Chan, A.W. and Nuefeld, R.J., Tuneable semi synthetic network alginate for absorptive encapsulation and controlled release of protein therapeutics. *Biomaterials*, 31, 9040–9047, 2010.
31. Silva, C.M., Riberio, A.J., Ferreira, D., Veiga, F., Insulin encapsulation in reinforced alginate microspheres prepared by internal gelation. *Eur. J. Pharm. Sci.*, 29, 148–159, 2006.
32. Wang, X., Wenk, E., Hu, X., Cast, G.R., Meinel, L., Wang, X., Li, C., Merkle, H., Kaplan, D.L., Silk coatings on PLGA and alginate microspheres for protein delivery. *Biomaterials*, 28, 4161–4169, 2007.
33. Lee, J., Tan, C.Y., Lee, S.K., Kim, Y.H., Lee, K.Y., Controlled delivery of heat shock protein using an injectable microsphere/hydrogel combination system for the treatment of myocardial infarction. *J. Control Release*, 137, 196–202, 2009.
34. Boateng, J.S., Matthews, K.H., Stevens, H.N.E., Eccleston, G.M., Wound healing dressings and drug delivery systems: A review. *J. Pharm. Sci.*, 97, 2892–2923, 2008.
35. Queen, D., Orsted, H., Sanada, H., Sussman, G., A dressing history. *Int. Wound J.*, 1, 59–77, 2004.
36. Balakrishnan, B., Mohantry, M., Fernandes, A.C., Mohanan, P.V., Jayakrishnan, A., Evaluation of the effect of incorporation of dibutyryl cyclic adenosine monophosphate in an *in situ* forming hydrogel would dressing based on oxidized alginate and gelatin. *Biomaterials*, 27, 1355–1361, 2006.

37. Connelly, J.T., Garcia, A.J., Leveston, M.E., Inhibition of *in vitro* chondrogenesis in RGD modified three-dimensional alginate gels. *Biomaterials*, 28, 4409–4417, 2007.
38. Bogdanowich-Knipp, S.J., Chakrabarti, S., Williams, T.D., Dillman, R.K., Siahaan, T.J., Solution stability of linear vs. cyclic RGD peptides. *J. Pept. Res.*, 53, 530–541, 1999.
39. Fischbach, C., Kong, H.J., Hsiong, S.X., Evangelista, M.B., Yuen, W., Mooney, D.J., Cancer cell angiogenic capability is regulated by 3D culture and integrin engagement. *Proc. Natl. Acad. Sci. USA*, 106, 399–404, 2009.
40. Carmeliet, P., Angiogenesis in life, disease and medicine. *Nature*, 438, 932–936, 2005.
41. Sun, Q.H., Silva, E.A., Wang, A.X., Fritton, J.C., Mooney, D.J., Schaffler, M.B., Grossman, P.M., Rajagopalan, S., Sustained release of multiple growth factors from injectable polymeric system as a novel therapeutic approach towards angiogenesis. *Pharm. Res.*, 27, 264–271, 2010.
42. Silva, E.A., Kim, E.S., Kon, H.J., Mooney, D.J., Material-based deployment enhances efficacy of endothelial progenitor cells. *Proc. Natl. Acad. Sci. USA*, 105, 14347–14352, 2008.
43. Basmanav, F.B., Kose, G.T., Hasirci, V., Sequential growth factor delivery from complexed microspheres for bone tissue engineering. *Biomaterials*, 29, 4195–4204, 2008.
44. Weir, M.D., Xu, H.H.K., Simon, C.G., Strong calcium phosphate cement-chitosan-mesh construct containing cell-encapsulating hydrogel beads for bore tissue engineering. *J. Biomed. Mater. Res. A*, 77, 487–496, 2006.
45. Lawson, M.A., Barralet, J.E., Wang, I., Shelton, R.M., Triffitt, J.T., Adhesion and growth of bone marrow stromal cells on modified alginate hydrogels. *Tissue Eng.*, 10, 1480–1491, 2004.
46. Thornton, A.J., Alsberg, E., Albertelli, M., Mooney, D.J., Shape-defining scaffolds for minimally invasive tissue engineering. *Transplantation*, 77, 1798–1803, 2004.
47. Park, H., Kang, S.W., Kim, B.S., Mooney, D.J., Lee, K.Y., Shear reversibly cross-linked alginate hydrogels for tissue engineering. *Macromol. Biosci.*, 9, 895–901, 2009.
48. Saxena, A.K., Marler, J., Benvenuto, M., Willital, G.H., Vacanti, J.P., Skeletal muscle tissue engineering using isolated myoblasts on synthetic biodegradable polymers: Preliminary studies. *Tissue Eng.*, 5, 525–532, 1999.
49. Levenberg, S., Rouwkema, J., Macdonald, M., Garfein, E.S., Kohane, D.S., Darland, D.C., Marini, R., van Blitterswijk, C.A., Mullingan, R.C., D'Amore, P.A., Langer, R., Engineering vascularized skeletal muscle tissue. *Nat. Biotechnol.*, 23, 879–884, 2005.
50. Hashimoto, T., Suzuki, Y., Suzuki, K., Nakashima, T., Tanihara, M., Ide, C., Peripheral nerve regeneration using non-tubular alginate gel cross-linked with covalent gel cross-linked with covalent bonds. *J. Mater. Sci. Mater. Med.*, 16, 503–509, 2005.

51. Katoaoka, K., Suzuki, Y., Kitada, M., Hashimoto, T., Chou, H., Bai, H.L., Ohta, M., Wu, S., Suzuki, K., Ide, C., Alginate enhances elongation of early regenerating axons in spinal cord of young rats. *Tissue Eng.*, 10, 493–504, 2004.
52. Selden, C. and Hodgson, H., Cellular therapies for liver replacement. *Transpl. Immunol.*, 12, 273–288, 2004.
53. Koizumi, T., Aoki, T., Kobayashi, Y., Yasuda, D., Izumida, Y., Jin, Z.H., Nishino, N., Shimizu, Y., Kato, H., Murari, N., Niiya, T., Enami, Y., Mitamura, K., Yamamoto, T., Kusano, M., Long-term maintenance of the drug transport activity in cryopreservation of microencapsulated rat hepatocytes. *Cell Transplant*, 16, 67–73, 2007.
54. Sakai, S., Ono, T., Ijima, H., Kawakami, K., *In vitro* and *in vivo* evaluation of alginate/sol-gel synthesized aminopropyl silicate/alginate membrane for bioartificial pancreas. *Biomaterials*, 23, 4177–4183, 2002.
55. Liu, K., Ding, H.J., Liu, J., Chen, Y., Zhao, X.Z., Shape-controlled production of biodegradable calcium alginate gel microparticles using a novel microfluidic device. *Langmuir*, 22, 9453–9457, 2006.
56. Gurikos, P., Raman, S.P., Weinrich, D., Fricke, M., Smirnova, I., A novel approach to alginate aerogels: Carbon dioxide induced gelation. 5, 7812–7818, 2015.

Part 2

ALGINATES IN BIOMEDICAL APPLICATIONS

5

Alginate in Biomedical Applications

Luiz Pereira da Costa

Postgraduate Program in Chemistry, Federal University of the State of Amazonas - UFAM, Coroado, Brazil

Abstract
In the last decades, alginates, multifunctional biopolymers, have increasingly drawn attention as attractive compounds for biomedical and pharmaceutical applications due to their unique physicochemical properties and versatile biological activities. The focus of the actual use of alginates is biological and pharmacological as a tool in drug formulation. The recent technological advancements with using alginates, issues related to alginates' suitability as a matrix for membranes, beads and three-dimensional tissue cultures, adjuvants of antibiotics, and antiviral agents in cell transplantation in disorder or neurodegenerative diseases treatment, and an update on the antimicrobial and antiviral therapy of alginate-based drugs are also highlighted.

Keywords: Biomedical applications, biocomposites, drug delivery, health, natural source

5.1 Introduction

Natural compounds and materials, biocompatible and nontoxic as polysaccharides, are an important source for utilization as an ingredient for formulations of biological applications both individually and in composition. Accordingly, they are widely used in drug delivery and other biomedical applications [1–5]. A bioactive component is covalently bound to the polysaccharide basis spontaneously or by the use of another binding substance through insertion. Polysaccharides have remarkable advantages because they are abundant in nature, e.g., obtained by different algae (alginate and

Email: lupeco7@hotmail.com

Shakeel Ahmed (ed.) Alginates, (81–94) © 2019 Scrivener Publishing LLC

carrageenan), other vegetable source (cellulose, pectin, hemicellulose, and guar gum), and microbial source (dextran, chitin, and xanthan gum), besides animal source (heparin, chondroitin, and hyaluronic acid) [5].

Due to their biocompatibility, biodegradability, hydrophilicity, technological properties, and great availability, polysaccharides are widely used as natural nontoxic polymers in different boughs of industry (biomedical, food, cosmetic, textile, construction material). Exopolysaccharides (such as pullulan and dextran) have recently taken up a significant position in biomedical industry. Alginate has an incomparable role in biomedicine, since its hydrogel form is satisfactorily applied in the immobilization of different types of enzymes, medication, biochemical compounds, and different kinds of microorganisms/cells (animal or herbal). Alginate microspheres and microcapsules have been used for self-control of biological molecules and active principle, as well as cell carriers and scaffolds for tissue engineering research. Special advantages are brought by the singleness of formulation of alginate hydrogel, among others the possibility of the manufacture production and scalability of micro- and nanoparticles.

In the areas of tissue engineering and controlled drug delivery of medical substances in the organism, these polysaccharides are dispersed for all the nature. The polysaccharides are found as a main composition of cell walls in some sea algae (thereabout 40% of dry mass), or as capsulated polysaccharides in determinate soil bacteria [6]. Usually, the species of brown algae explored for getting alginic acid, i.e., its salts (alginates), are *Ascophyllum nodosum*, *Laminaria hyperborean*, and *Macrocystis pyrifera*. The bacterial alginates are produced from *Azotobacter vinelandii*, as well as some species of *Pseudomonas*. Alginates can be found as salt mixtures of different cations (Mg^{2+}, Sr^{2+}, Ba^{2+}, and Na^{+}) present in salt seawater [7].

Abundant studies have demonstrated that chemical structure, molecular dimensions, and kinetics of gel forming have an important influence on alginates' functional properties such as porosity, viscosity, stability, biodegradability, immunologic features, and biocompatibility. Alginates have been included by the Food and Drug Administration (FDA) in the group of compounds considered generally safe for oral application. Alginates are nontoxic and biodegradable when applied orally, and they cause no immunoreactions, which is not the case when they are applied intravenously [8].

Among other properties, alginates have mucoadhesive characteristics, contributing to the transport of medical substances to mucosal tissues [9]. The results of other authors have reported that these polymers may be good mucoadhesive agents where a higher charge density would enable better adhesion [10]. Polyanionic polymers are more efficient as bioadhesives than polycationic and nonionic polymers. Therefore, alginates are very good mucoadhesive agents as anionic polymers with carboxyl groups. Alginates have greater mucoadhesive strength than polymers such as polystyrene, chitosan, carboxymethyl cellulose, and poly(lactic acid) [11]. The bioavailability and efficacy of a drug is higher as a result of better adhesivity of alginate particles onto the mucosal tissues, while the passage time for proteins is longer and the drug is localized on the absorptive surface.

The main disadvantages of alginates for hard tissue engineering are their low mechanical strength and poor controllability of the microinternal architecture because of their poor mechanical properties and excessive hydrophilicity [12]. To address these limitations, several researchers have combined alginates with other biomaterials (bioceramics and synthetic polymers) to produce biocomposites that provide both the biological advantages of alginates and greater mechanical strength because of the reinforcing material [12–14].

5.2 Chemical Structure and Properties of Alginate

Alginate is a naturally occurring linear polysaccharide abundant in various species of brown algae composed of β-D-mannuronic acid and α-L-guluronic acid units [15]. The molecular weight can vary between 10 and 1000 kDa depending on the source and production process. Upon adding multivalent cations, an alginate solution rapidly forms an ionotropical gel, which makes it extremely interesting to be applied in the biomedical field [16, 17]. To enable *in vivo* injection of a Ca^{2+}-cross-linked alginate hydrogel, the cross-linking rate should be reduced. Interestingly, polyols have already been applied to slow down the hydrogel formation. It has been anticipated that the polyols hinder the immediate complexation of Ca^{2+} by alginate.

Alginates extracted from different sources differ in M and G contents as well as the length of each block, and more than 200 different alginates are currently being manufactured [18].

Although alginate does not possess cell interactive properties, several authors have overcome this issue by coupling cell-interactive peptides (e.g., RGD) or growth factors (e.g., VEGF) to the alginate backbone [16]. In addition, alginate-based semi-interpenetrating polymer networks have been prepared possessing stimuli-responsive behavior. Wang and Wang [19] have developed a pH-sensitive superabsorbent hydrogel composed of sodium alginate-g-poly(sodium acrylate) and polyvinylpyrrolidone by free-radical solution polymerization using ammonium persulfate as an initiator and N,N-methylene-bisacrylamide as a cross-linker. As alginates form hydrogels under gentle conditions, the gels may be used to entrap cells and function as an extracellular matrix (ECM) with non-immunogenic and adjustable bioresorption properties [20].

5.3 Types of Interaction of Alginate

Alginate is an anionic copolymer consisting of (1→4)-linked β-D-mannuronic acid (M units) and its C-5 epimer α-L-guluronic acid (G units) with varying compositions and sequences and possesses unique complexation capability with divalent cations such as Ca^{2+} to form the so-called egg-box structure [21].

Under special conditions, alginate would depolymerize to oligosaccharides, monosaccharides, and decomposition products. The modifications in molecular structure and interactions can be verified by Fourier transform infrared spectroscopy spectra and magnetic nuclear resonance.

Alginates can be easily formed into diverse semisolid or solid structures under mild conditions because of their unique ability of sol/gel transition. Therefore, alginates are commonly used as viscosity increasing agents, thickeners, and suspension and emulsion stabilizers in food and pharmaceutical industry. The fact that alginate may undergo *in situ* gelation makes alginate materials promising tools for a wide range of applications, including injectable vehicles for tissue engineering or topical drug delivery systems.

Wang *et al.*, [22] discovered that the sulfated polymannuroguluronate (SPMG), the sulfated form of alginate, is a characteristic anti-AIDS drug candidate, as it caused the inhibition of HIV-1 infection mainly through the robust attachment of virus gp120 protein with CD4 molecules on the surface of T cells. Moreover, there is huge correlation between the size of SPMG oligosaccharides and their inhibitory significance that the octasaccharide will be the minimal active fragment preventing syncytium formation and reducing the P24 core antigen level in HIV-IIIB-infected T lymphoblastoid cell line, CEM cells [23, 24].

Modification or derivatization of alginates or chemical cografting could alter its solubility, hydrophobicity, and physicochemical and biological characteristics. Grafting and cross-linking are special chemical treatments by which modification of natural polymer can be done that contains abundance of polysaccharides.

In a study, Hussain *et al.*, have prepared a sustained-release gel containing ibuprofen with varying ratios of acrylamide monomer (AAm) and N,N-methylene bisacrylamide (MBA). They found that 4% w/w of AAm and 0.5% w/w of MBA would be able to generate transparent, thin jelly, and gradual increase in ratio (concentration-dependent) could produce more thick or opaque acrylamide gels. The compactness of the gel depends either on the concentration of monomer or cross-linker, or initiator, or free divalent cations, or a combination of all. The dispersed drug gets entrapped within these gel networks [25]. Ingber *et al.*, [26] prepared poly(N-vinyl-2-pyrrolidone) grafted sodium alginate hydrogels for the controlled release of indomethacin. The poly(N-vinyl-2-pyrrolidone) chains could improve swelling degree, entrapment efficiency, and cumulative release of indomethacin owing to the increase in hydrophilicity of the polymeric matrix.

Superabsorbents are loosely cross-linked hydrophilic polymers that can absorb and retain aqueous fluids up to thousands of times their own weight even under very high pressure. Superabsorbents received much attention and are widely used in many fields because of their superior characteristics to traditional absorbents, e.g., sponge, cotton, and pulp. Similar to the pristine sodium alginate hydrogels, most of anionic superabsorbents are swellable in neutral and basic aqueous media. However, most of anionic superabsorbents do not disintegrate in pH 7.4 phosphate buffer solutions, which makes them different from the pristine sodium alginate hydrogels [27]. Sodium alginate hydrogels collapse in acidic aqueous solutions, but swell in neutral and basic aqueous media. The pH sensitivity of sodium alginate hydrogels makes them ideal matrices for controlled drug delivery for oral administration. The drugs loaded in the drug delivery systems based on sodium alginate hydrogels cannot be released in the stomach but can be released in the intestinal tract at controlled speed, which are helpful to maintain the blood drug concentration, improve drug efficiency, and avoid side effects.

Oxidation reactions on –OH groups at C-2 and C-3 positions of the uronic units of sodium alginate are performed with sodium periodate, which leads, by rupture of carbon–carbon bond, to the formation of two aldehyde groups in each oxidized monomeric unit. Therefore, larger rotational freedom and new reactive groups along the backbone are obtained. Exclusion of light during oxidation is essential for the limitation of side reaction. It is possible to control the alginate degree of oxidation by varying the concentration of the oxidant. Gomez *et al.*, [28] synthesized oxidized alginates at room temperature during 24 h, in an aqueous solution of sodium alginate, using sodium periodate in the dark. Oxidation of alginate chain produced a decrease in the stiffness of the polymer by breaking the C2–C3 bond with a chain scission as a simultaneous reaction. It was observed that over 10% oxidation, no more gels were formed in excess of calcium even over their respective overlap concentration [28].

When alginate is sulfated, it will show high blood compatibility because of the structural similarity to that of heparin, which has been widely used for anticoagulant therapy. Huang *et al.*, [29] reported the alginate sulfates prepared from sodium alginate through reaction with $ClSO_3H$ in formamide. A simple procedure is summarized as follows: 10 g sodium alginate is added to the sulfating reagent containing 80-mL formamide and 20-mL $ClSO_3H$, and the mixture is preserved at 60°C for 4 h to give a brown solution. Two hundred milliliters of acetone is added to precipitate the solution, and the precipitate is redissolved in distilled water and its pH is adjusted to 10–11 by 0.1-mol/L NaOH; then the solution is dialyzed for 72 h and concentrated to give alginate sulfates. The *in vitro* coagulation assay of human plasma containing the sulfates indicated that alginate sulfates had considerably high anticoagulant activity, especially to the intrinsic coagulation pathway [29].

Esterification is often used as a simple method whereby alkyl groups are attached to molecules. This method was successfully used by researchers to modify native alginate, increasing its hydrophobic nature by the addition of alkyl groups to the backbone of the native alginate. Native alginate can be modified by direct esterification with several alcohols in the presence of a catalyst, and the alcohol is present in excess to ensure that the equilibrium is in favor of product formation. Yang *et al.*, [30] synthesized a novel water-soluble amphiphilic cholesteryl ester of alginate. The reaction between the carboxylic acid groups of protonated sodium alginate and the hydroxyl of cholesterol was carried out at room temperature for 24 h, after adding N,N-dicyclohexylcarbodiimide (DCC) as a coupling agent and 4-(N,N-dimethylamino) pyridine (DMAP) as a catalyst. Their research showed that the amphiphilic cholesteryl ester of alginate can self-assemble into the more stable and compact nano-aggregates through the intra- and intermolecular hydrophobic interactions between cholesteryl grafts in aqueous NaCl solution, compared with the parent sodium alginate [30].

The preparation of alginate derivatives via amidation has also been practiced in the last decades. Vallée *et al.*, [31] reported that new amphiphilic derivatives of sodium alginate were prepared by covalent attachment of dodecylamine onto the polysaccharide via amide linkages at different substitution ratios, using 2-chloro-1-methylpyridinium iodide (CMPI) as a coupling reagent. The derivatives are prepared as follows: Na^+ alginate is transformed into its acidic form, and then neutralized to pH 7 by TBA^+OH^-. The TBA-alginate salt is dissolved in DMF and stirred overnight to allow its complete dissolution. Then, CMPI (required amount) and an excess of dodecylamine are added at 0°C. Triethylamine is added into the solution at a concentration similar to that of dodecylamine. The solution is kept at

0°C for 45 min, and then left at room temperature for 20 h. Aqueous NaCl is added to the solution in order to exchange TBA^+ by Na^+ ions. Finally, the polymers are purified by precipitation in 7:1 EtOH–water, washed three times, and dried under diminished pressure [31].

Most of the research on alginates consisted in the chemical modification of the free groups (-OH and -COOH) in their structure, in order to modify their properties for specific applications. Some chemically modified alginates, such as propylene glycol alginate, have been used for some time in industries beyond others that have not hit the market. Although there are many works of synthesis of modified alginate compounds, other forms of study can be explored, as well as the scope of new initiatives for the scientific, technological, and innovational future.

5.4 Biomedical Application of Alginates

Sodium alginate is a particularly attractive material to form hydrogels for biomedical applications [32, 33]. They can be classified into different categories depending on various parameters including the preparation method, the charge, and the mechanical and structural characteristics. Hydrogels are extremely suitable for a variety of applications in the pharmaceutical and medical industry. According to this definition, hydrogels are water-swollen, cross-linked polymeric structures containing covalent bonds produced by the reaction of one or more comonomers [34], physical cross-links due to chain entanglements [2], association bonds including hydrogen bonds or strong van der Waals interactions between chains [35], or crystallites bringing together two or more macromolecular chains [36]. Furthermore, drugs can be incorporated into the matrices and can be released subsequently according to various release profiles depending on the hydrogel properties.

The recent technological advancements with using alginates, issues related to alginates' suitability as a matrix for three-dimensional tissue cultures, adjuvants of antibiotics, and antiviral agents in cell transplantation in diabetes or neurodegenerative diseases treatment, and an update on the antimicrobial and antiviral therapy of the alginate-based drugs are also highlighted. Alginates have been extensively evaluated as vaccine adjuvants or coadjuvants as these polymers were displayed to enhance bioavailability and immunogenicity of antigens after nasal and oral administration.

Alginates can be easily modified through chemical or physical cross-linking in order to form alginate hydrogels and improve physicochemical properties and/or biological activity. Many methods have been described for alginates cross-linking, which includes ionic cross-linking, covalent cross-linking, cell cross-linking, phase transition (thermal gelation), “click” reaction, and free radical polymerization [37].

Due to mechanical stability and proper viscoelastic behavior, alginates are also applied as structural supporting biomaterials for tissue (teeth, bone, and cartilage) reconstruction. Markstedt *et al.*, [38] report on a novel bioink composed of nanocellulose and alginate. Several structures representing cartilage tissue were 3D printed with high fidelity and stability. The rheological properties of the bioink provided excellent printability at room temperature and low pressure—essential conditions when 3D bioprinting with living cells. The aim of this study was to 3D bioprint human chondrocytes using a bioink consisting of cellulose nanofibrils and alginate [38].

Luo *et al.*, [12] suggested a biphasic organic–inorganic scaffold, which consisted of alginate and calcium phosphate cement. The fabricated composite scaffold showed significantly increased mechanical properties, compared with pure alginate, and its cytocompatibility was demonstrated by culturing human mesenchymal stem cells (hMSCs) for 3 weeks [12].

Micro- and mesoporous alginate materials have been elaborated for different applications such as parenteral scaffolds for tissue engineering, drug carrier and release of cell, and repair of cartilage, cardiac, bone, and liver. Moreover, porous alginate scaffolds have the capability to adsorb and absorb an elevated quantity of wound exudates and create conditions with beneficial effect for wound repair. Alginate sponges, based on the lyophilization of ionic or covalently cross-linked hydrogels, produced by the use of gas *in situ* or alternatively alginate precipitates with other oppositely charged polymers, have been widely used as carriers of drugs and biomolecules. One of the great difficulties of this type of production is the industrial staggering, since the characteristics obtained in several academic works were not reproduced in a larger scale.

Popa *et al.*, [39] prepared and characterized a new hydrogel composite based on natural origin polymers, namely, alginate and carrageenan, into different morphology and with adequate properties to sustain the viability of encapsulated cells, envisioning their application as cell delivery vehicles for tissue regeneration. In this study, several methodologies were tested to produce a wide range of systems based on carrageenan–alginate mixtures, with different properties in terms of biological and mechanical performance. The results obtained indicated that the different formulations, both in the form of beads and fibers, have considerable potential as cell-carrier materials for cell delivery in tissue engineering/regenerative medicine applications [39].

Huang *et al.*, [40] fabricated novel nanocomposite hydrogels based on polyelectrolytes alginate and chitin whiskers with biocompatibility based on the pH-induced charge shifting behavior of chitin whiskers. The experimental results indicated that their mechanical properties were significantly improved compared to alginate hydrogel, and the swelling trends were inhibited as a result of the strong electrostatic interactions between the chitin whiskers and alginate. The nanocomposite hydrogels exhibited certain crystallinity and hierarchical structure with nanoscale chitin whiskers, similar to the structure of the native ECM. Moreover, the nanocomposite hydrogels were successfully applied as bone scaffolds for MC3T3-E1 osteoblast cells, showing their excellent biocompatibility and low cytotoxicity [40].

When combined with varied nanostructures, freestanding ultrathin polymeric films demonstrate an increasing potential toward biomedical applications, including thermomechanical sensors, nanoactuators, chemical and biological sensors, and in drug delivery systems. In this context, Gil *et al.*, [41] prepared new multilayered thin films based on chitosan and alginate with magnetic nanoparticles (MNPs), the structural control along the thickness, using the layer-by-layer methodology. The *in vitro* biological performance studies showed that cells were more viable and adherent with higher proliferation rates when MNPs were included in the membranes. Our results suggested the potential of the developed magneto-active free-standing membranes for biomedical applications [41].

Pereira *et al.*, [42] prepared and characterized thin hydrogel films composed by alginate and *Aloe vera* hydrogel in different proportions (95:5, 85:15, and 75:25, v/v) and used calcium ions as cross-linking agents. *A. vera* gel has been attracting the interest of the scientific community due to its healing properties and anti-inflammatory, antiseptic, and antimicrobial activity, which can be particularly important for wound-healing applications. The incorporation of *A. vera* into the alginate films can contribute to preserving its therapeutic properties for a longer period of time, when compared with other formulations such as creams. This research work also suggests that alginate/*A. vera* films have suitable properties for wound healing and drug delivery applications [42].

Alginate microcapsules have been used for the encapsulation of living cells to address a variety of diseases. The alginate microcapsules are routinely produced using an ionic gelation procedure wherein alginate solution is dropped into a calcium chloride gelling solution.

Currently, two alginate-based products, DIABECELL® and NTCELL®, are in the advanced stage of clinical investigations. DIABECELL® implant consists of microencapsulated neonatal porcine islets capable of secreting insulin. The single system is designed to be delivered into the patient's abdomen during laparoscopy procedure. Each multilayer microcapsule comprises the inner core of alginate (M/G ratio 60:40) cross-linked with calcium chloride and coated with poly-L-ornithine-(PLO-) polycationic polymer responsible for strengthening the capsule wall. To reduce the risk of immunogenicity arisen from the presence of PLO, the additional outer layer prepared of alginate is present [43, 44].

NTCELL® with choroid plexus cells encapsulated within Immupel platform has been displayed to regenerate damaged tissue and significantly restore function in humans with Parkinson's disease. Following implantation into an impaired site within the brain of model animals, NTCELL® was found to promote the production of cerebrospinal fluid as well as nerve growth factors. In addition to Parkinson's disease, the product may have the potential to be utilized in a number of other neurodegenerative disorders, including Huntington's, Alzheimer's, or motor neuronal diseases [45].

5.5 Future Perspective of the Use and Biomedical Applications

Modified cell encapsulation technology, stem cell research, and different genic therapy strategies point to a rapid development of grafting techniques, as well as the use of scaffold 3D as support for tissue engineering with simultaneous use of the treatment of inflammatory processes of intelligent and progressive delivery.

Additionally, aginates have potential for application in the biomedical area, due to their high biocompatibility and bioabsorption for wound regeneration and artificial organs production, thus presenting a path of great scientific and technological growth.

On the other hand, a preparation of biopolymer nanomaterials points to the use of biological control systems for the biomarkers of specific targets in the treatment of cancer, among others.

Alginates have also been applied in nutritional and functional foods, thus demonstrating the great versatility of this natural polymer.

References

1. Nikolić, G.S. and Cakić, M.D., Analysis of bioactive oligosaccharide-metal complexes by modern FTIR spectroscopy: Copper complexes, in: *Fourier Transforms—New Analytical Approaches and FTIR Strategies*. pp. 15–44, In Tech., 2011.

2. Peppas, N., Bures, P., Leobandung, W., Ichikawa, H., Hydrogels in pharmaceutical formulations. *Eur. J. Pharm. Biopharm.*, 50, 27–46, 2000.
3. Coviello, T., Matricardi, P., Marianecci, C., Alhaique, F., Polysaccharide hydrogels for modified release formulations. *J. Control. Release*, 119, 5–24, 2007.
4. Cakic, M., Mitic, Z., Nikolic, G.S., Savic, I., Design and optimization of drugs used to treat copper deficiency—Review. *Drug Discov.*, 8, 1–11, 2013.
5. Cakić, M.D. and Nikolić, G.S., Iron complexes of pharmaceutical interest—Antianemics. *Chem. Ind.*, 57, 437–443, 2003.
6. George, M. and Abraham, T., Polyionic hydrocolloids for the intestinal delivery of protein drugs: Alginate and chitosan—Review. *J. Control. Release*, 114, 1–14, 2006.
7. Skjak-Braek, G., Grasdalen, H., Larsen, B., Monomer sequence and acetylation pattern in some bacterial alginates. *Carbohydr. Res.*, 154, 239–250, 1986.
8. Espevik, T., Otterlei, M., Skjak-Braek, G., Ryan, L., Wright, S.D., Sundan, A., The involvement of CD14 in stimulation of cytokine production by uronic acid polymers. *Eur. J. Immunol.*, 23, 255–261, 1993.
9. Gombotz, W.R. and Wee, S.F., Protein release from alginate matrices. *Adv. Drug Deliv. Rev.*, 31, 267–285, 1998.
10. Chickering, D.E. and Mathiowitz, E., Bioadhesive microspheres: A novel electrobalance-based method to study adhesive interactions between individual microspheres and intestinal mucosa. *J. Control. Release*, 34, 251–261, 1995.
11. Chang, H., Park, H., Kelly, P., Robinson, J., Bioadhesive polymers as platforms for oral controlled drug deliver synthesis and evaluation of some swelling, water insoluble bioadhesive polymers. *J. Pharm. Sci.*, 74, 399–405, 1985.
12. Luo, Y., Wu, C., Lode, A., Gelinsky, M., Hierarchical mesoporous bioactive glass/alginate composite scaffolds fabricated by three-dimensional plotting for bone tissue engineering. *Biofabrication*, 5, 1–10, 2013.
13. Lin, H.R. and Yeh, Y.J., Porous alginate/hydroxyapatite composite scaffolds for bone tissue engineering: Preparation, characterization, and *in vitro* studies. *J. Biomed. Mater. Res., B*, 71B, 52–65, 2004.
14. Liu, X. and Ma, P.X., Polymeric scaffolds for bone tissue engineering. *Ann. Biomed. Eng.*, 32, 477–486, 2004.
15. Augst, A.D., Kong, H.J., Mooney, D.J., Alginate hydrogels as biomaterials. *Macromol. Biosci.*, 6, 623–633, 2006.
16. Baldwin, A.D. and Kiick, K.L., Polysaccharide-modified synthetic polymeric biomaterials. *Biopolymers*, 94, 128–140, 2010.
17. Abbah, S.A., Lu, W.W., Chan, D., Cheung, K.M.C., Liu, W.G., Zhao, F., Li, Z.Y., Leong, J.C.Y., Luk, K.D.K., Osteogenic behavior of alginate encapsulated bone marrow stromal cells: An *in vitro* study. *J. Mater. Sci.: Mater. Med.*, 19, 2113–2119, 2008.

18. Tonnesen, H.H. and Karlsen, J., Alginate in drug delivery systems. *Drug. Dev. Ind. Pharm.*, 28, 621–630, 2002.
19. Wang, W.B. and Wang, A.Q., Synthesis and swelling properties of pH-sensitive semi-IPN superabsorbent hydrogels based on sodium alginate-g-poly (sodium acrylate) and polyvinylpyrrolidone. *Carbohydr. Polym.*, 80, 1028–1036, 2010.
20. Lee, K.Y. and Mooney, D., Alginate: Properties and biomedical applications. *J. Prog. Polym. Sci.*, 37, 106–126, 2012.
21. Donati, I., Morch, Y.A., Strand, B.L., Skjåk-Bræk, G., Paoletti, S., Effect of elongation of alternating sequences on swelling behavior and large deformation properties of natural alginate gels. *J. Phys. Chem. B*, 113, 12916–12922, 2009.
22. Wang, W., Wang, S.-X., Guan, H.-S., The antiviral activities and mechanisms of marine polysaccharides: An overview. *Marine Drugs*, 10, 2795–2816, 2012.
23. Ji, J., Wang, L., Wu, H., Luan, H., Bio-function summary of marine oligosaccharides. *Int. J. Biol.*, 3, 74–86, 2011.
24. Hu, X., Jiang, X., Hwang, H., Liu, S., Guan, H., Antitumour activities of alginate-derived oligosaccharides and their sulphated substitution derivatives. *Eur. J. Phycol.*, 39, 67–71, 2004.
25. Omidian, H., Rocca, J.G., Park, K., Elastic, superporous hydrogel hybrids of polyacrylamide and sodium alginate. *Macromol Biosci.*, 6, 703–710, 2006.
26. Ingber, D., Karp, S., Plopper, G., Hansen, L., Mooney, D., Chapter 2, in: *Physical Forces and the Mammalian Cell*, Academic Press, New York, 1993.
27. Zhang, J., Chen, H., Wang, A., Study on superabsorbent composite III. Swelling behaviors of polyacrylamide/attapulgite composite based on acidified attapulgite and organo-attapulgite. *Eur., Polym. J.*, 41, 2434–2442, 2005.
28. Gomez, C.G., Rinaudo, M., Villar, M.A., Oxidation of sodium alginate and characterization of the oxidized derivatives. *Carbohydr. Polym.*, 67, 296–304, 2007.
29. Huang, R.H., Du, Y.M., Yang, J.H., Preparation and *in vitro* anticoagulant activities of alginate sulfate and its quarterized derivatives. *Carbohydr. Polym.*, 52, 19–24, 2003.
30. Yang, L.Q., Zhang, B.F., Wen, L.Q., Amphiphilic cholesteryl grafted sodium alginate derivative: Synthesis and self-assembly in aqueous solution. *Carbohydr. Polym.*, 68, 218–225, 2007.
31. Vallée, F., Müller, C., Durand, A., Schimchowitsch, S., Dellacherie, E., Kelche, C., Cassel, J.C., Leonard, M., Synthesis and rheological properties of hydrogels based on amphiphilic alginate–amide derivatives. *Carbohydr. Res.*, 344, 223–228, 2009.
32. Isíklan, N., Inal, M., Yiğitoğlu, M., Synthesis and characterization of poly(N-vinyl-2-pyrrolidone) grafted sodium alginate hydrogel beads for the controlled release of indomethacin. *J. Appl. Polym. Sci.*, 110, 481–493, 2008.

33. Atala, A., Kim, W., Paige, K.T., Vacanti, C.A., Retik, A.B., Endoscopic treatment of vesicoureteral reflux with a chondrocyte-alginate suspension. *J. Urol.*, 152, 641–643, 1994.
34. Wichterle, O. and Lim, D., Hydrophilic gels for biological use. *Nature*, 185, 117–118, 1960.
35. Ovsianikov, A., Deiwick, A., Van, V.S., Dubruel, P., Moller, L., Drager, G., Chichkov, B., Laser fabrication of three-dimensional cad scaffolds from photosensitive gelatin for applications in tissue engineering. *Biomacromolecules*, 12, 1387–1408, 2011.
36. Gupta, P., Vermani, K., Garg, S., Hydrogels: From controlled release to pH-responsive drug delivery. *Drug Discov Today*, 7, 569–579, 2002.
37. Sun, J. and Tan, H., Alginate-based biomaterials for regenerative medicine applications. *Materials*, 6, 1285–1309, 2013.
38. Markstedt, K., Mantas, A., Tournier, I., Ávila, H.M., Hägg, D., Gatenholm, P., 3D Bioprinting human chondrocytes with nanocellulose–alginate bioink for cartilage tissue engineering applications. *Biomacromolecules*, 16, 1489–1496, 2015.
39. Popa, E.G., Gomes, M.E., Reis, R.L., Cell delivery systems using alginate–carrageenan hydrogel beads and fibers for regenerative medicine applications. *Biomacromolecules*, 12, 3952–3961, 2011.
40. Huang, Y., Yao, M., Zheng, X., Liang, X., Su, X., Zhang, Y., Lu, A., Zhang, L., Effects of chitin whiskers on physical properties and osteoblast culture of alginate based nanocomposite hydrogels. *Biomacromolecules*, 16, 3499–3507, 2015.
41. Gil, S., Silva, J.M., Mano, J.F., Magnetically multilayer polysaccharide membranes for biomedical applications. *Biomater. Sci. Eng.*, 1, 1016–1025, 2015.
42. Pereira, R., Carvalho, A., Vaz, D.C., Gil, M.H., Mendes, A., Bártolo, P., Development of novel alginate based hydrogel films for wound healing applications. *Int. J. Biol. Macrom.*, 52, 221–230, 2013.
43. Hillberg, A.L., Kathirgamanathan, K., Lam, J.B., Law, L.Y., Garkavenko, O., Elliott, R.B., Improving alginate-poly-L-ornithine-alginate capsule biocompatibility through genipin cross-linking. *J. Biol. Mater. Res. B: Appl. Biomater.*, 101, 258–268, 2013.
44. Qi, M., Transplantation of encapsulated pancreatic islets as a treatment for patients with type 1 diabetes mellitus. *Adv. Med.*, 1–15, 2014.
45. Irving, S., Gillespie, L., Richardson, R., Rowe, D., Fallon, J.B., Wise, A.K., Electroacoustic stimulation: Now and into the future. *BioMed Res. Int.*, 2014, 1–17, 2014.

6

Alginates in Pharmaceutical and Biomedical Application: A Critique

Vivek Dave*, Kajal Tak, Chavi Gupta, Kanika Verma and Swapnil Sharma

Department of Pharmacy, Banasthali Vidhyapith, Rajasthan, India

Abstract

Alginate is one of the most widely used polymers in the pharmaceutical, cosmetic, food, and biomedical industry. One of the reasons for its widespread use is its biocompatibility with body fluids, which make it suitable for human use. The present chapter mainly focuses on wide application of alginate usage in pharmaceutical and biomedical industry. Various studies have found that alginate is suitable in biomedical applications such as 3-D tissue culture, antibiotics, antiviral agents in cell transplant in diabetes, and various neurodegenerative diseases. Alginate helps in wound dressings as it maintains a moist microenvironment and minimizes bacterial infection at wound site. In pharmaceutical field, it is used as a tablet binder, disintegrant, and release modifying agents, and also possesses mucoadhesive behavior and has a unique ability to undergo *in situ* gelation, which makes it useful in a wide range of injectable vehicles, for tissue engineering and topical drug delivery system, taste masking, etc. While in biomedical field, it is used for reduction of secretion of inflammation cytokines, cell carrier with mechanical stability and relative permeability, immobilization of bacteria, etc. In the physiological medium, alginate biodegrades so there is no chance of toxicity with alginates. There are many other properties that are possessed by alginates that make it a promising tool for many industries.

Keywords: Alginate, biocatalysts, biomedical, glycol, glycosidic, microsphere, polymer

6.1 Introduction

There are various natural biomaterials that are being used for many years; one of these are alginates. Alginate was first discovered by British chemist

**Corresponding author*: vivekdave1984@gmail.com

Shakeel Ahmed (ed.) Alginates, (95–124) © 2019 Scrivener Publishing LLC

E.C.C. Stanford in 1881 [7, 11]. It consists of a group of anionic polysaccharides, which is obtained from the cell wall of brown algae and various species of *Ascophyllum*, *Durvillaea*, *Ecklonia*, *Laminaria*, *Lessonia*, *Macrocystis*, *Sargassum*, and *Turbinaria* [8]. It is also known as algin, alginate, or alginic acid belonging to polysaccharide category of the biomaterial, carrying anionic charge on it. On the basis of two sugars—mannuronic (M) and guluronic (G) acid—alginates are built and both sugars belong to the category of urinate [9–11]. Alginates are a biopolymer of linear chain made up of 1,4 linked β-D-mannuronic acid and 1,4 α-1-gluconic acid group arranged in homogeneous (poly M, poly G) or heterogeneous (MG) block-like pattern [13]. According to the initial source, the composition and sequence of G and M block in alginates vary [12–15]. Alginate molecule contains different proportion of M, G, and MG blocks [18] so it can be regarded as a block copolymer varying with the seaweed source. Different types of alginate contribute different properties due to the difference in the proportion of blocks present in them such as alginates, which possess high G block found to have good gel strength, while alginates with high proportion of M block have immune-stimulating behavior [16, 17]. There are different derivatives of alginates such as sodium alginate, propylene glycol alginate, and ammonium alginate, which possess different properties [19]. Natural origin-based biodegradable films have been widely discovered for biomedical purpose such as structural supporting biomaterial for tissue reconstruction including wound healing due to its biodegradability, biocompatibility, and similarity in human tissue properties [1–3, 20]. There are various other applications of alginate in the field of pharmaceutical, biomedical, cosmetic, textile, and food industry. The salts of alginate are considered safe, biocompatible, and nonimmunogenic. In this section, basic information about the alginate is provided, along with its characteristic behavior and different applications [21–24].

Specifically, this chapter deals with basic information about the alginate, along with its characteristic behavior and application in various fields involving pharmaceutical industries, food industries, and cosmetic industries and in other fields such as in textile printing, welding rods, paper industries, etc. [25–27].

6.2 Structure of Alginate

Alginate is an unbranched binary copolymer or we can say alginate is a true block copolymer composed of homopolymeric region of M and G, called as M and G block [4–7], which have no regular repeating units [28]. These blocks (M and G) provide unique features to the alginates; for example,

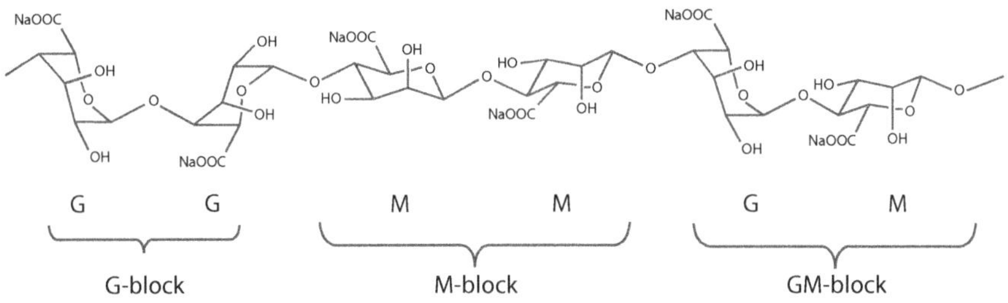

Figure 6.1 Structure of alginate showing all four glycosidic linkages, i.e., linkage in MM, GG, MG, and GM group.

G block has a gelling property. But the M content alginate was more immunogenic and more potent in inducing cytokine production as compared to G alginates. There are various characterization studies performed to gain knowledge about the confirmation of this uronate group. X-ray diffraction study was performed to observe the confirmation of mannuronate- and guluronate-rich alginate in homopolymeric blocks. It was found that guluronate possesses ${}^{1}C_{4}$ confirmation and mannuronate possesses ${}^{4}C_{1}$ confirmation. According to the arrangement of these blocks, the viscosity of alginate varies. The order of their viscosity is

$$MG < MM < GG$$

Alginates possess all four glycosidic linkages in their structure: (1) diequatorial linkage in MM group, (2) diaxial linkage in GG group, (3) equatorial axial linkage in MG group, and (4) axial equatorial linkage in GM group, as shown in Figure 6.1 [29].

6.3 Different Types of Alginates Used in Pharmaceutical Industries

Commercially, alginate is present in its different salt forms. Various salts of alginates are

1. Sodium alginate
2. Ammonium alginate
3. Potassium alginate
4. Calcium alginate
5. Magnesium alginate
6. Propylene glycol alginate

6.4 Properties of Alginate

Alginate appears in the form of filamentous, granular, and powder form with white to yellowish-brown color [31]. The molecular formula of alginate is $C_{14}H_{22}O_{13}$ [32–35]. The molecular weight of the alginate varies from 33,000 to 40,000 gm/mol. Residues of M and G and their length vary in different types of alginate [36–39]. Alginate is found to be insoluble in water and organic solution, but its monovalent salts and esters are water soluble. Alginates having heterogeneous structure are soluble at less pH compared to poly M and poly G alginic molecules [40, 41], which get precipitated in this condition. The viscosity of the alginates depends on the M and G block present in it [42–46]. The property of the alginates changes according to physical, chemical, and environmental changes. The order of stability of different salts of alginate is given as

Sodium > Ammonium > Alginic acid

Thus, it is found that the salt of the alginate is more stable as compared to pure alginate [47].

6.5 Pathway for the Biosynthesis of Alginate

The pathway of alginate biosynthesis is generally divided into four broad categories:

(i) Synthesis of precursor substrate,
(ii) Polymerization and cytoplasmic membrane transfer,
(iii) Periplasmic transfer and modification, and
(iv) Export through the outer membrane.

For the production of sodium alginate, firstly, the brown algae and seaweeds were chopped; then, the seaweeds were milled, washed with acid, and then extracted with the help of alkali like Na_2Co_3; and they were clarified, filtered, and precipitated with the help of calcium salt (mixture of calcium alginate + alginic acid formed) [45–48].

Then, the mixture was washed to obtain the pure form and then neutralized with the help of sodium carbonate, potassium carbonate, ammonium carbonate, magnesium carbonate, and calcium carbonate so as to obtain the desirable alginate [30, 49].

The procedure for the extraction of sodium alginate was explained in Figures 6.2 and 6.3 [50–53].

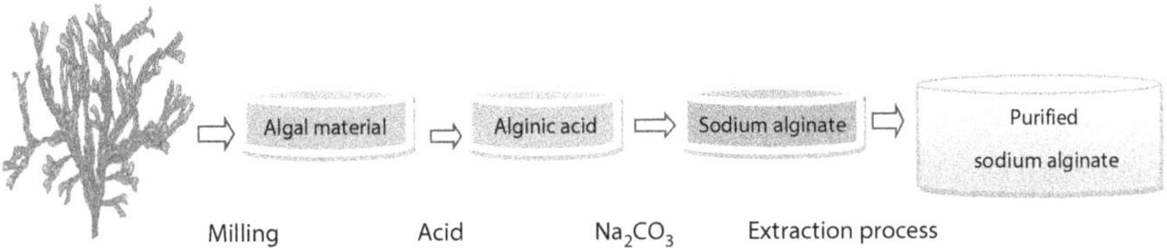

Figure 6.2 Sodium alginate extraction from brown algae.

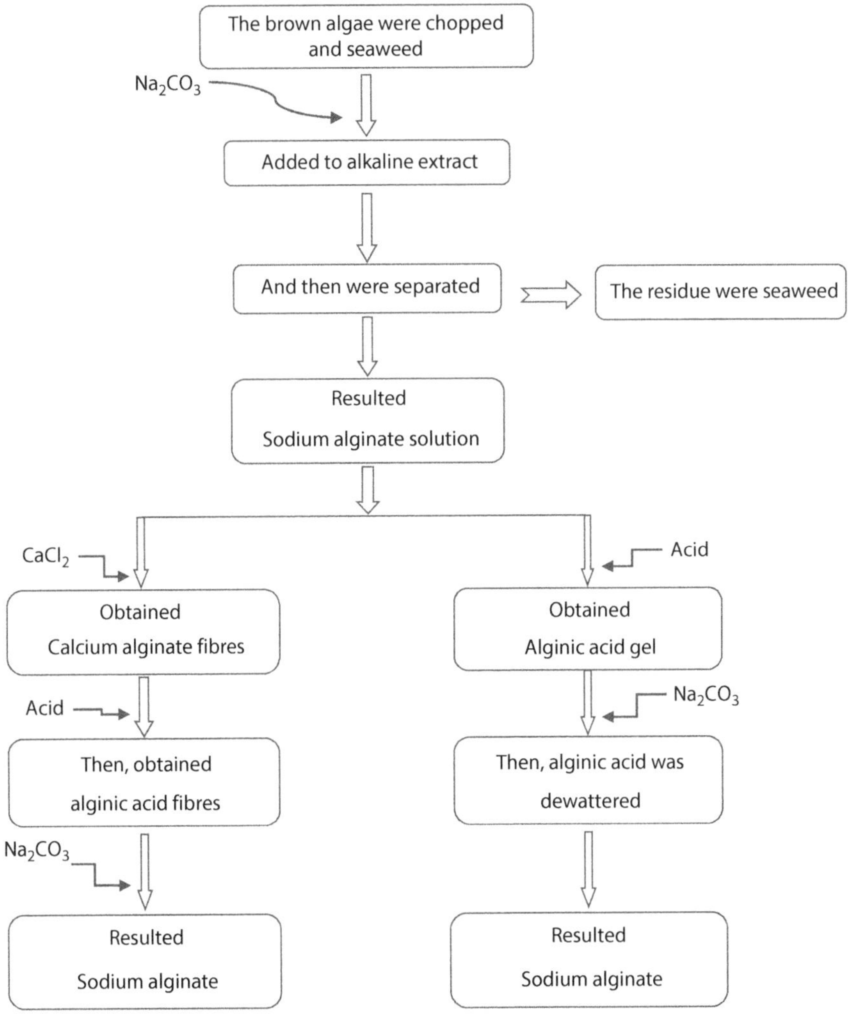

Figure 6.3 Flow chart for the production of sodium alginate.

6.6 Regulatory Consideration of Alginate

According to US Pharmacopoeia, alginate is explained as a polymannuronate even on the presence of glucuronate, and the building block structure of alginate is now well known [54]. Before the use of alginate in the pharmaceutical and biotechnology field, its regulatory aspect must be addressed, mainly regarding long-term safety, toxicity [55–59], reproducibility of the product, and its characterization and functionality, which includes product specification, product stability, and validation of analytical methods. The use of alginate in human studies must require a regulatory approval by the FDA [60]. Drug master file (DMF) contains all the information regarding the manufacturing, specification, and safety of the product. Regulatory considerations regarding alginate in different pharmacopoeia are discussed in Table 6.1 [61, 62].

6.7 Applications

1. Pharmaceutical field—The application of alginate in industry is based on its ability to retain water and swelling, viscosifying, and stabilizing properties.

Dosage form—The presence of free carboxyl group exhibits mucoadhesive property and allows it to interact with the mucin, which makes it a good excipient for buccal, nasal, ocular, and gastrointestinal dosage form. This interaction is due to the hydrogen and electrostatic bonding. pH has a great impact on the mucoadhesive character as only an ionized carboxyl group is capable of interacting with mucin.

In tablet—Alginate acts as a binding or adhesion agent in tablet so as to hold the powder together or in granular form, but it should be added during the granulation process, not after the granulation. Depending on the concentration of alginate added, it acts as a disintegrant agent and helps in rapid release of the drug. Thus, the tablets prepared with alginate have more mechanical strength as compared to other tablets that are prepared from starch. From different types of alginate, sodium alginate was used to prepare a matrix tablet by direct compression method or wet granulation method [63]. Interaction of sodium alginate with calcium salts results in formation of water insoluble gels which allows encapsulation of variety of drugs in substantial amount. Upon coming in contact with water, an *in situ* gel is formed and the drug molecule diffuses slowly from this gel [64]. Thus, sodium alginate helps to provide a sustained and controlled release

Table 6.1 Characteristics of sodium alginate recommended by the United States Pharmacopeia (USP), European Pharmacopeia (Eur. Ph.), and Indian Pharmacopeia (IP).

Parameter	**US 32-NF 27**	**Eur. Ph. 8.0**	**IP**
Identification	+	+	+
Loss on drying	≤15.0%	≤15.0%	≤15.0%
Microbial limits	≤200 cfu/g	TAMC: ≤1000 cfu/g TYMC: ≤100 cfu/g	1.0 g free from *Escherichia coli*, 10.0 g test free from *Salmonellae*
Appearance of solid product	n.d.	White or pale yellowish-brown powder	White or pale yellowish-brown powder
Solubility	n.d.	Slowly soluble in water, practically insoluble in ethanol 96%	n.d.
Content	n.d.	90.8%–106.0% of dried basis	n.d.

(Continued)

Table 6.1 Characteristics of sodium alginate recommended by the United States Pharmacopeia (USP), European Pharmacopeia (Eur. Ph.), and Indian Pharmacopeia (IP). (*Continued*)

Parameter	**US 32-NF 27**	**Eur. Ph. 8.0**	**IP**
Lead	Not more than 0.001%	n.d.	>20 ppm
Packaging and storage	Preserved in tight containers	n.d.	n.d.
Appearance of solution	n.d.	Not more opalescent than reference formazin suspension in water and not more intensely colored than intensity 6 of the range of reference solutions of the most appropriate color	Not more opalescent than reference formazin suspension in water and not more intensely colored than intensity 6 of the range of reference solutions of the most appropriate color
Calcium	n.d.	≤1.5%	n.d.
Total ash	18.0%–27.0%	n.d.	n.d.
Heavy metals	≤0.004%	≤20 ppm	40.0 ppm
Arsenic	≤1.5 ppm	n.d.	n.d.
Chlorides	n.d.	≤1.0%	≤1.0%
Sulfated ash	n.d.	30.0%–36.0%	30.0–36.0%
Absence of specified microorganisms	*Salmonella* sp., *Escherichia coli*	*Salmonella* sp., *Escherichia coli*	*Salmonella* sp., *Escherichia coli*

of the drug [65, 66]. At higher concentration of alginate, alginate gel will be most effective in retarding the drug release [67].

In suspension—Alginate acts as a gelling agent and thus increases the viscosity of the formulation and reduces the sedimentation rate, which improves the overall stability of the product. In case of emulsion, alginate, to a small extent, helps to prevent the separation of oil and water phase. Alginate suspension was used in controlling postprandial esophageal acid exposure in gastroesophageal reflux disease (GERD) patients [68].

Hydrogel—Alginates form a 3-D hydrophobic polymer network in aqueous solution and maintain its structure so it is used in the making of hydrogel, which is used in wound dressing. It possesses properties such as smoothness, high water retention, elasticity, and malleability and provides moisture, which protects the wound from desiccation. The cost-effectiveness of the alginate provides good relationship with the clinical efficacy and manufacturing cost. Alginate hydrogel was used to provide localized delivery of the drug, mainly the low-molecular-weight drugs, and provides a controlled release of the drug with minimum exposure to other tissue. However, the main problem with alginate hydrogel was that it loses its mechanical strength over time, but this problem can be overcome by modifying the alginate polysaccharide with stable covalent cross-linkers such as RGD-containing peptides [69].

Role of alginate in gastrointestinal dosage form—Alginate was used in various GI formulations to modify the GI transit time of drugs, especially the drugs that have poor bioavailability from the lower regions of the digestive tract. Alginate was given in the form of buoyant capsule containing drug substance along with a hydrocolloid gelling agent, i.e., HPMC, to provide a controlled and sustained release of basic drug [71]. In the stomach, as the water penetrates into the capsule, it causes the hydration of the HPMC layer and forms the gel layer. At low pH, alginate is converted into alginic acid; as the gel layer gets eroded, the drug gets dissolved in the gel and diffuses to the surrounding environment and the dosage form is emptied from the stomach after the loss of buoyancy; the gelled structure becomes more porous as the alginic acid is converted into more soluble salt, and from the porous structure, release of drug occurs very slowly [70]. Alginate was used in various raft formulations to treat heartburn and esophagitis [72]. As sodium alginate enters the stomach, it will form a colloidal gel and a raft is created with the bubbles of carbon dioxide in the gel; then the gel will float on the surface of gastric content and drug diffuses slowly from the gel [73].

Role of alginate in nasal drug delivery—Nasal drug delivery provides certain advantages such as rapid absorption of drugs, improved patient

Table 6.2 List of pharmaceutical products based on ALG.

Product	Main ingredients	Description	Indications	Reference
	Dermal application			
Saf-Gel_ gel	Carbomer, propylene glycol sodium/calcium alginate	Provides moist environment at wound surface	Dry and sloughy necrotic wounds, pressure and venous ulcers, second-degree burns, cuts, abrasions and skin tear, noninfected diabetic foot ulcers	Sussman and Bates-Jensen [96]
Flaminal Forte_gel	Hydrated alginates polymers in a polyethylene glycol (PEG) matrix with a biologic enzyme system based on glucose oxidase and lactoperoxidase stabilized by guaiacol	Dissolution of dry scab and necrotic material, absorption of lysed material and bacteria by alginates in hydrated form	Leg and diabetic ulcers, pressure sores, complex grazes, burns, oncology, and wounds dermatosurgery	Rashaan *et al.*, [97]
Purilon gel	Carboxy methylcellulose, calcium alginate	Provides moist environment at wound surface	Dry and sloughy necrotic wounds, pressure and venous ulcers, second-degree burns, cuts, abrasions and skin tear, noninfected diabetic foot ulcers	Cal'o and Khutoryanskiy [98]

(*Continued*)

Table 6.2 List of pharmaceutical products based on ALG. (*Continued*)

Product	Main ingredients	Description	Indications	Reference
SeaSorb_dressing	Calcium alginate	Creates moist environment at wound surface, conversion soft fibers to wet gel	Heavily exuding wounds including leg and pressure ulcers, diabetic ulcers, and second-degree burns, cavity wounds	Ausili *et al.*, [100]
Fibracol Plus_ dressing [32]	90% collagen and 10% calcium alginate	Provides moist environment at wound surface, tissue granulation, epithelialization, and healing	Exuding wounds including full-thickness and partial-thickness wounds; pressure ulcers; venous ulcers; ulcers caused by mixed vascular etiologies; diabetic ulcers; second-degree burns	Bale *et al.*, [101]
Tromboguard_ dressing	Two-layer dressing built from hydrophilic polyurethane sponges and biologically active layer containing chitosan, sodium alginate, calcium alginate, and silver cations	Strong hemostatic and antibacterial activity	Control of bleeding traumatic and postoperative wounds	Bale *et al.*, [101]

(*Continued*)

Table 6.2 List of pharmaceutical products based on ALG. (*Continued*)

Product	Main ingredients	Description	Indications	Reference
Guardix-SG_	Sodium alginate, poloxamer, calcium chloride	Creates thermosensitive viscous gel in contact with body temperature and forms a mechanical barrier that separates injured tissues	In spine and thyroid surgeries to reduction of the incidence postoperative adhesions	Sohn *et al.*, [103], Park *et al.*, [104]
Hyalogran_dressing	Ester of hyaluronic acid (HA) and sodium alginate	Exudate absorbs and transforms to soft gel; removes necrotic tissue	Variety of exuding wounds including leg ulcers, pressure sores, ischemic and diabetic wounds, particularly those that are covered with slough and necrotic tissue or areas that are difficult to dress	Carella *et al.*, [99]
Algivon_dressing	Calcium alginate dressing impregnated with Manuka honey	Binds of exudate, regeneration	Sloughy, necrotic, and malodorous wound	Porter and Kelly [102]
	Periodontal application			
Progenix putty_, Progenix plus_injection	Demineralized bone matrix in type-1 bovine collagen and sodium alginate	Regeneration, complementation of bone losses; periodontal diseases	Bony voids or gaps of the skeletal system	Gruskin *et al.*, [106]

(*Continued*)

Table 6.2 List of pharmaceutical products based on ALG. (*Continued*)

Product	Main ingredients	Description	Indications	Reference
Emdogain_ gel	Enamel matrix derivative (EMD), propylene glycol alginate	Regeneration, periodontal diseases, paradontosis	1-, 2-, and 3-wall intrabony defects, class II mandibular furcation defects with minimal interproximal bone loss, recession defects	AlMachot *et al.*, [107], Sculean *et al.*, [109]
Natalsid_ suppositories	Sodium alginate	Anti-inflammatory local action	Chronic hemorrhoids, proctosigmoiditis, and chronic anal fissures, after surgical interventions in the area of the rectum	Abramowitz *et al.*, [105]
	Arthroscopic application			
Chondro Art 3D_ injection	Autologous chondrocytes situated on a hydrogel scaffold built from connection of alginate and agarose	Increase production and growth of cartilage	Degenerative diseases of joints and backbones (osteoarthrosis, osteochondrosis)	Barzegari and Saei [110]

(*Continued*)

Table 6.2 List of pharmaceutical products based on ALG. (*Continued*)

Product	**Main ingredients**	**Description**	**Indications**	**Reference**
	Oral administration			
Algicid_ suspension/ tablets	500-mg sodium alginate, 100-mg potassium bicarbonate per 5 ml/ per 1 tablet	Creates a mechanical barrier between the stomach and the esophagus, which prevents the reflux, recurrent symptoms of respiratory, choking, dysphagia, heartburn, belching, irritability	Children and infants from the first days of life reflux treatment	Jakaria *et al.*, [112]
Gastrotuss® baby syrup	Magnesium alginate, simethicone, fructose, xanthan gum, honey, D-panthenol, fluid extracts of *Althaea officinalis*, *Papaver rhoeas,* zinc oxide, sodium bicarbonate, sodium	Creates a mechanical barrier between the stomach and the esophagus, which prevents the reflux, recurrent symptoms of respiratory, choking, dysphagia, heartburn, belching, irritability	Children and infants from the first days of life reflux treatment	Ummarino *et al.*, [111]

(*Continued*)

Table 6.2 List of pharmaceutical products based on ALG. (*Continued*)

Product	Main ingredients	Description	Indications	Reference
Gaviscon double action tablets	250-mg sodium alginate, 106.5-mg sodium bicarbonate, and 187.5-mg calcium carbonate per tablet	Creates a mechanical barrier between the stomach and the esophagus, which prevents the reflux, recurrent symptoms of respiratory, choking, dysphagia, heartburn, belching, irritability	Children and infants from the first days of life reflux treatment	Thomas *et al.*, [114]
Gaviscon double action liquid	250-mg sodium alginate, 106.5-mg sodium bicarbonate, and 162.5-mg calcium carbonate per 5 ml	Creates a mechanical barrier between the stomach and the esophagus, which prevents the reflux, recurrent symptoms of respiratory, choking, dysphagia, heartburn, belching, irritability	Children and infants from the first days of life reflux treatment	De Ruigh *et al.*, [113]

compliance, ease of administration, fast onset of action, and avoiding irritation of GI membrane, but the major drawback is nasal mucociliary clearance, which prevents sustained nasal drug administration [74, 75]. Thus, to retain the drug in the nasal cavity, mucoadhesive polymer is required to form microparticle, which prevents nasal mucociliary clearance [76]. Alginate has excellent mucoadhesive property, which forms an intimate contact with the mucus layer, and the release of microparticle from gel layer occurs slowly and thus retains the drug in the nasal cavity for a prolonged time [77].

As antibacterial, antiviral, antifungal—Alginate shows antiviral, antibacterial, and antifungal property [78–79]. Negative charge on the alginate was found to interact with the outer surface of bacterial cell, which leads to disruption and intracellular substance leakage, function of membrane decrease due to formation of viscous alginate layer around the bacterial cell, which prevents the nutrition intake [80, 81]. The antibacterial action is thought to be due to chelating property, which is responsible for modulating toxin and microbial growth and production of other factor that is crucial for stability of microbe. Alginate shows antibacterial activity against a number of species including *Pseudomonas*, *Escherichia*, *Proteus*, etc. [82, 83]. The antiviral property of alginate is due to that it interacts with positively charged host cell and prevents the contact of virus with the host cell due to which viral material could not enter the cell. Another mechanism by which alginate shows antiviral property is that it forms a physical barrier for the penetration of virus. It shows antiviral property against a number of viruses including *Herpesviridae*, *Togaviridae*, *Rhabdoviridae*, etc. [84–86].

Vaccine adjuvant—Alginate possesses the property of increasing bioavailability and immunogenicity of the antigen; it is widely used in the vaccine adjuvant or coadjuvant. Alginates prolong the release of antigen and thus enhance the immunogenicity. Also, the alginate particle does not agglomerate to any organ of the body. Vaccines containing alginate as adjuvant are more efficacious in producing the antibodies as compared to other adjuvants. Alginate was used to encapsulate hepatitis B antigen to stimulate the local and systemic immune response followed after oral vaccination [91]. Alginate is nontoxic, nonirritant, and biodegradable and does not produce any toxic effects, so alginate is considered as a suitable polymer for the delivery of vaccines [92].

Anti-diabetic effect—Hypoglycemic activity of the alginate is due to the association with reduced α-amylase activity; intestinal enzyme hydrolyzes the bond present between glucose residues in carbohydrate polymer [87, 88]. Alginate is capable of inhibiting the glucose transporters and also glucose intestinal absorption rate and thus attenuates the postprandial glycemic response. Sodium alginate in small dose lowers the postprandial rise in blood glucose and serum insulin [89, 90].

Wound healing property—Alginate dressing can be useful for moderately to heavily exuding wounds. As compared to various cotton gauzes, alginates are more popular in treating the wounds because of its capability of absorbing large volume of exudate and provide a moist environment for wound healing. The sodium ions of the wound fluid convert the cross-linked sodium alginate to viscous sodium alginate liquid, which helps to protect and soothe the wound. Moreover, it was observed that calcium alginate activates human macrophages to produce TNFα (tissue necrosis factor α), which helps in providing the inflammatory signals and also helps in the wound healing process [93]. When alginate dressings are placed on wounds, they are firstly dry, but upon absorbing exudate, they become larger and more gel-like, which helps in clearing the wounds and protects them from harmful bacteria and helps in new skin growth by keeping the wound area moist [94, 95].

2. Biomedical field—Over the last few decades, alginate was used to deliver various biomolecules such as protein, cell, and DNA.

Protein delivery—Alginate was used to deliver protein drugs. When proteins are incorporated in the formulation containing alginate, then the denaturation and degradation of protein in the gastric environment are minimized [117]. As the alginate gels are hydrophilic and porous in nature, the release of protein occurs very rapidly. For the sustained delivery of protein, ionically cross-linked alginate formulation was prepared. Due to their capability to get assembled at neutral pH and low temperature, they can easily be used in encapsulation and delivery of proteins. Alginate gel was used to encapsulate various growth factors such as vascular endothelial growth factor (VEGF) and basic fibroblast growth factor (bFGF) to provide localized and sustained delivery, and the release rate of protein from the gel can be manipulated by altering the degradation rate of the gel by using partially oxidized alginate; in this way, the release of protein depends on degradation reaction [115–118].

To protect the insulin from the gastric environment and to achieve its sustained delivery, insulin-loaded alginate microspheres were prepared by mixing alginate with various anionic polymers (e.g., CAP, dextran sulfate, etc.) and further coating with chitosan was done [119]. Sometimes, alginate microsphere was coated with *Bombyx mori* silk fibroin, which provides more stable shell and acts as a strong diffusion barrier for the encapsulated protein [120]. Further study was performed on combination of PLGA microspheres and alginate hydrogel to provide sustained release of bovine serum albumin (BSA), a model protein. A combination was prepared by encapsulation of suspension of PLGA microsphere into alginate before ionic cross-linking [115, 121].

Alginate in cell culture system—Alginate was considered as an excellent material for cell and tissue culture and was used as a 3-D-based system because of its capability of forming the gel and has good mechanical strength and forms a bioadhesive bond with cell [122]. Alginate gels were used as a model for mammalian cell culture where they serve as a 2-D or physiologically relevant 3-D culture system. They are used to explain the influence of 3-D culture on cancer cell signaling to recruit the blood vessels and on tumor vascularization [123]. Alginate gels have been used in the transplantation of stem cell so that they can be explored in bone tissue engineering. They can be used in immobilization of particular cells, mainly pancreatic islet cells. Alginate can be used to stimulate the immune response.

For the cell culture, two alginate-based preparations are available in the market, i.e., AlgiMatrix® and NovaMatrix® [124]. AlgiMatrix® is lyophilized sponge preparation and the cell grown on this culture resembles with the normal cell morphology and behavior. So, it is used in many fields such as drug development, cancer research, tissue engineering, etc. [125]. NovaMatrix® is a sterile foam preparation used in cell immobilization under physiological conditions [124].

Alginate gel modified with RGD peptide has been used as a substrate for *in vitro* cell culture and to control the phenotype of interacting myoblast, osteoblast, ovarian follicle, and bone marrow stromal cell, and further, the modified gel enhances the adhesion and proliferation of myoblast [126]. The number of cells adheres to the gel and their growth rate on the gel was dependent on the density of RGD in the gel. Alginate gel modified with (glycine)n-arginine-glycine aspartic acid-serine-proline (GnRGDSP) have been used in the growth of human fibroblast, but at least 12 glycine units are necessary for the cell adhesion [108, 127].

For the enhancement of osteogenic differentiation of stem cells, alginate gel was modified with the cyclic RGD peptide, and they are more resistant to proteolysis and have more binding affinity and selectivity as compared to linear RGD peptide; they are also used to enhance the stem cell differentiation and tissue regeneration [128, 129].

But the major limitation with 3-D culture system is that it is difficult to analyze and quantify cell interaction. This problem was overcome by the development of FRET technique. This technique helps to understand the relationship between the cell receptor and ligand binding and provides information on arrangement of cell [130].

6.7.1 Other Applications

Cosmetics—Alginate was used in many cosmetic preparations because it acts as a thickening agent and helps in retaining moisture. In lipstick, alginate helps to retain the color of the lipstick by forming a gel network. Alginate was used as a consistency agent in various cosmetic products and also helps in forming moisture-retaining surface film on the skin and thus prevents the problem of dry skin. Alginate is used in instant cold emulsion-type cream where it leaves the skin soft and smooth because of the water-binding capacity of alginate. At higher concentration, alginate helps in stabilizing the oil phase in emulsifier-free cosmetics. In shampoos, alginate is used as a bubble stabilizer and thickener. In toothpaste, it is used as a binding agent [131].

Food—Because of the thickening property of alginate, it is useful in sauces and syrups and as a topping in ice creams. In certain fruit drinks, sodium alginate was added to prevent the sedimentation of pulp and to keep the drink in the suspension form. In ice creams, alginates act as a stabilizer and prevent the formation of ice crystal during freezing and provide a smooth product and reduce the rate of melting of ice cream. Alginate prevents the separation of oil and water phase from the water-in-oil emulsions such as mayonnaise. In yoghurt, addition of alginate improves its texture, body, and sheen. Alginate was added in the chocolate milk so as to keep the cocoa in the suspension form. In beer, a very low concentration of alginate was added to keep the foam on the top of the newly poured glass for a long time and thus improves the subject's compliance. To preserve the frozen fish, calcium alginate films and coatings have been used. The oils in oily fish are very prone to rancidity through oxidation, even on immediate freezing and when stored at low temperature, but when the fish is frozen in calcium alginate jelly, the oils get protected from the rancidity through oxidation. Because of the gelling property of the alginate, it was used in the production of artificial cherries. Edible desert jellies can be prepared from alginate and they are referred to as instant jellies because they are prepared simply by mixing the powder either with milk or water without any heating. Alginate gels were used in restructured food products like meats. Restructured meat was prepared by taking the various meat pieces and binding them together with the help of alginate powder and then shaping them to resemble usual cuts of meats. In food industry, only propylene glycol alginate was used [132–134].

Immobilized biocatalyst—Different conversions and chemical synthesis, which are commercialized, are carried out, the best by using enzymes or active whole cell that are biocatalysts. These enzymes are used for the conversion of glucose to fructose, and for the production of L-amino acids, which are used in foods, after hydrolysis of penicillin G for the synthesis of new penicillins, whole cells are used in order to convert starch to ethanol (for beer brewing), and for the continuous production of yoghurt. The concentrated biocatalysts are used, when these processes had to be carried out at a moderate to large scale, and for reuse, these should be recoverable from the process [135].

By "immobilizing" these enzymes/cells, the process can be achieved by their entrapment in a material that allows the penetration by a converted or changed substance. Target enzymes is usually isolated in pure form and are useful in catalyzing specific conversions. These conversions can also be achieved via use of whole cells and have been found economical. An additional advantage of immobilization is that the cells last longer. The suspended cells could have better activity for just 1–2 days, whereas immobilized cells could last for 30 days. The first materials to be used for immobilization are the beads, which are made with calcium alginate. The cells are fully suspended in a solution containing sodium alginate, and this solution is further added drop by drop to a calcium chloride solution. The beads are resultantly formed in the same way as previously described for artificial cherries. While in use, they are tightly packed in a column and a solution of a substance which is to be converted is allowed to be added drop by drop into a column to flow through beads bed containing immobilized biocatalyst in the cell. The product comes out at the bottom through the conversion process. In order to immobilize yeast cells, a simple example is to flow a solution of sugar on through the beads, and this sugar is converted to alcohol.

6.8 Conclusion

In summary, specific property, mucoadhesive property, swelling capacity, and ability of sol/gel alginate have gained a place in the developed advance drug delivery systems. These multifunctionalized, natural polymers have been studied in the design of microsized particulate systems for manageable release, targeted drug delivery, and biomedical application (as a matrix for 3-D tissue cultures, adjuvant of antiviral agents and antibiotics, or in transplantation of cell in diabetes and treatment of neurodegenerative diseases). In addition, more absorbent alginate-based hydrogels with viscoelastic properties and mechanical stability are used as an application for wound dressing. To sum up, this chapter highlighted alginates, possible modifications in their chemical structure, and their application in the field of pharmaceuticals at the lab and commercial levels.

References

1. Huebsch, N. and Mooney, D.J., Inspiration and application in the evolution of biomaterials. *Nature*, 462, 426–432, 2009.
2. Williams, D.F., On the nature of biomaterials. *Biomaterials*, 30, 5897–5909, 2009.
3. Gombotz, W.R. and Wee, S.F., Protein release from alginate matrices. *Adv. Drug Delivery Rev.*, 31, 267–285, 1998.
4. Langer, R. and Vacanti, J.P., Tissue engineering. *Science*, 260, 920–926, 1993.
5. Lee, K.Y. and Mooney, D.J., Hydrogels for tissue engineering. *Chem. Rev.*, 101, 1869–1879, 2001.
6. Smidsrod, O. and Skjak-Bræk, G., Alginate as immobilization matrix for cells. *Trend Biotechnol.*, 8, 71–78, 1990.
7. Rinaudo, M., Main properties and current applications of some polysaccharides as biomaterials. *Polym. Int.*, 57, 397–430, 2008.
8. Qin, Y., Alginate fibers: An overview of the production processes and applications in wound management. *Polym. Int.*, 57, 171–180, 2008.
9. Remminghorst, U. and Rehm, B.H.A., Bacterial alginates: From biosynthesis to applications. *Biotechnol. Lett.*, 28, 1701–1712, 2006.
10. Thu, H.-E., Zulfakar, M.H., Ng, S.-F., Alginate based bilayer hydrocolloid films as potential slow-release modern wound dressing. *Int. J. Pharm.*, 434, 375, 2012.
11. Tonnesen, H.H. and Karlsen, J., Alginate in drug delivery systems. *Drug Dev. Ind. Pharm.*, 28, 621–630, 2002.
12. Hay, I.D., Rehman, Z.U., Ghafoor, A., Rehm, B.H.A., Bacterial biosynthesis of alginates. *J. Chem. Technol. Biotechnol.*, 85, 752–759, 2010.
13. Rinaudo, M., On the abnormal exponent's a_η and a_D in Mark-Houwink type equations for wormlike chain polysaccharides. *Polym. Bull.*, 27, 585–589, 1992.
14. LeRoux, M.A., Guilak, F., Setton, L.A., Compressive and shear properties of alginate gel: Effects of sodium ions and alginate concentration. *J. Biomed. Mater. Res.*, 47, 46–53, 1999.
15. Kong, H.J., Smith, M.K., Mooney, D.J., Designing alginate hydrogels to maintain viability of immobilized cells. *Biomaterials*, 24, 4023–4029, 2003.
16. Kong, H.J., Lee, K.Y., Mooney, D.J., Decoupling the dependence of rheological/mechanical properties of hydrogels from solids concentration. *Polymer*, 43, 6239–6246, 2002.
17. Otterlei, M., Ostgaard, K., Skjakbraek, G., Smidsrod, O., Soonshiong, P., Espevik, T., Induction of cytokine production from human monocytes stimulated with alginate. *J. Immunother.*, 10, 286–291, 1991.
18. Zimmermann, U., Klock, G., Federlin, K., Haning, K., Kowaslski, M., Bretzel, R.G., Horcher, A., Entenmann, H., Siebers, U., Zekorn, T., Production of mitogen contamination free alginates with variable ratios of mannuronic to guluronic acid by free flow electrophoresis. *Electrophoresis*, 13, 269–274, 1992.

19. Orive, G., Ponce, S., Hernandez, R.M., Gascon, A.R., Igartua, M., Pedraz, J.L., Biocompatibility of microcapsules for cell immobilization elaborated with different type of alginates. *Biomaterials*, 23, 3825–3831, 2002.
20. Lee, J. and Lee, K.Y., Local and sustained vascular endothelial growth factor delivery for angiogenesis using an injectable system. *Pharm. Res.*, 26, 1739–1744, 2009.
21. Pelletier, S., Hubert, P., Payan, E., Marchal, P., Choplin, L., Dellacherie, E., Amphiphilic derivatives of sodium alginate and hyaluronate for cartilage repair: Rheological properties. *J. Biomed. Mater. Res.*, 54, 102–108, 2001.
22. Sionkowska, A., Current research on the blends of natural and synthetic polymers as new biomaterials: Review. *Progr. Polym. Sci.*, 36, 1254, 2011.
23. Abdelrahman, T. and Newton, H., Wound dressings: Principles and practice. *Surgery*, 29, 491, 2011.
24. An, Q.-D., Zhang, G.-L., Wu, H.-T. *et al.*, Alginate-deriving oligosaccharide production by alginase from newly isolated *Flavobacterium* sp. LXA and its potential application in protection against pathogens. *J. Appl. Microbiol.*, 106, 1, 161–170, 2009.
25. Kong, H.J., Kaigler, D., Kim, K., Mooney, D.J., Controlling rigidity and degradation of alginate hydrogels via molecular weight distribution. *Biomacromolecules*, 5, 1720–1727, 2004.
26. Chan, A.W. and Neufeld, R.J., Tuneable semi-synthetic network alginate for absorptive encapsulation and controlled release of protein therapeutics. *Biomaterials*, 31, 9040–9047, 2010.
27. Rowley, J.A., Madlambayan, G., Mooney, D.J., Alginate hydrogels as synthetic extracellular matrix materials. *Biomaterials*, 20, 45–53, 1999.
28. Wang, L., Shelton, R.M., Cooper, P.R., Lawson, M., Triffitt, J.T., Barralet, J.E., Evaluation of sodium alginate for bone marrow cell tissue engineering. *Biomaterials*, 24, 3475–3481, 2003.
29. Gombotz, W.R. and Wee, S.F., Protein release from alginate matrices. *Adv. Drug Delivery Rev.*, 31, 267–285, 1998.
30. D.E. Clark and H.C. Green, Alginic acid and process of making same. US Patent. 2036922, 1936.
31. Haug, A., Fractionation of alginic acid. *Acta Chem. Scand.*, 13, 601–603, 1959.
32. Hay, I.D., Rehman, Z.U., Ghafoor, A., Rehm, B.H.A., Bacterial biosynthesis of alginates. *J. Chem. Technol. Biotechnol.*, 85, 752–759, 2010.
33. Vallee, F., Muller, C., Durand, A., Schimchowitsch, S., Dellacherie, E., Kelche, C., Cassel, J.C., Leonard, M., Synthesis and rheological properties of hydrogels based on amphiphilic alginate-amide derivatives. *Carbohydr. Res.*, 344, 223–228, 2009.
34. Augst, A.D., Kong, H.J., Mooney, D.J., Alginate hydrogels as biomaterials. *Macromol. Biosci.*, 6, 623–633, 2006.
35. Rashaan, Z.M., Krijnen, P., van den Akker-van Marle, M.E. *et al.*, Clinical effectiveness, quality of life and cost-effectiveness of Flaminal_ versus

Flamazine_ in the treatment of partial thickness burns: Study protocol for a randomized controlled trial. *Trials*, 17, article122, 9 pages, 2016.
36. Eiselt, P., Lee, K.Y., Mooney, D.J., Rigidity of two-component hydrogels prepared from alginate and poly (ethylene glycol)-diamines. *Macromolecules*, 32, 5561–5566, 1999.
37. Cal′o, E. and Khutoryanskiy, V.V., Biomedical applications of hydrogels: A review of patents and commercial products. *Eur. Polym. J.*, 65, 252–267, 2015.
38. Sussman, C. and Bates-Jensen, B., Management of the wound environment with dressings and topical agents, in: *Wound Care a Collaborative Practice Manual for Health Professionals*, 3rd edition, p. 254, Lippincott Williams & Wilkins, 2007.
39. Carella, S., Maruccia, M., Fino, P., Onesti, G., An atypical case of Henoch-Shonlein purpura in a young patient: Treatment of the skin lesions with hyaluronic acid-based dressings. *In Vivo*, 27, 1, 147–151, 2013.
40. Maiti, S., Singha, K., Ray, S., Dey, P., Sa, B., Adipic acid dihydrazide treated partially oxidized alginate beads for sustained oral delivery of flurbiprofen. *Pharm. Develop. Technol.*, 14, 461–470, 2009.
41. Lucinda-Silva, R.M., Salgado, H.R.N., Evangelista, R.C., Alginate-chitosan systems: *In vitro* controlled release of triamcinolone and *in vivo* gastrointestinal transit. *Carbohydr. Polym.*, 81, 260–268, 2010.
42. Eiselt, P., Lee, K.Y., Mooney, D.J., Rigidity of two-component hydrogels prepared from alginate and poly (ethylene glycol)-diamines. *Macromolecules*, 32, 5561–5566, 1999.
43. Lee, K.Y., Rowley, J.A., Eiselt, P., Moy, E.M., Bouhadir, K.H., Mooney, D.J., Controlling mechanical and swelling properties of alginate hydrogels independently by cross-linker type and cross-linking density. *Macromolecules*, 33, 4291–4294, 2000.
44. Lu, M.Z., Lan, H.L., Wang, F.F., Chang, S.J., Wang, Y.J., Cell encapsulation with alginate and α-phenoxy cinnamylidene-acetylated poly (allylamine). *Biotechnol. Bioeng.*, 70, 479–483, 2000.
45. Roy, D., Cambre, J.N., Sumerlin, B.S., Future perspectives and recent advances in stimuli responsive materials. *Progr. Polym. Sci.*, 35, 278–301, 2010.
46. Sachan, N.K. *et al.*, *J. Pharm. Res.*, 2, 8, 1191–1199, 2009.
47. Chan, A.W. and Neufeld, R.J., Tuneable semi-synthetic network alginate for absorptive encapsulation and controlled release of protein therapeutics. *Biomaterials*, 31, 34, 9040–9047, 2010.
48. Kanczler, J.M., Ginty, P.J. *et al.*, The effect of the delivery of vascular endothelial growth factor and bone morphogenic protein2 to osteoprogenitor cell populations on bone formation. *Biomaterials*, 31, 6, 1242–1250, 2010.
49. Park, D.J., Choi, B.H. *et al.*, Injectable bone using chitosan-alginate gel/mesenchymal stem cells/BMP-2 composites. *J. Cranio-Maxillofacial Surg.*, 33, 1, 50–54, 2005.

50. Lin, H.R. and Yeh, Y.J., Porous alginate/hydroxyapatite composite scaffolds for bone tissue engineering: Preparation, characterization, and *in vitro* studies. *J. Biomed. Mater. Res. B Appl. Biomater.*, 71, 1, 52–65, 2004.
51. Rhim, J.W., Physical and mechanical properties of water resistant sodium alginate films. *LWT-Food Sci. Technol.*, 37, 3, 323–330, 2004.
52. Tonnesen., H.H. and Karlsen., J., Alginate in drug delivery systems. *Drug Dev. Ind. Pharm.*, 28, 6, 621–630, 2002.
53. Ertesvag., H. and Valla, S., Biosynthesis and applications of alginates. *Polym. Degrade. Stability*, 59, 1–3, 85–91, 1998.
54. Johnson, F.A., Craig, D.Q.M., Mercer, A.D., Characterization of the block structure and molecular weight of sodium alginates. *J. Pharm. Pharmacol.*, 49, 7, 639–64, 1997.
55. Mc Neely, W.H. and Pettitt, D.J., *Algin, Industrial Gums; Polysaccharides and Their Derivatives*, 2nd Ed, R.L. Whistler, J.N. Be Miller, (Eds.), pp. 1–50, Academic, New York, 1973.
56. Smidsrod, O. and Skjak-Braek, G., Alginate as immobilization matrix for cells. *Trends Biotech.*, 8, 71–78, 1990.
57. Linker, A. and Jones, R.S., A polysaccharide resembling alginic acid from *Pseudomonas* micro-organism. *Nature*, 204, 4954, 187–188, 1964.
58. Rehm, B.H. and Valla, S., Bacterial Alginates: Biosynthesis and applications. *Appl. Microbiol. Biotechnol.*, 48, 3, 281–288, 1997.
59. Bernkop-Schnürch, A., Kast, C.E., Richter, M.F., Improvement in the mucoadhesive properties of alginate by the covalent attachment of cysteine. *J. Control Release*, 71, 3, 277–285, 2001.
60. Sachan, K.N., Pushkar, S., Jha, A., Bhattacharya, A., Sodium alginate: The wonder polymer for controlled drug delivery. *J. Pharm. Res.*, 2, 8, 1191–1199, 2009.
61. Kuo, C.K. and Ma, X., Ionically cross-linked alginate hydrogels as scaffold for tissue engineering part 1. Structure, gelation rate and mechanical properties. *Biomaterials*, 22, 6, 511–521, 2001.
62. Thu, H.-E., Zulfakar, M.H., Ng, S.-F., Alginate based bilayer hydrocolloid films as potential slow-release modern wound dressing. *Int. J. Pharm.*, 434, 375, 2012.
63. Giunchedi, P., Gavini, E., Moretti, M.D.L., Pirisino, G., Evaluation of alginate compressed matrixes as prolonged drug delivery systems. *AAPS Pharm. Sci. Tech.*, 1, 3, 31–36, 2000.
64. Liew, C.V., Chan, L.W., Ching, A.L., Heng, P.W.S., Evaluation of sodium alginate as drug release modifier in matrix tablets. *Int. J. Pharm.*, 309, 25–37, 2006.
65. Nokhodchi, A. and Tailor, A., *In situ* cross-linking of sodium alginate with calcium and aluminum ions to sustained the release of theophylline from polymeric matrices. *Il Farmaco*, 59, 999–1004, 2004.

66. Bayomi, M.A., Al-Suwayeh, S.A., El-Heiw, A.M., Excipient–excipient interaction in the design of sustained-release theophylline tablets: *In vitro* and *in vivo* evaluation. *Drug Dev. Ind. Pharm.*, 27, 6, 499–506, 2001.
67. Sriamornsak, P., Thirawong, N., Korkerd, K., Swelling, erosion and release behavior of alginate-based matrix tablets. *Eur. J. Pharm. Biopharm.*, 66, 435–450, 2007.
68. De Ruigh, A., Roman, S., Chen, J., Pandolfino, J.E., Kahrilas, P.J., Gaviscon Double Action Liquid (antacid & alginate)is more effective than antacid in controlling post-prandial esophageal acid exposure in GERD patients: A double-blind cross-over study. *Aliment Pharmacol. Ther.*, 40, 5, 531–537, 2014.
69. Rowley, J.A., Madlambayan, G., Mooney, D.J., Alginate hydrogels as synthetic extracellular matrix materials. *Biomaterials*, 20, 45–53, 1999.
70. Prinderre, P., Sauzet, C., Fuxen, C., Advance in gastro retentive drug-delivery systems. *Expert Opinion Drug Deliv*, 8, 9, 1189–1203, 2011.
71. Dennis, A., Timmins, P. *et al.*, Buoyant controlled release powder formulation., 1992. US Patent 5,169,638 https://patents.google.com/patent/US5169638A/en.
72. Mandel, K., Daggy, B. *et al.*, Of heartburn and acid reflux. *Aliment Pharmacol. Ther.*, 14, 669–690, 2000.
73. Hampson, F.C., Farndale, A., Strugala, V., Sykes, J., Jolliffe, I.G., Dettmar, P.W., Alginate rafts and their characterisation. *Int. J. Pharm.*, 294, 1–2, 137–147, 2005.
74. Mygind, N. and Dahl, R., Anatomy, physiology and function of the nasal cavities in health and disease. *Adv. Drug Deliv. Rev.*, 29, 1, 3–12, 1998.
75. Romeo, V., DeMeireles, J. *et al.*, Effects of physicochemical properties and other factors on systemic nasal drug delivery. *Adv. Drug Deliv. Rev.*, 29, 1–2, 89, 1998.
76. Illum, L., Jørgensen, H. *et al.*, Bioadhesive microspheres as a potential nasal drug delivery system. *Int. J. Pharm.*, 39, 3, 189–199, 1987.
77. Stokke, B.T., Draget, K.I. *et al.*, Small-angle X-ray scattering and rheological characterization of alginate gels. 1. Ca alginate gels. *Macromolecules*, 33, 5, 1853–1863, 2000.
78. Ngo, D.-H. and Kim, S.-K., Sulfated polysaccharides as bioactive agents from marine algae. *Int. J. Biol. Macromol.*, 62, 70–75, 2013.
79. Lee, J.-B., Takeshita, A., Hayashi, K., Hayashi, T., Structures and antiviral activities of polysaccharides from *Sargassum trichophyllum*. *Carbohydr. Polym.*, 86, 2, 995–999, 2011.
80. Yan, G.L., Guo, Y.M., Yuan, J.M., Liu, D., Zhang, B.K., Sodium alginate oligosaccharides from brown algae inhibit *Salmonella enteritidis* colonization in broiler chickens. *Poultry Sci.*, 90, 7, 1441–1448, 2011.
81. Benavides, S., Villalobos-Carvajal, R., Reyes, J.E., Physical, mechanical and antibacterial properties of alginate film: Effect of the cross-linking degree and oregano essential oil concentration. *J. Food Eng.*, 110, 2, 232–239, 2012.

82. Khan, S., Tøndervik, A., Sletta, H. *et al.*, Overcoming drug resistance with alginate oligosaccharides able to potentiate the action of selected antibiotics. *Antimicrob. Agents Chemother.*, 56, 10, 5134–5141, 2012.
83. Pritchard, M.F., Powell, L.C., Menzies, G.E. *et al.*, A new class of safe oligosaccharide polymer therapy to modify the mucus barrier of chronic respiratory disease. *Mol. Pharm.*, 13, 3, 863–872, 2016.
84. Sano, Y., Antiviral activity of alginate against infection by tobacco mosaic virus. *Carbohydr. Polym.*, 38, 2, 183–186, 1999.
85. Witvrouw, M. and De Clercq, E., Sulfated polysaccharides extracted from sea algae as potential antiviral drugs. *Gen. Pharmacol.*, 29, 4, 497–511, 1997.
86. Meiyu, G., Fuchuan, L., Xianliang, X., Jing, L., Zuowei, Y., Huashi, G., The potential molecular target of marine sulfated poly mannuroguluronate interfering with HIV 1 entry. Interaction between SPMG and HIV-1 rgp120 and CD4 molecule. *Antiviral Res.*, 59, 2, 127–135, 2003.
87. Yavorska, N., Sodium alginate—A potential tool for weight management: Effect on subjective appetite, food intake, and glycemic and insulin regulation. *J. Undergraduate Life Sci.*, 6, 1, 66–69, 2012.
88. Williams, J.A., Lai, C.S., Corwin, H. *et al.*, Inclusion of guargum and alginate into a crispy bar improves postprandial glycemia in human. *J. Nutr.*, 134, 4, 886–889, 2004.
89. Wolf, B.W., Lai, C.-S., Kipnes, M.S. *et al.*, Glycemic and insulin emic responses of non-diabetic healthy adult subjects to an experimental acid-induced viscosity complex incorporated into a glucose beverage. *Nutrition*, 18, 7–8, 621–626, 2002.
90. Torsdottir, I., Alpsten, M., Holm, G., Sandberg, A.-S., Tolli, J., A small dose of soluble alginate–fiber affect postprandial glycemia and gastric emptying in humans with diabetes. *J. Nutr.*, 121, 6, 795–799, 1991.
91. Borges, Tavares, J., deSousa, A., Borchard, G., Junginger, H.E., Cordeiro-da-Silva, A., Evaluation of the immune response following a short oral vaccination schedule with hepatitis B antigen encapsulated into alginate-coated chitosan nanoparticles. *Eur. J. Pharm. Sci.*, 32, 4–5, 278–290, 2007.
92. Sarei, F., Dounighi, N.M., Zolfagharian, H., Khaki, P., Bidhendi, S.M., Alginate nanoparticles as a promising adjuvant the model brown alga. *PLoS One*, 9, 5, Article ID e96470, 2014.
93. Queen, D., Orsted, H. *et al.*, A dressing history. *Int. Wound J.*, 1, 1, 59–77, 2004.
94. Balakrishnan, B., Mohanty, M. *et al.*, Evaluation of the effect of incorporation of di butyryl cyclic adenosine monophosphate in an *in situ*-forming hydrogel wound dressing based on oxidized alginate and gelatin. *Biomaterials*, 27, 8, 1355–1361, 2006.
95. Rabbany, S.Y., Pastore, J. *et al.*, Continuous delivery of stromal cell-derived factor-1 from alginate scaffolds accelerates wound healing. *Cell Transplant*, 19, 4, 399–408, 2010.
96. Sussman, C. and Bates-Jensen, B., Management of the wound environment with dressings and topical agents, in: *Wound Care a Collaborative Practice*

Manual for Health Professionals, 3rd edition, p. 254, Lippincott Williams & Wilkins, 2007.
97. Rashaan, Z.M., Krijnen, P., van den Akker-van Marle, M.E. *et al.*, Clinical effectiveness, quality of life and cost-effectiveness of Flaminal_ versus Flamazine_ in the treatment of partial thickness burns: Study protocol for a randomized controlled trial. *Trials*, 17, article 122, 9 pages, 2016.
98. Cal'o, E. and Khutoryanskiy, V.V., Biomedical applications of hydrogels: A review of patents and commercial products. *Eur. Polym. J.*, 65, 252–267, 2015.
99. Carella, S., Maruccia, M., Fino, P., Onesti, M.G., A typical case of Henoch-Shonlein purpura in a young patient: Treatment of the skin lesions with hyaluronic acid-based dressings. *In Vivo*, 27, 1, 147–151, 2013.
100. Ausili, E., Paolucci, V., Triarico, S. *et al.*, Treatment of pressure sores in spina bifida patients with calcium alginate and foam dressings. *Eur. Rev. Med. Pharmacol. Sci.*, 17, 12, 1642–1647, 2013.
101. Bale, S., Baker, N., Crook, H., Rayman, A., Rayman, G., Harding, K.G., Exploring the use of an alginate dressing for diabetic foot ulcers. *J. Wound Care*, 10, 3, 81–84, 2001.
102. Porter, M. and Kelly, J., Pressure ulcer treatment in a patient with spina bifida. *Nursing Standard*, 28, 35, 60–69, 2014.
103. Sohn, E.J., Ahn, H.B., Roh, M.S., Ryu, W.Y., Kwon, Y.H., Efficacy of temperature-sensitive guardix-sg for adhesiolysis in experimentally induced eyelid adhesion in rabbits. *Ophthalmic Plastic Reconstruct. Surg.*, 29, 6, 458–463, 2013.
104. Park, S.O., Han, J., Minn, K.W., Jin, U.S., Prevention of capsular contracture with Guardix-SG_ after silicone implant insertion. *Aesthetic Plastic Surg.*, 37, 3, 543–548, 2013.
105. Abramowitz, L., Weyandt, G.H., Havlickova, B. *et al.*, The diagnosis and management of haemorrhoidal disease from a global perspective. *Aliment Pharmacol. Ther.*, 31, supplement 1, 1–58, 2010.
106. Gruskin, E., Doll, B.A., Futrell, F.W., Schmitz, J.P., Hollinger, J.O., Demineralized bone matrix in bone repair: History and use. *Adv. Drug Deliv. Rev.*, 64, 12, 1063–1077, 2012.
107. AlMachot, E., Hoffmann, T., Lorenz, K., Khalili, I., Noack, B., Clinical outcomes after treatment of periodontal intrabony defects with nanocrystalline hydroxyapatite (Ostim) or enamel matrix derivatives (Emdogain): A randomized controlled clinical trial. *BioMed Res. Inte.*, 2014, Article ID786353, 9 pages, 2014.
108. Yan, X.Z., Rathe, F., Gilissen, C. *et al.*, The effect of enamel matrix derivative (Emdogain_) on gene expression profiles of human primary alveolar bone cells. *J. Tissue Eng. Regenerative Med.*, 8, 6, 463–472, 2014.
109. Sculean, A., Auschill, T.M., Donos, N., Brecx, M., Arweiler, N.B., Effect of an enamel matrix protein derivative (Emdogain) on *ex vivo* dental plaque vitality. *J. Clin. Periodontol.*, 28, 11, 1074–1078, 2001.
110. Barzegari, A. and Saei, A.A., An update to space biomedical research: Tissue engineering in microgravity bioreactors. *BioImpacts*, 2, 1, 23–32, 2012.

111. Ummarino, D., Miele, E., Martinelli, M. *et al.*, Effect of magnesium alginate plus simethicone on gastro esophageal reflux in infants. *J. Pediatr. Gastroenterol. Nutr.*, 60, 2, 230–235, 2015.
112. Jakaria, M.D., Zaman, R., Parvez, M. *et al.*, Comparative study among the different formulation of antacid tablets by using acid-base neutralization reaction. *Global J. Pharmacol.*, 9, 3, 278–281, 2015.
113. De Ruigh, A., Roman, S., Chen, J., Pandolfino, J.E., Kahrilas, P.J., Gaviscon Double Action Liquid (antacid & alginate) is more effective than antacid in controlling post-prandial oesophageal acid exposure in GERD patients: A double-blind crossover study. *Aliment Pharmacol. Ther.*, 40, 5, 531–537, 2014.
114. Thomas, E., Wade, A., Crawford, G., Jenner, B., Levinson, N., Wilkinson, Randomized clinical trial: Relief of upper gastrointestinal symptoms by an acid pocket-targeting alginate–antacid (Gaviscon Double Action)—A double-blind, placebo controlled, pilot study in gastro-oesophageal reflux disease. *Aliment Pharmacol. Ther.*, 39, 6, 595–602, 2014.
115. Lee, K.Y., Peters, M.C., Mooney, D.J., Comparison of vascular endothelial growth factor and basic fibroblast growth factor on angiogenesis in SCID mice. *J. Controlled Release*, 87, 49–56, 2003.
116. Xie, H., Wang, Z., Zhang, L. *et al.*, Development of an angiogenesis promoting microvesicle-alginate polycaprolactone composite graft for bone tissue engineering applications, *Peer J.*, 1–19, 2016.
117. Chan, A.W. and Neufeld, R.J., Tuneable semi-synthetic network alginate for absorptive encapsulation and controlled release of protein therapeutics. *Biomaterials*, 31, 34, 9040–9047, 2010.
118. Lee, K.Y., Peters, M.C. *et al.*, Comparison of vascular endothelial growth factor and basic fibroblast growth factor on angiogenesis in SCID mice. *J. Controlled Release*, 87, 1, 49–56, 2003.
119. Wang, X., Wenk, E. *et al.*, Silk coatings on PLGA and alginate microspheres for protein delivery. *Biomaterials*, 28, 28, 4161–4169, 2007.
120. Lee, J., Tan, C.Y. *et al.*, Controlled delivery of heat shock protein using an injectable microsphere/hydrogel combination system for the treatment of myocardial infarction. *J. Controlled Release*, 137, 3, 196–202, 2009.
121. Lee, J. and Lee, K.Y., Injectable microsphere/hydrogel combination systems for localized protein delivery. *Macromol. Biosci.*, 9, 671–676, 2009.
122. Sun, J. and Tan, H., Alginate-based biomaterials for regenerative medicine applications. *Materials*, 6, 4, 1285–1309, 2013.
123. Kong, H.J., Polte, T.R. *et al.*, FRET measurements of cell traction forces and nano-scale clustering of adhesion ligands varied by substrate stiffness. *Proc. Natl. Acad. Sci. USA*, 102, 12, 4300–4305, 2005.
124. Andersen, T., Auk-Emblem, P., Dornish, M., 3D cell culture in alginate hydrogels. *Microarrays*, 4, 2, 133–161, 2015.

125. Godugu, C., Patel, A.R., Desai, U., Andey, T., Sams, A., Singh, M., Algi Matrix based 3D cell culture system as an *in-vitro* tumor model for anticancer studies. *PLoS One*, 8, 1.
126. Rowley, J.A., Madlambayan, G., Mooney, D.J., Alginate hydrogels as synthetic extracellular matrix materials. *Biomaterials*, 20, 45–53, 1999.
127. Comisar, W.A., Hsiong, S.X., Kong, H.J., Mooney, D.J., Linderman, J.J., Multi-scale modeling to predict ligand presentation within RGD nano patterned hydrogels. *Biomaterials*, 27, 2322–2329, 2006.
128. Kumagai, H., Tajima, M., Ueno, Y., Giga-Hama, Y., Ohba, M., Effect of cyclic RGD peptide on cell adhesion and tumor metastasis. *Biochem. Biophys. Res. Commun.*, 177, 74–82, 1991.
129. Liu, S., Radiolabeled multimeric cyclic RGD peptides as integrin αvβ3 targeted radiotracers for tumor imaging. *Mol. Pharm.*, 3, 472–487, 2006.
130. Kong, H.J., Polte, T.R., Alsberg, E., Mooney, D.J., FRET measurements of cell-traction forces and nano-scale clustering of adhesion ligands varied by substrate stiffness. *Proc. Natl. Acad. Sci. USA.*, 102, 4300–4305, 2005.
131. Ahmed, A.B.A., Adel, M., Karimi, P., Peidayesh, M., Chapter ten—Pharmaceutical, cosmeceutical, and traditional applications of marine carbohydrates. *Adv. Food Nutr. Res.*, 73, 197–220, 2014.
132. Capela, P., Hay, T.K.C., Shah, N.P., Effect of cryo protectants, prebiotics and microencapsulation on survival of probiotic organisms in yoghurt and freeze-dried yoghurt. *Food Res. Int.*, 39, 203–211, 2006.
133. Mancini, F. and McHugh, T.H., Fruit-alginate interactions in novel restructured products. *Nearing*, 44, 152–157, 2000.
134. Mancini, F., Montanari, L., Peressini, D., Fantozzi, P., Influence of alginate concentration and molecular weight on functional properties of mayonnaise. *LWT Food Sci. Technol.*, 35, 517–525, 2002.
135. Illans, A., Immobilized biocatalyst. *Comprehensive Biotechnol.*, 1, 25–39, 2011.

7

Alginates in Evolution of Restorative Dentistry

S.C. Onwubu[1*], P.S. Mdluli[2], S. Singh[3] and Y. Ngombane[1]

[1]Department of Dental Sciences, Durban University of Technology, South Africa
[2]Department of Chemistry, Durban University of Technology, South Africa
[3]Discipline of Dentistry, University of KwaZulu-Natal, South Africa

Abstract
Over the last decades, alginate biomaterial has gained sufficient attention among researchers for its inherent properties. These properties, such as the lack of toxicity and biocompatibility, have endeared it to various pharmaceutical and biomedical applications. Since its introduction in restorative dentistry in the early 1930s, alginate has remained the material of choice for impression taking due to its low cost, ease of use, and ability to reproduce sufficient degree of the oral cavity and its related tissues. The conventional alginate used in dentistry, however, has poor dimensional stability and low tear strength. In recent years, several modifications and advancement in alginate material have led to the development of alginate with an improved dimensional stability, dust-free, and color-changing alginate attributes, etc. This chapter foregrounds the evolution of alginate impression materials with an emphasis in the art of impression taking using alginate material.

Keywords: Alginate, biomaterial, impression, restorative dentistry

7.1 Introduction

The history of dental materials indicates that restorative dentistry has witnessed unprecedented advancements, particularly with respect to replacement of teeth for partially or completely edentulous jaws [1, 2]. Much of this was largely due to the development of new and smart biomaterials [3, 4], which has allowed direct and accurate replication of the oral tissues

**Corresponding author*: profstan4christ@yahoo.com

Shakeel Ahmed (ed.) Alginates, (125–140) © 2019 Scrivener Publishing LLC

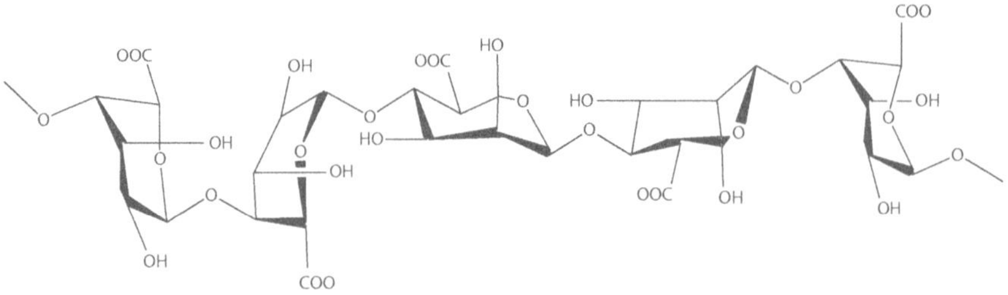

Figure 7.1 Chemical structure of alginate (adopted from Skaugrud, Hagen [11]).

and cavity [5]. Among these materials, the use of alginates has significantly gained much attention—owing to their biocompatibility [6], low toxicity [2], and simplicity in their use with minimal equipment requirement [7].

Alginates are linear biopolymers consisting of 1,4-linked β-D-mannuronic acid (M) and 1,4-α-L-guluronic acid (G) residues that are arranged in homogenous (poly-G, poly-M) or heterogenous (MG) block-like patterns (Figure 7.1). Owing to this unique chemical structure, alginate has successfully been used in various biomedical and pharmaceutical applications. For instance, alginate hydrogels have found applications most notably in wound healing [8], drug delivery, and tissue engineering [2]. More so, alginates are also used in textile printing, paper industry, and food industry [9, 10]. In the context of dentistry, Srivastava *et al.*, [9] noted that alginates are widely used for recording the impressions of edentulous arches, making dental models for diagnostic purposes, and for duplications of casts. However, these materials have long been known for their inferior ability to record fine details and to resist tearing when used in partial denture cases. Hence, in the present chapter, the evolution of alginate and its modification in dentistry are discussed with particular emphasis on the techniques of alginate impression applications.

7.2 Method of Alginate Extraction

Ever since the discovery of alginate, several attempts have been explored to extract alginates in commercial quantities for both pharmaceutical and biomedical applications. Studies [11, 12] reported that commercial alginates are mainly produced from farmed brown seaweed, mainly from genera such as *Macrocystic pyrifera*, *Ascophyllum nodosum*, and *Laminaria hyperborean* [6, 13]. The constituents of these seaweeds are structurally made up of alginate, playing similar roles analogue to cellulose found in terrestrial plants [10, 14, 15]. The seminal work of E.C.C. Stanford, a British chemist, in 1881, historically extracted alginate from brown seaweed. He observed that when a mineral acid is added to alginate, it precipitates into a gelatinous substance that hardens upon drying. Forty years after Stanford discovered alginate, S. William Wilding patented the use of alginate as a dental impression material when he discovered that alginate in water forms a viscous solution or sol, which is subsequently converted to gel using a calcium salt (Figure 7.2).

Interestingly, the process of extracting alginates from these seaweeds is not complicated, but requires a series of steps [12]. As illustrated in Figure 7.3, this multistep usually begins with treating dried raw materials with diluted mineral acid, which is subsequently purified. The obtained alginic acid is then converted to water-soluble sodium salt in the presence of calcium carbonates. While most commercial alginates are traditionally produced from marine source, Remminghorst and Rehm [16] point out that alternative production by microbial fermentation has been recently explored in order to produce alginate with more defined properties. Notwithstanding the methods of extraction, alginates intended for pharmaceutical and biomedical applications are tailor-made by series of purification and microfiltration steps to serve the desired functions [11].

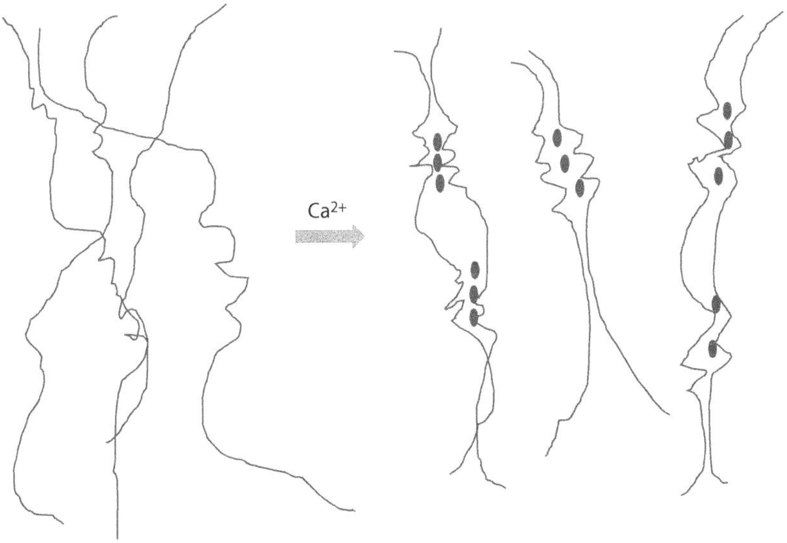

Figure 7.2 Cross-linking of alginate G blocks with calcium.

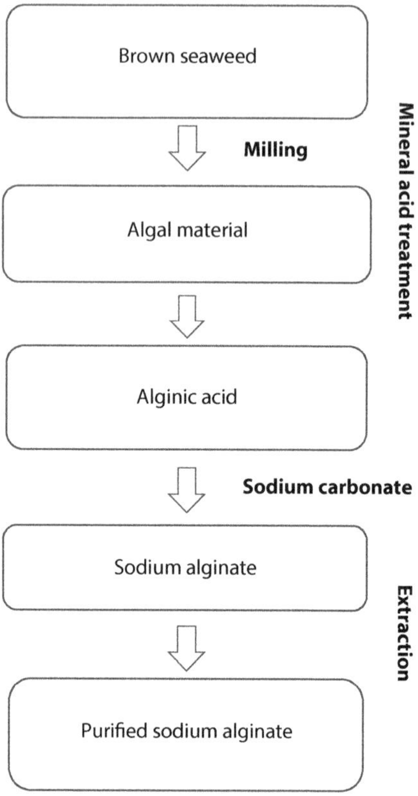

Figure 7.3 Method of sodium alginate extraction from brown seaweed (adapted from Sachan, Pushkar [6]).

7.3 Evolution of Alginate in Restorative Dentistry

Although alginate impression material was first used in the 1930s, the introduction of alginate as an alternative impression material in dentistry was, however, attributed to the scarcity of agar impression material during the Second World War [9]. Unlike agar, alginate is an elastic, irreversible hydrocolloid impression material [17]. Giordano 2nd [18] and Craig and Robert [19] alleged that the hydrophilic nature of irreversible hydrocolloids enables alginate to capture accurate impressions in the presence of blood and saliva. In addition, alginate has an ability to sufficiently record details of tooth undercut with sufficient elastic recovery [20].

7.3.1 Problems with Conventional Alginate

Concerning, and as reported by Kaur *et al.*, [5], conventional alginate's ability to reproduce fine detail was relatively limited compared with other impression materials such as elastomeric material. Their work supports other studies [19, 21] who noted that alginate is not highly accurate for use in fixed partial dentures. The above authors, however, acknowledged the suitability and accuracy of alginate for partial framework impressions. Moreover, literature [7, 22, 23] documents that conventional alginate has poor dimensional stability. For example, casts must be poured within 10–12 minutes of impression making or distortion becomes a major issue [5]. Furthermore, Craig and Robert [19] and McCabe and Walls [20] reported that the conventional alginate has a poor tear strength. While alginate impressions may capture subgingival contours and anatomy, McCabe and Walls [20] observed that the impression tears upon removal. In addition, Anusavice [24] noted that conventional alginate contains silica dust in the form of diatomaceous earth and other components such as cadmium and lead among its composition, which are considered toxic when inhaled.

7.3.2 Current Trends and Modification of Alginate

Despite the above-mentioned drawbacks, since its introduction, alginate has remained the commonest impression material used in restorative dentistry. This is mainly attributed to its ease of use and cost-effectiveness when compared against other impression materials [22]. Hence, it was highly sensible and critical to modify alginate in order to enhance its continuous suitability for clinical performance. For instance, Kaur *et al.*, [5] revealed that various advancements in alginate modifications have been introduced in dentistry. Corroborating with them, Srivastava *et al.*, [9] moot that the composition of alginate has been altered in recent years by modifying its composition with other materials. This was done in an attempt to improve its properties and clinical performance. Notably, there have been a variety of alginate impression materials in the market, with some varying in their properties, setting, and pouring time [22, 25]. As different brands of alginate are commercially available in the markets, it is essential to thoroughly research the properties of these products in order to establish their inherent properties. This section therefore examines the recent progress in alginate modification and applications in restorative dentistry.

7.3.2.1 *Extended Pour Time Alginate*

The dimensional stability of alginate is an important and well-debated discourse among clinicians. Given credence to the debate, Powers and Sakaguchi [25] and Rosenstiel *et al.*, [26] recommend that alginate impression be poured immediately after impression taking. Donovan and Chee [27], in their own view, suggest that alginate impression must be poured within a few minutes after removal from the mouth. They, however, caution against wrapping the impression in a wet paper towel. This, and according to Nassar *et al.*, [22], was to prevent inhibition of water, which may later lead to dimensional changes, since it is not possible to determine the amount of water absorbed by the impression. Consequently, Kaur *et al.*, [5] proposed that alginate impression should be poured immediately or until 12 minutes in 100% humid environments at room temperature. They maintain that if alginate was poured within 15 minutes, it could still be used as a final impression material; otherwise, if not poured within the time limit, dimensional changes will occur.

Drawing from the above, it is evident that pouring time of alginate is a critical parameter that could influence its dimensional stability as well as the clinical performance. Significantly, Nassar *et al.*, [22] reveal that industries have developed an improved alginate material with extended pour time. With this in mind, several other studies [28–30] claimed that alginate impression material can now be left unpoured for a period of time based on the specifications of the manufacturer, wrapped in a damp paper towel or sealed in a plastic bag. Table 7.1 highlights some commercial available alginate impression materials with an improved extended pour time. While noticeable improvement in the extended pouring time has been observed, laboratory studies failed to confirm some of the extraordinary claims made by manufacturers.

7.3.2.2 *Dust-Free Alginates*

Generally, in clinical practice, alginate is mixed with cold water to extend the manipulation and mixing time. Srivastava *et al.*, [9] reported that conventional alginate formulations that contain only soluble alginate and calcium salt produce sticky calcium alginate gel of devoid strength—which is not desirable for dental impression. To overcome this, inert fillers in the form of diatomaceous earth are added to improve its consistency while reducing the tackiness. Concerning, Kaur *et al.*, [5] voiced that due to the silica content of diatomaceous earth, prolonged inhalation may cause silicosis. In attempting to minimize the inhalation of diatomaceous dust during

Table 7.1 Studies on the effect of extended pour time on dimensional stability of the modified alginate.

Studies	Name of brands	Manufacturer	Storage time	Results/recommendations
Aalaei *et al.*, [7]	Alginplus	Major, Prodotti Dentari S.P.a, Moncalieri, Italy	12 minutes, 24 hours, and 120 hours	Shows dimensional stability for 24 hours. Significant dimensional changes observed for 120 hours
Aalaei *et al.*, [7]	Hydrogum5	Zhermack S.p.a., Badia Polesine, Rovigo, Italy	12 minutes, 24 hours, and 120 hours	Shows dimensional stability at 24 hours. Significant dimensional changes observed at 120 hours
Todd *et al.*, [23]	Kromopan	Lascord, Florence, Italy	10 minutes, 24 hours, and 100 hours	Shows significant dimensional changes at 24 and 100 hours
Todd *et al.*, [23]	Triphasix	Parkell, Edgewood, NY	10 minutes, 24 hours, and 100 hours	Shows significant dimensional changes at 24 and 100 hours
Imbery *et al.*, [29]	Cavex color change		Immediate, 1 d, 2 d, 3 d, 4 d, and 5 d	Shows accuracy till day 5 d
Walker, Burckhard [30]	Jeltrate Plus, Alginmax		30 minutes, 48 hours, and 100 hours	Shows stability within 30 minutes

manipulations, alginate is coated with de-dusting agents that agglomerate the powder to a denser form. According to Kaur *et al.*, [5], materials used as de-dusting agents include glycerine, glycol, polyethylene glycol, and polypropylene glycol. Other notable de-dusting materials are squalene, decane, dudecane, and isoparaffin [31].

In recent years, attempts have been made to partially substitute diatomaceous earth in the alginate with sepiolite or tetrafluoroethylene. Sepiolite is a natural mineral fiber containing magnesium silicate. Kaur *et al.*, [5] alleged that when 20% of sepiolite was added to alginate, it traps the particles, thereby reducing dust generation. On the other hand, tetrafluoroethylene traps alginate particles by forming a cobweb-like structure due to stresses generated during manipulations [24].

7.3.2.3 Infection-Free Alginates

As a general rule, before impression is sent from the dental office to the laboratory, it is usually disinfected to prevent the transfer of diseases from the clinics to the laboratory [32, 33]. Concerning, however, disinfection of alginates either by immersion or spray techniques was found to cause changes in dimensional stability [5]. In an attempt to prevent dimensional inaccuracies associated with the disinfection process, recent dental alginate compositions have been modified by the incorporation of disinfectant materials. Kaur *et al.*, [5] noted that water-soluble antimicrobial agents such as quaternary ammonium compounds, bisquanidine compounds, chlorhexidine, and didecyldimethyl ammonium chloride are generally employed since they do not alter the inherent properties of the alginates. Another notable advancement in the alginate modification is the incorporation of disinfectants in the form of microcapsules, which will release the disinfectant on mixing with the liquids [9].

7.3.2.4 High Viscosity Alginates

Conventional alginate powder is reported to have slow permeation speed of water, as it takes a prolonged time period to form a wet mass with water during mixing. The modification of alginates by the incorporation of hydrophobic materials such as polyoxyethylene alkyl phenyl ether, polyoxyethylene/polyoxypropylene alkyl ether, and polyoxyethylene alkyl ether reportedly enhances the permeation speed of alginates [34]. In addition, the incorporation of polysaccharide materials such as carrageenan, pullulan, curdlan, xanthan gum, gellan gum, pectin, konjak, glucomannan, xyloglucan, guar gum, gum Arabic, and locust bean gum has been reported to improve alginate viscosity as well as the prevention of its deterioration during storage [9].

7.3.2.5 *Alginates in Two Pastes Form*

Traditionally, alginate is supplied as powder to be mixed with water. The amount of the powder alginate is determined by the manufacturer's specifications. Conventional alginates, however, have the tendency to release dusts during manipulations [24]. Equally concerning, Kaur *et al.*, [5] noted that the inconsistency of dispersing the right amount of powder, separation of the alginate compositions, as well as contamination of the alginate powder during storages pose serious setbacks to its use for dental impressions. In attempting to address the above-mentioned drawbacks, alginates in the form of two pastes were introduced [17]. The newly modified two pastes form of alginates has two parts, namely: (1) the base pastes and (2) the reactor or catalyst.

The base paste is made up of soluble alginate, water, and fillers such as carrageenan, pullulan, xanthene gellan gum, guar gum, and gum Arabic. According to Srivastava *et al.*, [9], the listed fillers mainly function to prevent the separation of components of alginate paste. On the other hand, the catalyst contains calcium salt mixed with a viscous liquid that is nonreactive toward calcium salts such as liquid paraffin, fatty acids, or aliphatic alcohol in the form of a paste. More so, polybutene is added to stabilize the reactor paste while a basic material such as magnesium hydroxide is used as a pH-stabilizing agent. The advantage of two-paste alginate over the conventional form of alginate is that it allows convenient mixing of the alginates, either manually or with mechanical mixing units [9].

7.3.2.6 *Tray Adhesive Alginates*

In dental practice, impression trays are used to load the alginate after mixing to form a paste. The poor retention of conventional alginates to impression trays has necessitated the use of perforated trays. In recent years, manufacturers have introduced tray adhesive materials in the form of liquid and sprays. These materials are reported to contain conjugate of polymers and solvent such as polyamide or diethylenetriamine polymer, ester gum, and rosin in isopropyl alcohol [9]. A combination of isopropyl with ethyl acetate is also utilized as tray adhesives [5]. The application of tray adhesives is reported to improve the bond strength between the alginate material and the trays [35].

7.4 The Art of Impression Taking Using Alginates

While alginate impression material has remained one of the most important materials in restorative dentistry, it is, however, far from being the ideal material for taking impressions. The previous sections have highlighted some of the drawbacks of alginate and the need for modification of alginates. As stated earlier, the modification of alginates created tailored products that have overcome most of the initial drawbacks associated with conventional alginates. Despite these, accurate impression taking using alginates still poses a difficult challenge even for skilled and experienced clinician. For instance, and as reported by Christensen [36], the premature contacts on indirect restorations are a consequence of inaccurate opposing arch impressions from alginate materials. Echoing further, the earlier report of Ashley *et al.*, [37] accentuates that failure of alginate impression was largely due to poor operator technique more than the inherent limitation of the alginate impression material. As such, a clinical reliable impression using alginate material is only feasible when the art of impression-taking protocol is strictly adhered to. Essentially, Nandini *et al.*, [17] outlined some of the required protocols for an effective impression taking using alginate material. These protocols include the following stages:

- Selection of appropriate impression trays,
- Mixing and loading of alginate impression material,
- Preparing the oral cavity,
- Taking the impression,
- Removal of the impression, and
- Cast production.

7.4.1 Selection of Impression Trays

Since the human oral cavity (mouth) differs from one person to another, it is highly pertinent to select the correct trays for individual patients that will match the dental arch. Another factor that is noteworthy when selecting impression trays is their ability to retain the alginate impression material. Consequently, a perforated stock tray is recommended since alginates have poor retention abilities on trays. However, recent modification introduces alginate adhesives that have helped counter the poor retention properties of alginates. Hence, the use of alginate adhesive is highly recommended, particularly to overcome the displacing forces that occur when impression is withdrawn from the oral cavity [17].

For some patients with unusual dental arches such as those with high palatal vault, there is a need for modification of trays prior to use for impression taking. Materials like wax, tracing stick, impression compounds, or heavy-bodied silicone are often considered for tray extension in the sulcus borders. Therefore, and adhering to the advice of Nandini *et al.*, [17], as a precautionary measure, the dental arch must be examined carefully before taking impression. They elaborated further that, if there are existing fixed partial dentures, it is advisable to block out the pontics (artificial tooth/teeth) with wax to prevent tearing of the impression material in these regions.

7.4.2 Mixing and Loading Alginates

Generally, commercial alginate used for dental impressions is usually supplied in a container containing the powder alginate. A scoop is provided by the manufacturer for accurate measurement of the powder, and a cylindrical plastic measuring cylinder needed to gauge the water proportion. However, caution must be taken against the use of water with high amount of mineral, as it would invariably affect the accuracy as well as the setting time of the alginate [17]. In such cases, demineralized or distilled water will be much more appropriate to use [38]. Mixing is often initiated by adding measured quantity of water into a clean bowl; thereafter, an accurate amount of powder is added. Mixing then commences by stirring vigorously using a spatula until a creamy consistency is achieved. In order to control the time required for manipulation, the use of cold water is recommended while the setting time of the mixture can be manipulated by varying the temperature of the water [17].

In terms of loading the mixed alginate, care must be taken to ensure that the desired amount of alginate is loaded onto the selected tray. It is highly advisable that the tray be filled with the impression material up to the borders. More so, excess material around the periphery should be removed with spatula. In addition, the surface of the alginate should be smoothened with a wet gloved finger [37]. While these protocols are much more suited for alginates in powder form, recent development has introduced two-paste form of alginate, which is more convenient to mix. The advantages and benefits of the two-paste form of alginates have been discussed (see Section 7.3.2.5).

7.4.3 Preparation of the Oral Cavity before Impression Taking

One of the drawbacks of alginate impression material mentioned in this chapter is its tendency to tear upon removal. Hence, to prevent alginate from tearing upon impression removal, the oral cavity (mouth) must be adequately prepared. Firstly, in order to minimize air-blows, the occlusal surface of the teeth must be blown with an air string to remove debris and saliva [17]. Secondly, adequate care must be taken to ensure that the mouth is not completely dried before taking the impression with alginate material. This is very important, because when the surface is dried, alginate easily forms chemical bonds with hydroxyapatite crystals of the enamel, causing tear upon removal [38]. Hence, it is highly advisable that that teeth should not be left to try completely [17]. Furthermore, the presence of air bubbles in the mouth will create voids when the impression has been taken. This, in turn, will negatively affect the accurate impression of the dental arch. To avoid this, it is required that the patient rinses with water and mouthwash mixture. This will help eliminate mucin (heavily glycosylated proteins), which lowers the surface tension, thereby eliminating air bubbles [17].

Additionally, for impression needed for removable prostheses such as dentures, impression of the sulci is significantly important. As such, Nandini *et al.*, [17] suggested that pre-packing of the sulci, particularly of the lower lingual, upper labial, and humular notches in the buccal areas with alginate impression material, will help improve the clinical accuracy of the impression.

7.4.4 Impression Taking Using Alginate Material

The way and manner alginate impression is taken could make or mar the accurate impression of the dental arch. In order to ensure that detailed and accurate impression is achieved, the mixed alginate should be rubbed onto the occlusal surfaces with a gloved finger, thus filling the occlusal grooves. Nandini *et al.*, [17] hinted that impression tray is positioned in the mouth by retracting the patient's lips on one side using a gloved finger or mouth mirror, while on the other side, the tray is rotated in the mouth. Critically, the tray must be placed centrally in the mouth and held in place with light pressure. More so, it is highly advisable that when the tray is seated in the mouth, pressure should be relieved and tray held lightly in place to prevent unseating. Nandini *et al.*, [17] reported that alginate impression materials begin setting from the tooth surface to the impression tray; hence, pressure will cause impression to set under strain. The consequence of this is that, on removing from the mouth, these strains are released, resulting to distortion and inaccurate cast. Equally important, removal of the impression before it completely sets will also cause similar strain and distortion.

In an attempt to give more quality assurance in the setting time of alginate, manufacturers have introduced color-changing alginates that indicate each phase of the alginate setting time. The color-changing alginate contains pigments (Cresol Red, naphtholphthalein, Tropaeolin OOO, Thymol) that react with pH during setting reactions [39]. The advantage of the color-changing alginate is that operators can avoid the impression being removed from the mouth before setting time is completed. It also puts a sensitive patient's mind at ease considering that the change in color is an indicative of the material being ready for removal.

7.4.5 Removal and Inspection of Alginate Material

Upon setting, care must be taken on removal of the impression from the mouth to avoid deformation. The optimal technique when removing the impression from the patient's mouth is to press on the side of alginate overflow from the labial vestibule. Nandini *et al.*, [17] advocated that the impression must be removed with a firm, quick snap. For a mandibular arch removal, it is advisable that the operator avoid using the handle of the tray as level. This is because the alginate is easily loosened from the tray with the leverages. More so, with respect to the removal of the maxillary arch impression, it is highly recommended that the operator's index finger on both hands be placed in the buccal sulci to break the seal [17].

Once the impression is removed from the mouth, it should be observed for any possible defects under good lighting condition and subsequently rinsed under running tap water to remove saliva or blood. With the modified form of alginate available in the market, impression can be covered in a damp towel/napkin to prevent syneresis (loss of moisture). It is not recommended to place the impression in water, as this could result to inhibition of water, thereby causing distortion and inaccurate casts. In addition, it is also important that the impression is trimmed to remove excess alginate material before sending it to the dental laboratory. While it is recommended that the cast be poured within 30 minutes to avoid dimensional instability, essentially, as highlighted in Table 7.1, modified forms of alginates can now be stored up to 5 days without any significant changes in its dimensional properties.

7.4.6 Effects of Cast Production Techniques

A dental cast is produced by filling the alginate impression of the dental arch with a semisolid material product [39]. While correct impression-taking techniques are important for accurate duplication of the dental arch, the cast production from the impression is much more critical. More so, a poorly poured cast will invariably negate the quality of fabricated prosthesis. As such, it is advisable that conventional alginate impression material be poured immediately or within 10–12 minutes (see Section 7.3.2.1).

Generally, dental gypsum is the material of choice commonly used for alginate impression. Prior to alginate impression pouring, it is highly recommended that the surface of the alginate impression be sprinkled with dental gypsum and the stone be dispersed with minute amounts of water [17]. This will help to remove any saliva or blood residues. The sprinkled gypsum must be carefully rinsed out and excess water removed after 1–2 minutes. Mixing of the dental gypsum with water then follows.

It is highly recommended that the manufacturer's instructions of powder and water be adhered to. Although mixing of dental gypsum and water can be done manually using a clean bowl and spatula, it is, however, difficult to prevent air bubbles in the mix [17]. Hence, a vacuum mixer at a rate of 300–400 rpm for 30–45 seconds is highly recommended. Mixing should be homogenous with a creamy constituency. Pouring of the mixed dental gypsum should start with a small portion, carefully vibrating it until the deepest areas of dental arch are filled. Larger portions are to be poured gradually until the margin is filled. Excess casts can be trimmed with a trimming machine to have the desired shape. Care must be taken to avoid excessive trimming of the dental arch. Equally important, the cast must be removed from the impression upon setting and must not be left overnight.

7.5 Conclusions

While alginate remains the most common impression material used in restorative dentistry, it is, however, far from being the ideal impression material. In fact, alginate impression has poor dimensional stability, has low tear strength, and hence cannot be stored for more than 5 days or repeatedly reused. Despite these, alginate is easily the material of choice in restorative dentistry due to its low cost, ease of use, nontoxicity, and ability to reproduce sufficient details of the oral cavity. Recent modifications of alginate are noted to have overcome some of the early drawbacks associated with the use of alginate impression materials. This chapter has highlighted some of the improved properties of alginate material. The ability for alginate impression material to reproduce accurate and correct impressions of the dental arch will largely depend on the operators complying with the manufacturer's instructions. Hence, it is highly recommended that operators adhere strictly to the art of impression taking as this could make or mar the quality of the final casts produced.

References

1. Vaderhobli, R.M., Advances in dental materials. *Dental Clin.*, 55, 619–25, 2011.
2. Lee, K.Y. and Mooney, D.J., Alginate: Properties and biomedical applications. *Progr. Polym. Sci.*, 37, 106–26, 2012.
3. Ratner, B.D. and Bryant, S.J., Biomaterials: Where we have been and where we are going. *Annu. Rev. Biomed. Eng.*, 6, 41–75, 2004.

4. Badami, V. and Ahuja, B., Biosmart materials: Breaking new ground in dentistry. *Sci. World J.*, 2014, 2014.
5. Kaur, G., Jain, P., Uppal, M., Sikka, R., Alginate impression material: From then till now. *Indicators*, 18, 20, 2012.
6. Sachan, N.K., Pushkar, S., Jha, A., Bhattcharya, A., Sodium alginate: The wonder polymer for controlled drug delivery. *J. Pharm. Res.*, 2, 1191–9, 2009.
7. Aalaei, S., Ganj-Khanloo, R., Gholami, F., Effect of storage period on dimensional stability of alginplus and hydrogum 5. *J. Dent. (Tehran, Iran)*, 14, 31, 2017.
8. Queen, D., Orsted, H., Sanada, H., Sussman, G., A dressing history. *Int. Wound J.*, 1, 59–77, 2004.
9. Srivastava, A., Aaisa, J., Kumar, T., Ginjupalli, K., Upadhya, P., Alginates: A review of compositional aspects for dental applications. *Trends Biomater. Artif. Organs*, 26, 2012.
10. Hay, I.D., Rehman, Z.U., Moradali, M.F., Wang, Y., Rehm, B.H., Microbial alginate production, modification and its applications. *Microb. Biotechnol.*, 6, 637–50, 2013.
11. Skaugrud, Ø., Hagen, A., Borgersen, B., Dornish, M., Biomedical and pharmaceutical applications of alginate and chitosan. *Biotechnol. Genet. Eng. Rev.*, 16, 23–40, 1999.
12. Szekalska, M., Puciłowska, A., Szymańska, E., Ciosek, P., Winnicka, K., Alginate: Current use and future perspectives in pharmaceutical and biomedical applications. *Int. J. Polym. Sci.*, 2016, 2016.
13. Repka, M.A. and Singh, A., *Alginic Acid*, 6th ed., Pharmaceutical Press, London, UK, 2009.
14. Draget, K.I., Smidsrød, O., Skjåk-Bræk, G., Alginates from algae. *Biopolymers Online*, 2005.
15. Donati, I. and Paoletti, S., Material properties of alginates, in: *Alginates: Biology and Applications*, B.H.A. Rehm (Ed.), pp. 1–53, Springer Berlin Heidelberg, Berlin, Heidelberg, Germany, 2009.
16. Remminghorst, U. and Rehm, B.H., *In vitro* alginate polymerization and the functional role of Alg8 in alginate production by *Pseudomonas aeruginosa. Appl. Environ. Microbiol.*, 72, 298–305, 2006.
17. Nandini, V.V., Venkatesh, K.V., Nair, K.C., Alginate impressions: A practical perspective. *J. Conserv Dent: JCD*, 11, 37, 2008.
18. Giordano, R., 2nd, Impression materials: Basic properties. *Gen. Dent.*, 48, 510, 2000.
19. Craig, R.G. and Robert, G., *Restorative Dental Materials*, Elsevier, 2002.
20. McCabe, J. and Walls, A., Elastic impression materials: Hydrocolloids, in: *Applied Dental Materials*, 9th ed, pp. 154–62, Blackwell, Ames, Iowa, 2008.
21. Phoenix, R.D. and Rodney, D., *Stewart's clinical removable partial prosthodontics*, 3rd ed, Quintessence, 2002.
22. Nassar, U., Aziz, T., Flores-Mir, C., Dimensional stability of irreversible hydrocolloid impression materials as a function of pouring time: A systematic review. *J. Prosthet. Dent.*, 106, 126–33, 2011.

23. Todd, J.A., Oesterle, L.J., Newman, S.M., Shellhart, W.C., Dimensional changes of extended-pour alginate impression materials. *Am. J. Orthod. Dentofacial Orthop.*, 143, S55–S63, 2013.
24. Anusavice, K.J., Elastic impression materials alginate, in: *Phillips' Science of Dental Materials*, 11th ed., Saunders, Philadelphia, 2003.
25. Powers, J.M. and Sakaguchi, R.L., Impression materials, in: *Restorative Dental Materials*, pp. 272–9, St. Louis Mosby, 2006.
26. Rosenstiel, S.F., Land, M.F., Fujimoto, J., Diagnostic casts and related procedures, in: *Contemporary Fixed Prosthodontics*, 4th ed., pp. 44–5, St. Louis Mosby, 2006.
27. Donovan, T.E. and Chee, W.W., A review of contemporary impression materials and techniques. *Dent. Clin. North Am.*, 48, 445–70, 2004.
28. Alcan, T., Ceylanoğlu, C., Baysal, B., The relationship between digital model accuracy and time-dependent deformation of alginate impressions. *Angle Orthod.*, 79, 30–6, 2009.
29. Imbery, T.A., Nehring, J., Janus, C., Moon, P.C., Accuracy and dimensional stability of extended-pour and conventional alginate impression materials. *J. Am. Dent Assoc.*, 141, 32–9, 2010.
30. Walker, M.P., Burckhard, J., Mitts, D.A., Williams, K.B., Dimensional change over time of extended-storage alginate impression materials. *Angle Orthod.*, 80, 1110–5, 2010.
31. P. Schwabe and R. Voigt, Dust-free alginate impression materials. Google Patents, 1987.
32. Rweyendela, I., Patel, M., Owen, C., Disinfection of irreversible hydrocolloid impression material with chlorinated compounds: Scientific. *South Afr. Dent. J.*, 64, 208–12, 2009.
33. Surna, R., Junevicius, J., Rutkauskas, E., *In vitro* investigation of the integration depth of oral fluids and disinfectants into alginate impressions. *Stomatologija*, 11, 129–34, 2009.
34. H. Kamohara and H. Morita, Dental alginate impression material composition. Google Patents, 2010.
35. Woortman, R., Hermans, J., Feilzer, A., Effect of alginate adhesives on the bond strength of alginate impression material to stainless steel. *J Dent Res*, B203–B, 2003. Int Amer Assoc Dental Research ADR/AADR Alexandria, VA 22314-3406 USA.
36. Christensen, G.J., Making fixed prostheses that are not too high. *J Am Dent Assoc.*, 137, 96–8, 2006.
37. Ashley, M., McCullagh, A., Sweet, C., Making a good impression: (a 'how to' paper on dental alginate). *Dent Update.*, 32, 169–70, 72, 74–5, 2005.
38. Rubel, B.S., Impression materials: A comparative review of impression materials most commonly used in restorative dentistry. *Dent Clin.*, 51, 629–42, 2007.
39. Kulzer, H., *Guidelines for taking perfect impression*, 2014.

8

Alginates in Drug Delivery

Srijita Basumallick

Asutosh College under Calcutta University 92, Kolkata, India

Abstract
Alginate is a biocompatible biopolymer and is widely used as a drug delivery vehicle in different forms. In this chapter, we will discuss the applications of different cross-linked networks of alginates, their microspheres, and hydrogel in relation to drug encapsulation and delivery processes with a brief introduction of chemistry and pharmaceutical chemistry of alginates. Biomedical applications of alginate–chitosan composites with special reference to their unique hydrogel-forming ability have been discussed. In conclusion, we have highlighted the promising features of alginates as an ideal biomaterial for drug delivery.

Keywords: Biocompatible, alginate–chitosan composites, microspheres, hydrogels, cross-linked alginates

8.1 Introduction

Drug delivery materials are selected and designed to overcome the defensive mechanism of our immune system, which resists introduction of foreign materials within its border. Thus, biocompatible materials and materials with natural origin-fulfilling criteria [1–4] of an ideal delivery system should be the first choice. Bio-friendly and biodegradable materials with drug-encapsulating and -release properties are a basic requirement for this purpose. Targeted delivery and controlled release are the most important criteria for designing material for drug delivery. A drug designer addresses many of these criteria in selecting an ideal material for drug delivery. In spite of hundreds of challenges like nature of drugs, their doses, and side effects, delivery materials are selected to minimize the interactions with

Email: srijitabasumallick@gmail.com

Shakeel Ahmed (ed.) Alginates, (141–152) © 2019 Scrivener Publishing LLC

the core. In this chapter, we focus our discussions on uses of alginates as a drug delivery material, its potentiality, and versatile applications.

8.2 Chemistry of Alginates

Alginate, also known as alginic acid, is an anionic polysaccharide obtained from cell walls of brown seaweed.

Alginates are found in three different forms of linear polymers of D-mannuronic acid (M) or L-guluronic acid (G) residues or copolymer of D-mannuronic acid (M) and L-guluronic acid (G) (as shown in Figure 8.1). D-mannuronic acid (M) and L-guluronic acid (G) residues are epimers (D-mannuronic acid converted to L-guluronic by an enzyme) and only differ at C5. Bacterial alginates are additionally O-acetylated on the two and/or three positions of the D-mannuronic acid residues. Three types of polymers of alginates are possible: GGGG, MMMM, and GMGM structures. As suggested by the name in GGGG polymer, all monomers are G alginic acid and it is a homopolymer, whereas in MMMM polymer, all monomers are M alginic acid and it is also a homopolymer. Only in the case of GMGM do the alternate groups of G and M residue form a copolymer structure. Alginates may be of a wide range of average molecular weights (50–100,000 residues) for appropriate application.

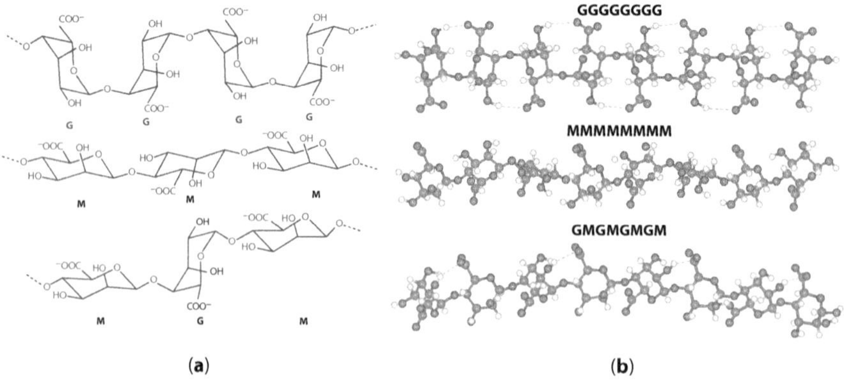

Figure 8.1 (a) Different forms of linear polymers of alginate [1]. (b) L-guluronic acid (G) residues forming hydrogen bonds and D-mannuronic acid (M) is non-hydrogen bonded (web source: http://www1.lsbu.ac.uk/water/alginate.html).

8.2.1 Hydrogel Formation by Alginates

Hydrogels of alginates are widely used as drug delivery vehicles, in addition to their use as tissue-repairing medicine. These gels are 3D networks of highly hydrophilic polymer segments held together by cross-linking with high water content. The cross-linking may be of physical nature like ionic interactions or chemical nature formed by covalent bond formation among the active functional groups of the polymer segments and added cross-linking molecules. Physicochemical properties of hydrogel depend on the nature of cross-linking type and cross-linking density, in addition to the molecular weight and chemical composition of the polymers. Hydrogels of biopolymers or semi-biopolymers are generally biocompatible, as they have structural similarity with the sugar-based constituent of the biological system.

8.2.1.1 Preparation of Hydrogel

8.2.1.1.1 Divalent and Trivalent Ions as Ionic Cross-Linker

Alginate is an anionic polymer coating $-COO^-$ groups in its skeleton, whereas chitosan is a cationic polymer with $-NH_3^+$ groups in its skeleton. Thus, alginate segments can be cross-linked by a polyvalent cation whereas chitosan by polyvalent anion like phosphate ion. Alginate forms gel from its aqueous solution when treated with $CaCl_2$ solution where Ca++ ion acts as an ionic cross-linker [1]. Generally, $CaCl_2$ is moderately soluble in water, leading to rapid cross-linking reaction, which may result in the formation of a stiff gel. This problem may be overcome by using sparingly soluble calcium salts such as calcium sulfate ($CaSO_4$) and calcium carbonate ($CaCO_3$). Formation of soft gel of alginates also depends on the chemical structure of alginates; alginates with more G units [2] result in the formation of softer gel. Ionically cross-linked gel is chemically less stable than those of a covalently cross-linked gel. This is probably due to the replacement of Ca^{2+} from the gel matrix by Na^+ present in the biofluid. The replaced Ca^{2+}, if accumulated in excess, may have some adverse biological effect [3].

8.2.1.1.2 Alginate Gel Preparation by Covalent Bond Formation with Poly Ethylene Glycol or Similar Polyfunctional Molecules

The basic principle of covalent bond formation between the -OH groups of glycol and acid groups of alginic acid is simple ester formation reaction between acid and alcohol with release of water molecules. The water molecules can migrate through the gel and do not provide any additional stress on the gel [4]. This makes the gel an ideal biomaterial for drug encapsulation. The physical properties of the gel may be regulated by varying the molecular weight of the alginate and that of the cross-linker [5] molecule. The cross-linking molecule is also a hydrophilic group containing macromolecules to restore the water retention property of the gel.

The condensation reaction is generally carried out under mild conditions using an acid or alkali as a catalyst. Sometimes, the polymerization reaction is initiated by light and known as photo cross-linking process [6, 7].

8.3 Pharmaceutical and Biomedical Chemistry of Alginates

Alginate is nontoxic and can be tailor-made for particular pharmaceutical and biomedical applications. It may be used as a stimulant (known as excipient) that can only modify drug release or drug absorption rate. Alginates are mostly used as a thickening, gel-forming, and stabilizing agent. For example, sodium alginate is used as a binding and disintegrating agent in tablets and as a suspending and thickening agent in water-miscible gels, lotions, and creams; it also acts as a stabilizer of emulsions [8–10].

Uses of alginate as polymeric- or alginate-controlled drug delivery systems [11] is more common. In controlled drug release, the drug releases slowly, reducing down the toxic effect of concentrated drug (toxic drug level). This also allows target-specific delivery (i.e., drug reaches target before degradation and full drug potency is used up), and bioavailability of the drug at the target site becomes optimum.

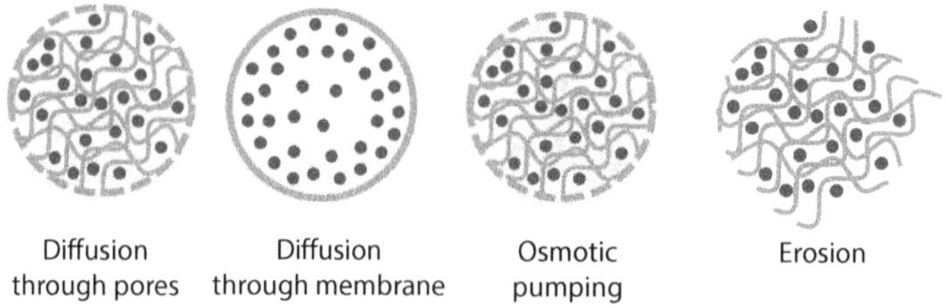

Figure 8.2 Different mechanisms of drug release (reproduced with permission [9]).

Polymeric-controlled systems [8, 9] can be separated into three distinct types:

(i) Polymer coatings dissolve slower than the drug. The drug comes to the aqueous environment only after degradation or dissolution of the coatings. This is also known as degradation-controlled release of the drug. Degradation-controlled drug delivery may occur in several ways like physical degradation and chemical degradation. Physical degradation of drug-encapsulated vesicle may occur by frictional interaction with another surface known as adhesion. It may take place by abrasion, i.e., some material is lost due to press, or by fatigue where the surface is weakened by application of load. Cavitations and corrosions may also rupture the drug-encapsulated surface. Chemical degradation of cross-linked network of polymer or breaking of drug–polymer link in drug–polymer conjugates occurs through hydrolysis reaction. Alginate is effectively used in chemical degradation route of drug delivery as alginates undergo acid-catalyzed hydrolysis, which depends on pH and temperature. Apart from degradation-controlled release, diffusion control and osmotic pressure-regulated drug release are also important, particularly when the drug is encapsulated within alginate vesicles.

(ii) Diffusion-controlled. Here, the rate of diffusion of the drug through the polymer membrane controls its release to the biofluid.

(iii) Osmotic pressure-regulated release. Here, water molecules pass through the semipermeable membrane of the encapsulating polymeric membrane due to osmotic pressure difference, and dissolve the drug, which gets out through pores.

8.3.1 Factors Governing Drug Encapsulation and Drug Delivery Processes

8.3.1.1 Delivery and Encapsulation of Small Drugs

Alginate microsphere or conventional Ca^{2+} cross-linked alginate matrix acts as polymeric vessels [12, 13] for drug encapsulation. In both cases, drug and polymer molecule coexist side by side, and swelling of polymer matrix or polymer capsule due to water permeation leads to enlargement of pores to allow drug or protein molecules to diffuse out of the matrix or reservoir.

In acidic conditions, such as in the stomach, swelling of the calcium alginate beads [14] does not occur. Under this situation, a drug is released by diffusion rather than by degradation. In neutral conditions (e.g., intestine), the calcium alginate beads swell and the drug releases by degradation process.

As we have mentioned that guluronic acid (G) of alginate coordinates with calcium ion during ionic cross-linking and gel forms are stiff, drugs cannot be released by erosion. On the other hand, with increasing mannuronic acid (M) content, gels become softer. It may be noted that H bonding within the polymer matrix makes it stiffer. Thus, this extra hydrogen bonding along the axis for GGGG polymer makes it more rigid. Rigidity of alginate polymer matrix may be altered by cross-linking with glutaraldehyde. This reduces swelling of alginate bead as is reported with the water-soluble drug nimesulide. Cross-linking of alginate with iron salt is also known. Iron cross-linked alginic acid has been shown to be effective in controlling the drug release rate of a number of drugs.

Aggregates of alginic acid also reduce net drug release rate. It has been found that ethyl cellulose or hydroxyl-propyl cellulose aggregates with alginic acid decrease the drug release rate due to aggregate structure.

Among other polymeric systems, calcium alginate microspheres with various copolymers are interesting. Drug release profiles have been found to depend on the nature of the copolymer. Composites of pectinate and alginate make the coating more brittle. Alginate forms complexes with chitosan [15], a cationic polymer. Chitosan–alginate composite tablets [16] have been used to reduce erosion of the tablet. Alginate beads may be coated with chitosan by spray dry technique. Water-soluble drugs are easily encapsulated and released from the composite tablets. Controlled release is observed [17] from an aggregate as the pores are blocked due to aggregation. Slow release rate may be achieved by putting an extra polymer cover onto polymer-coated drug. The effect of alginate on drug release rate from liposome has been studied; here, alginate acts as a coating agent [18]. In this case, the drug inside the liposome is released at a slower rate. In addition, alginates have an antioxidative activity that stabilizes the liposome.

The rate of diffusion of drug [19] through polymer membrane depends on swelling of the polymer matrix and interaction of drug molecules with polymer membrane. Among different types of interactions, cationic drug interacts with anionic $-CO_2^-$ group of alginate, which reduces diffusion rate for positively charged drug molecules.

Other interactions may be dominating; for example, the drug gentamicin sulfate interacts with mannuronic acid residues of alginates via nonionic interactions. In such a case, a higher mannuronic acid content in the alginate polymer would lead to a higher binding capacity for the drug molecules.

In general, the molecular weight of the alginates does not affect the drug release rate [20]. But in some cases, the release rate depends on the molecular weight; for example, the release rate of pindolol depends on the alginate molecular weight.

It has been found that the composition ratio of drug and alginate and calcium chloride concentration used in preparation of the tablet affect the drug release rate. For example, the release rate of nicardipine from alginate particles prepared in a ratio of 1:1 has been found to be less than that from 1:2 formulations.

It is reported that while the curing time of alginate bead is of minor importance, a long gelling time is favorable for obtaining control release rate. The cross-linker type and its concentration play an effective role on the drug release rate. For example, calcium alginate beads displayed a longer release time compared to that of alginate beads prepared using Ba^{++} and Sr^{++} ions. Variation in bead size may be used to optimize the release time.

Alginate–chitosan composite beads show release behavior dependent on pH, primarily because of the fact that protonation of the amine groups of chitosan improves its solubility. The inter-polymeric complex between alginate and chitosan exists in a gel form in acidic pH [21]. At neutral pH, the viscous complex swells and the gel slowly disintegrates, releasing the drug. Apart from pH, the release rate is also a function of the degree of cross-linking between the polymers.

Alginate and its composites with another biopolymer of opposite charge may be used to prepare compressed tablets. A compressed alginate tablet [22] will have a compact structure compared to a gel bead, and the drug release rate from such a compact tablet is reduced further. While water-soluble drugs are released primarily by diffusion mechanism, less soluble drugs are released by erosion mechanism. In a drug–alginate blended tablet, it has been shown that drugs of high water solubility are released at a faster rate, whereas the opposite effect is observed for drugs with poor water solubility.

Alginate beads prepared by simple gelation method show slow release for cationic drugs than that of anionic drugs because of the presence of the negative charges on alginate polymer. But when cellulose acetate is used as a gelling agent, the release rate of basic drug has been found independent of pH [23]. The release rate of acetaminophen from tablets containing spray-dried lactose–alginate particles was found to be slower than that of a sodium alginate tablet. This may be due to the fact that the spray-dried particles [24] have a much smaller particle size.

When a drug is coated with alginate and compressed, the drug surface becomes hydrophilic. For a lipophilic drug, it can be incorporated into alginate microspheres by the use of oil-in-water emulsion technique. For example, the drug may be dissolved in soybean oil shake with aqueous alginate microspheres to obtain immobilized drug containing oil microdroplets.

Encapsulation and release of hydrophobic drugs from alginate gel are difficult. To overcome this difficulty, alginate is grafted with poly-caprolactone (PCL) and cross-linked with calcium ions to form amphiphilic gel bead. Ophylline, a model hydrophobic drug, may be encapsulated in such an amphiphilic gel. It may be mentioned that the hydrophobic chain length of PCL controls the drug encapsulation efficiency.

Carbon nano tube (CNT)-incorporated alginate gel shows sustainable release of the ophylline. The addition of CNT [25] enhances the mechanical stability of gels, without affecting the structure and morphology and no significant cytotoxicity is reported.

Chitosan is an acetyl derivative of natural polymer chitin [26, 27] containing 80% d-glucosamine and 20% N-acetyl-d-glucosamine. A multi-component composite particle consisting of alginate, chitosan, and triamcinolone has been prepared [28] for sustainable release. The same method has been extended to encapsulate metronidazole into chitosan–alginate composites [28].

8.3.1.2 Macromolecular Drug Delivery by Alginates

So far, we have discussed encapsulation and delivery of hydrophilic, hydrophobic, and lipophilic small drug molecules by alginate and its composites. Now, we shall discuss some features of macromolecular drug delivery by alginate and its derivatives. Alginate and its composite are potential materials for encapsulation and delivery of enzymatic (protein) drugs. These drugs are susceptible to denaturation under heat, acidic, and alkaline conditions. These drugs can be easily encapsulated into alginate-based [29] composites under relatively mild conditions, which protect the drug from denaturation. Apart from various enzymatic drugs, macromolecular drugs from recombinant DNA are also known to deliver by alginates. Different strategies are adopted to regulate the release rate of macromolecular drugs from alginate gels. In general, the release rate of such drugs from alginate gels is rapid due to the porous and hydrophilic nature of the gels. Enzymes such as lysozyme and chymotrypsin can form physical cross-linking with sodium alginate and are thus encapsulated within the gel. Enzymes with low encapsulation efficiency release at a faster rate from alginate gels. Insulin-loaded alginate [30] gel prepared by blending alginate with chitosan-coating protects insulin from denaturation even at gastric pH [31]. Hydrogel containing bovine serum albumin (BSA) [29] dispersed in poly(lactic-co-glycolic acid) (PLGA) microspheres is an example of combined gel and microsphere vehicle systems. A similar system is the hydrogel of heat shock protein fused with a transcriptional activator in a vesicle.

8.4 Conclusions

In this chapter, potential biomedical applications of alginates in the areas of drug encapsulation and drug delivery have been discussed. Biocompatibility, mild gelation conditions, and easy modifications of its chemical structure to prepare alginate derivatives with new properties have been highlighted throughout this chapter. Factors governing encapsulation and delivery of different types of drugs, such as hydrophilic, amphiphilic, and hydrophobic drugs, have been presented. Encapsulation and delivery of macromolecular drugs like enzymatic drugs by alginates and their composites with other biomaterials have been discussed. It has been shown that alginate–chitosan composites are prospective biomaterials for the development of safe drug delivery system.

Acknowledgments

The author is thankful to the Head of the Department (HOD), Chemistry and Principal, Asutosh College under Calcutta University for their academic encouragement.

References

1. Hanne, H.T. and Jan, K., Alginate in drug delivery systems. *Drug Dev. Ind. Pharm.*, 28, 6, 621–630, 2002.
2. Kuen, Y.L. and David, J.M., Alginate: Properties and biomedical applications. *Progr. Polym. Sci.*, 37, 106–126, 2012.
3. Drury, J.L., Dennis, R.G., Mooney, D.J., The tensile properties of alginate hydrogels. *Biomaterials*, 25, 3187–99, 2004.

4. Zhao, X.H., Huebsch, N., Mooney, D.J., Suo, Z.G., Stress-relaxation behaviour in gels with ionic and covalent cross-links. *J. Appl. Phys.*, 107, 063509, 1–5, 2010.
5. Eiselt, P., Lee, K.Y., Mooney, D.J., Rigidity of two-component hydrogels prepared from alginate and poly(ethylene glycol)-diamines. *Macromolecules*, 32, 5561–6, 1999.
6. Smeds, K.A. and Grinstaff, M.W., Photo cross-linkable polysaccharides for *in situ* hydrogel formation. *J. Biomed. Mater. Res.*, 54, 115–21, 2001.
7. Tanaka, H. and Sato, Y., Photosensitivity of polyvinylesters of substituted cinnamylidene acetic acids. *J. Polym. Sci.*, 10, 3279–87, 1972.
8. George, M. and Abraham, T.E., Polyionic hydrocolloids for the intestinal delivery of protein drugs: Alginate and chitosan–A review. *J. Control Release.*, 114, 1, 1–14, 2006.
9. Susanne, F., Marie, W., Mats, R., Anders, A., The mechanisms of drug release in poly (lactic-co-glycolic acid)-based drug delivery systems-A review. *Int. J. Pharm.*, 415, 1–2, 34–52, 2011.
10. Lee, K.Y., Bouhadir, K.H., Mooney, D.J., Controlled degradation of hydrogels using multifunctional cross-linking molecules. *Biomaterials*, 25, 2461–6, 2004.
11. Al-Musa, S., Fara, D.A., Badwan, A.A., Evaluation of parameters involved in preparation and release of drug loaded in cross-linked matrices of alginate. *J. Control Rel.*, 57, 223–232, 1999.
12. Filipovic-Grcic, J., Maysinger, D., Zorc, B., Jalsenjak, I., Macromolecular prodrugs., I.V., alginate-chitosan microspheres of PHEA-L-dopa adduct. *Int. J. Pharm.*, 116, 39–44, 1995.
13. Chanp, L.W., Heng, W.S., Wan, L.S.C.J., Effect of cellulose derivatives on alginate micro spheres prepared by emulsification. *Microencaps*, 14, 5, 545–555, 1997.
14. Kulkarni, A.R., Soppimath, K.S., Aralaguooi, M.I., Aminabhavi, T.M., Rudzinski, W.E., Preparation of cross-linked sodium alginate microparticles using glutaraldehyde in methanol. *Drug Dev. Ind. Pharm.*, 26, 1121–1124, 2000.
15. Hari, P.R., Chandi, T., Sharma, C.P., Chitosan/calcium alginate microcapsules for intestinal delivery of nitrofurantoin. *J. Microencaps.*, 13, 319–329, 1996.
16. Shiraishi, S., Arahira, M., Imai, T., Otagiri, M., Enhancement of dissolution rates of several drugs by low-molecular chitosan and alginate. *Chem. Pharm. Bull.*, 38, 185–187, 1990.
17. Takeuchi, H., Yasuji, T., Yamamoto, H., Kawashima, Y., Spray-dried lactose composite particles containing an ion complex of alginate-chitosan for designing a dry-coated tablet having a time-controlled releasing function. *Pharm. Res.*, 17, 94–99, 2000.

18. Takagi, I., Nakashima, H., Takagi, M., Yotsuyanagi, T., Ikeda, K., Application of alginate gel as a vehicle for liposomes. II. Erosion of alginate gel beads and the release of loaded liposomes. *Chem. Pharm. Bull.*, 45, 389–393, 1997.
19. Aslani, P. and Kennedy, R.A., Studies on diffusion in alginate gels. I. Effect of cross-linking with calcium or zinc ions on diffusion of acetaminophen. *J. Control Rel.*, 42, 75–82, 1996.
20. Lee, K.Y., Rowley, J.A., Eiselt, P., Moy, E.M., Bouhadir, K.H., Mooney, D.J., Controlling mechanical and swelling properties of alginate hydrogels independently by cross-linker type and cross-linking density. *Macromolecules*, 33, 4291–4, 2000.
21. Park, H.Y., Choi, C.R., Kim, J.H., Kim, W.S., Effect of pH on drug release from polysaccharide tablets. *Drug Del.*, 5, 13–18, 1998.
22. Pillay, V., Dangor, C.M., Govender, T., Moopanar, K.R., Hurbans, N., Drug release modulation from cross-linked calcium alginate microdiscs, 2: Swelling, compression, and stability of the hydrodynamically-sensitive calcium alginate matrix and the associated drug release mechanisms. *Drug Del.*, 5, 35–46, 1998.
23. Rao, V.M., Engh, K., Qiu, Y., Design of pH-independent controlled release matrix tablets for acidic drugs. *Int. J. Pharm.*, 252, 1–2, 81–6, 2003.
24. Takeuchi, H., Yasuji, T., Hino, T., Yamamoto, H., Kawashima, Y., Spray-dried composite particles of lactose and sodium alginate for direct tabletting and controlled releasing. *Int. J. Pharm.*, 174, 91–100, 1998.
25. Zhang, X.L., Hui, Z.Y., Wan, D.X., Huang, H.T., Huang, J., Yuan, H., Yu, J.H., Alginate microsphere filled with carbon nanotube as drug carrier. *Int. J. Biol. Macromol.*, 47, 389–95, 2010.
26. Rinaudo, M., Chitin and chitosan: Properties and applications. *Progr. Polym. Sci.*, 31, 603–32, 2006.
27. Sandford, P.A., Steinnes, A. *et al.*, Biomedical application of high-purity chitosan, in: *Water soluble polymers: Synthesis, solution properties, and applications*, S.W. Shalaby, C.L. McCormick, G.B. Butler (Eds.), pp. 430–45, American Chemical Society, Washington, DC, 1991.
28. Murata, Y., Maeda, T., Miyamoto, E., Kawashima, S., Preparation of chitosan-reinforced alginate gel beads—Effects of chitosan on gel matrix erosion. *Int. J. Pharm.*, 96, 139–145, 1993.
29. Wells, L.A. and Sheardown, H., Extended release of high pI proteins from alginate microspheres via a novel encapsulation technique. *Eur. J. Pharm. Biopharm.*, 65, 329–35, 2007.
30. Silva, C.M., Ribeiro, A.J., Ferreira, D., Veiga, F., Insulin encapsulation in reinforced alginate microspheres prepared by internal gelation. *Eur. J. Pharm. Sci.*, 29, 148–59, 2006.
31. Lee, K.Y. and Yuk, S.H., Polymeric protein delivery systems. *Progr. Polym. Sci.*, 32, 669–97, 2007.

9

Alginate in Wound Care

Satyaranjan Bairagi and S. Wazed Ali*

Department of Textile Technology
Indian Institute of Technology, Delhi, New Delhi, India

Abstract
Wound is a broken part of the skin due to physical or chemical actions such as friction, rubbing, temperature, etc. Suitable wound management system is necessary for proper wound healing. Commercially available wound-dressing materials like wool cotton, cotton gauge, and different synthetic dressings have some problems in terms of wound-healing rate, the moist environment at the wound, and comfort of the patients. Different existing problems can be overcome by using natural biopolymers (alginate, chitosan, collagen, etc.) based materials for wound dressings. Natural biopolymers have various advantageous properties, including low cost, easy to synthesize, and biocompatibility. Among the different biopolymers, alginate is one of the potential candidates for wound-dressing materials. Alginate is generally categorized as a naturally synthesized polyanionic copolymer, extracted from marine kelps, mainly the brown sea algae. Its advantageous properties (e.g., viscosity, thermostability, sol–gel transformation, hydrophilicity, swelling capacity, and drug release) give rise to different medical applications such as tissue engineering and drug delivery, particularly in wound care. In this chapter, synthesis of alginate, physicochemical properties, biomedical applications of the alginate, and particularly wound care application in the various forms of alginate-based wound dressings have been discussed in various sections.

Keywords: Alginate, biopolymer, biomedical applications, wound dressing, wound care, alginate fiber, alginate hydrogel

**Corresponding author*: wazed@textile.iitd.ac.in

Shakeel Ahmed (ed.) Alginates, (153–178) © 2019 Scrivener Publishing LLC

9.1 Introduction

Wound has been defined as a break or a defect in the screen due to physical or thermal destruction. There are mainly two types of wounds in terms of wound repair procedure. One type of wound is acute wound, which can be cured permanently with the minimum time periods, mainly within 8–12 weeks [1]. Acute wounds used to occur due to mechanical injury by means of abrasion, tears, knives, gun shots, etc. or chemical injury with an exposure to corrosive chemicals or burn injury by electricity, radiation, thermal stimulus, etc. [2]. Chronic wounds, on the other hand, are cured very slowly and it takes more than 12 weeks to heal permanently [3]. Chronic wounds fail to heal under diabetes and malignancies. So, the wound-healing system has an important role to give the solution for the treatment of various wounds. Wound healing is a biological process in which cell growth and regeneration take place. In the wound-healing process, different stages work independently and also overlap each other. Mainly five stages, i.e., hemostasis, inflammation, migration, proliferation, and maturation phases, occur during the wound-healing process [4].

In recent years, biomaterials have drawn a great interest toward biomedical applications like wound healing, as compared to their synthetic counterpart. Materials extracted from biological sources, for instance, wood, seaweeds, etc., have been greatly used in cell and tissue regeneration process. Different types of biomaterials such as alginate, chitosan, etc. have been used in the recent past in the wound-healing treatment of different acute and chronic wounds. Among different biomaterials, alginate has been found as a promising material in wound healing because of its various beneficial physicochemical properties. It is a polyanionic polysaccharide-based biopolymer and is extracted from brown seaweeds. Its good compatibility with the biological system and eco-friendly nature have given rise to the application of wound healing. Also, it has a good ion exchange and gelation properties, which are very important parts of the wound-healing process. Herein, the different biomedical applications of the naturally extracted alginate biopolymer have been described with special emphasis on its wound-healing application.

9.2 Sources and Synthesis of Alginate

Alginate is a natural biopolymer that can be extracted from the sea algae as well as bacteria. It occurs as a structural component in the case of marine brown algae and molecular polysaccharides in some bacteria. Among the two sources, the algae source has a great interest in comparison to the bacteria source. The different algae species are *Laminaria hyperborea*, *Ascophyllum nodosum*, and *Macrocystis pyrifera* [5, 6].

Alginate is generally synthesized in alkaline condition. The brown algae collected from the sea is dried first and is finally treated with different chemicals to get impurities-free fresh alginate powder in the form of acid or salt [6]. In the synthesis of alginate biopolymer (Figure 9.1), fructose-6-phosphate is produced through the gluconeogenesis process, wherein first, acetyl-CoA is generated by the oxidation of carbon materials, and finally, this CoA converts into fructose-6-phosphate. Then, the produced fructose-6-phosphate goes through different biosynthetic transformations to convert GDP–mannuronic acid. Generally, the biosynthetic process involves four steps, i.e., (i) synthesis of the GDP–mannuronic acid, (ii) cytoplasmic membrane transfer and formation of the polymannuronic acid, (iii) periplasmic transfer and modification, and (iv) transfer through the outer membrane. During the modification step as mentioned in step 3, polymannuronic acid is acetylated by different transacetylases at the O-2 or O-3 places. Then, some of the nonacetylated M deposits are converted into G parts. Finally, through transmembrane porins, cell alginate is unconfined [7–9]. Generally, α-L-guluronic acid (G) and β-D-mannuronic

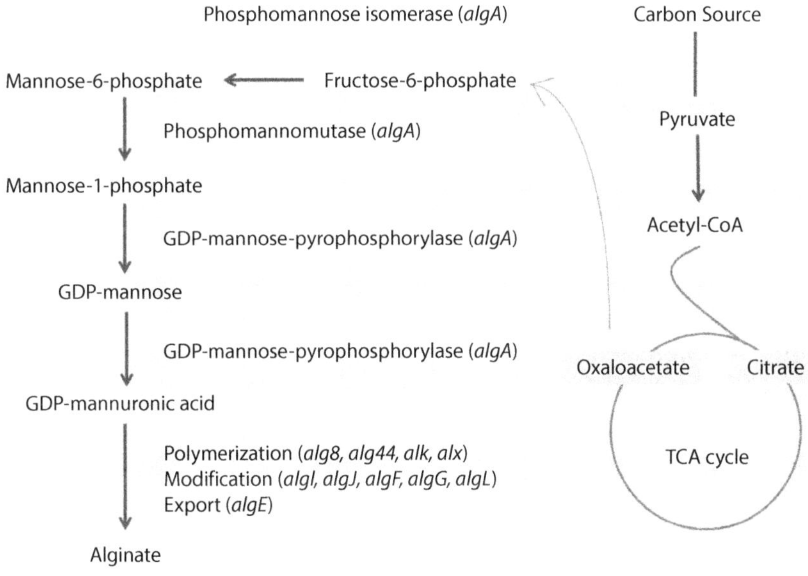

Figure 9.1 Bacterial alginate biosynthesis pathway [11].

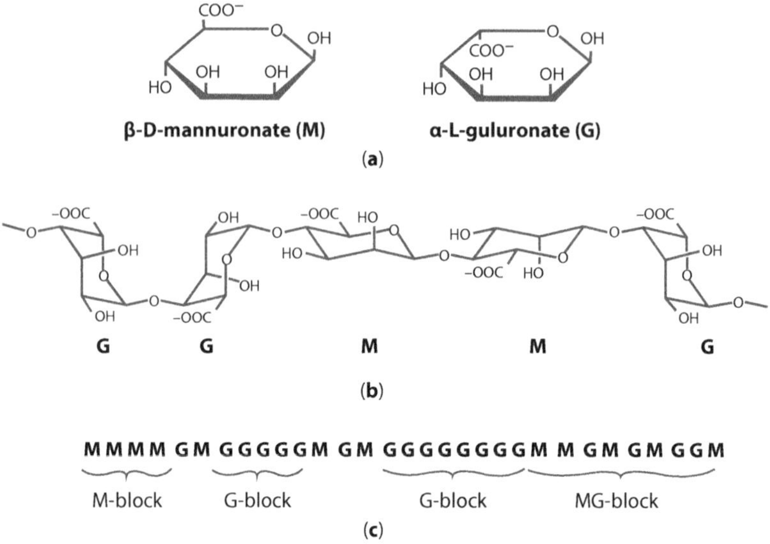

Figure 9.2 Structural characteristics of alginates: (a) alginate monomers, (b) chain conformation, and (c) block distribution [12].

acid are two basic linear blocks, which are linked together by 1–4 linkages in the alginate biopolymer, as shown in Figure 9.2 [10].

9.3 Physicochemical Properties of the Alginate Biopolymer

The physical properties of the alginate biopolymer depend upon the chemical structure, molecular weight, length of the polymer chain, and concentration of the alginate. The gel-to-sol transformation property of the alginate is varied with a concentration of the multivalent cations (Ca^+) [13, 14]. The viscosity of the alginate solution changes with the concentration and number of monomers in the chain. The guluronic acid-based alginate has more solubility in the water as compared to the mannuronic acid-based alginate. Another important physical property of alginate is its thermal stability. In general, between 0 and 100 °C temperature ranges, the alginate gel is formed. The noncovalent polymer segments help to protect deformation of the alginate at below the boiling temperature of the water. But the stability of the alginate gel reduces with increasing temperature. The most vital property of alginate is the sol-to-gel transformation property. The alginate solution is transferred from sol state to gel state by using

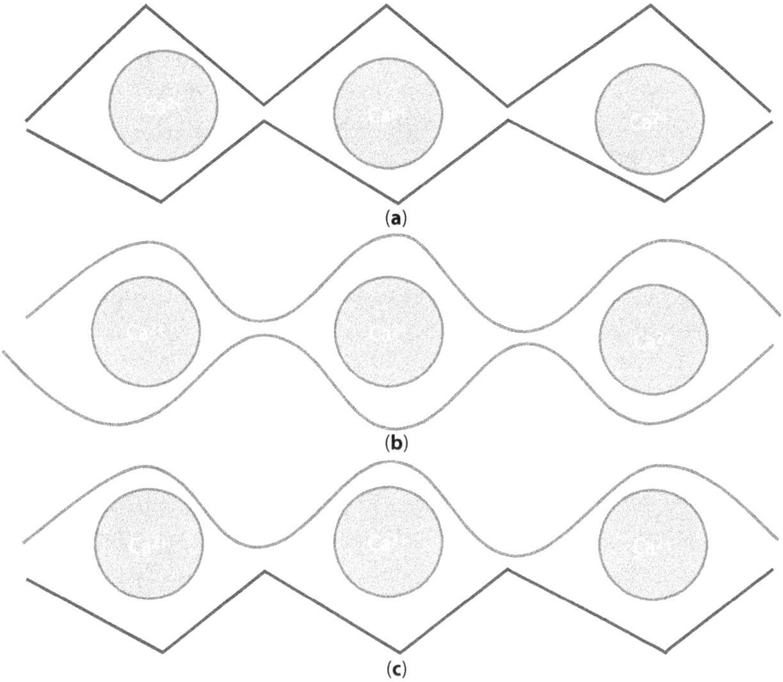

Figure 9.3 Possible junction points in alginates. (a) GG/GG junctions, (b) MG/MG junctions, and (c) GG/MG junctions [15].

the cationic cross-linking agent at the aqueous medium. The G blocks in the synthesized alginate biopolymer are the main component to create the gel form of the alginate in the presence of divalent cations. The MG blocks also help to form weak bonds during gelation process. There are different divalent cations such as Pb > Cu > Cd > Ba > Sr > Ca > Co, Ni, Zn > Mn as per the affinity of the alginate with the same. Moreover, the most usable cation is Ca^+ to help in gel formation of the alginate. Figure 9.3 shows a schematic representation of the different possible junctions with the Ca^+ into the alginate biopolymer during gelation process.

9.4 Biomedical Applications of Alginate

Over the last few years, researchers have drawn most attention to research on the different natural biopolymers, particularly alginate. Alginate has different suitable properties, including biocompatibility and suitability of sol-to-gel transformation capability in the presence of cationic cross-linking agents, which promotes its various biomedical applications.

Alginate has been used in different biomedical applications such as drug delivery, protein delivery, wound dressing, cell culture, tissue engineering (blood vessels, bone, cartilage, muscle, nerve, pancreas, and liver), formation of dental impression materials, and gas preventive formulations. Cross-linked alginate (Na or Ca alginate) has been used to make an oral tablet because of its suitable gel formation property. This acceptable gel-forming ability of alginate permits to protect the different mild amalgams from acidic conditions inside the stomach by its buffering action [16, 17]. In this chapter, only the wound care application by using alginate has been elaborated in the next sections.

9.4.1 Alginate in Wound Care

Wound is a broken part of the skin due to some physical and thermal actions, as mentioned earlier. Wound can also be explained as the disruption of the normal anatomic activities of the cells in the presence of different actions (friction, temperature, etc.). It is mainly categorized into two types on the basis of wound repairing process, i.e., acute and chronic wounds. An acceptable wound management system is necessary to treat the wounds. Wound dressing is a biological process to regenerate and promote growth of cells and tissues. In the nineteenth century, synthetic or natural bandages, wool cotton, lint, and gauzes have been used in wound management applications because of their absorbance capability. The function of wound dressing is to protect the wound from infection by allowing evaporation of the exudates through the dressing and creating a barrier against different external harmful microorganisms [18]. Nowadays, wound could be healed in a better and successful manner by creating a warm, moist environment around the wound. Epithelial cells can be removed from the surface of the wound so that oxygen circulation around the wound area would increase. Therefore, regeneration of the cells and tissues will be improved in the presence of warm, moist environment [19]. Different factors, such as the type of wound, chemical and physical properties of the wound-dressing materials, and the condition of the patient in terms of health, influence the wound management process. Therefore, it is important to understand the all above-mentioned factors before starting the wound dressing process for a given type of wound. There are various natural carbohydrate-based materials like neutral (e.g., cellulose, dextran, and (1→3)-β-D-glucans), basic (e.g., chitin and chitosan), acidic (e.g., alginic acid and hyaluronic acid), and sulfated polysaccharides (e.g., heparin, chondroitin, dermatan, and keratan sulfates) from which wound-dressing materials can be developed for wound care application. Among them, the alginate-based material has different advantages like higher gilation characteristic, higher hydrophilicity, etc. over other natural materials [20]. Alginate dressing has been made by cross-linking of an alginate solution with cationic ions to form a gel-like structure. Then, a porous sheet has been developed by freeze-drying the gel structure of the alginate. The alginate dressing in dry form can absorb wound exudates from the wound bed and change into the gel form again. Therefore, alginate dressing maintains a moist condition at around the wound bed by supplying the water and also helps to regenerate and grow the cells without any bacterial infection. Commercially available different alginate-based dressings are shown in Table 9.1. Alginate-based dressing materials can be developed in different forms such as (i) pure alginate polymer-based wound dressing, (ii) intercellular mediators incorporated alginate polymer-based wound dressing, (iii) zinc/alginate- and silver/alginate-based wound dressing, (iv) chitosan/alginate- and collagen/alginate-based wound dressing, (v) alginate fiber-based wound dressing, and (vi) alginate hydrogel-based wound dressing. Above-mentioned forms of the alginate-based wound dressing have been explained in the next section of this chapter.

Table 9.1 Alginate-based wound dressings commercially available in the market [21].

Product	Manufacturer
AlgiDERM	Bard
AlgiSite	Smith & Nephew, Inc.
Algosteril	Johnson & Johnson
CarraSorb H	Carrington
CURASORB	Kendall
CURASORB Zinc	
Dermacea	Sherwood-Davis & Geck
FyBron	B. Braun
Gentell	Gentell
Hyperion Advanced Alginate Dressing	Hyperion Medical, Inc.
KALTOSTAT	ConvaTec
KALGINATE	DeRoyal
Maxorb	Medline
PolyMem	Ferris Mfg.
Restore	Hollister
SORBSAN	Dow Hickam
SeaSorb	Coloplast Sween Corp.
Tegagen HG	3M Health Care
Tegagen HI	

9.4.1.1 *Pure Alginate Polymer-Based Wound Dressing*

Sayaq *et al.*, [22] performed a comparative performance study on alginate-based wound dressings compared to a control dressing. These wound dressings have been applied on 92 patients to treat the full-thickness pressure ulcer disease. With alginate dressing, 40% reduction in wound area has been explored in 74% of the patients, while 42% of the patients responded properly for dextranomer dressing. Four and eight weeks have been taken for alginate-based wound dressing and dextranomer dressing, respectively. The wound reduction surface area in the case of alginate-based dressing has been found to be 2.39 cm^2, wherein the same was recorded for 0.27 cm^2 in the case of dextranomer dressings. Therefore, they concluded that alginate-based wound dressing is the best option in the case of wound care compared to other wound dressings.

9.4.1.2 *Intercellular Mediators Incorporated Alginate Polymer-Based Wound Dressing*

Alginate-based wound-dressing materials have drawn great attention in wound care applications. Different intercellular mediators have been used in the alginate-based matrix for suitable proliferation and migrations of wound cells. Balakrishnan *et al.*, [23] studied dibutyryl cyclic adenosine monophosphate (DBcAMP)-incorporated alginate-based wound-dressing materials. DBcAMP lipophilic analog of cyclic adenosine monophosphate (cAMP) and cAMP have been used as a strong regulator of human keratinocyte proliferation. DBcAMP was released very slowly from the alginate matrix during application in the wound dressing, and gradually, it was found to release firstly at a later time. Researchers observed that a complete re-epithelialization of wounds has been taking place within 10 days in the case of DBcAMP-incorporated alginate-based wound-dressing materials, wherein 15 days was required for the complete re-epithelialization in the case of control wounds. They have mainly evaluated the wound-healing capacity of the DBcAMP-incorporated alginate-based dressing on full-thickness wounds in a rat model. In another study, the cytocompatibility and proangiogenesis function in the wound has also been investigated by strontium (an inorganic angiogenesis factor)-incorporated silk fibroin (SF)/sodium alginate (SA)-based blended films [24]. The silk fibroin has some beautiful properties in terms of medical treatment, mainly in wound healing. It has good biocompatibility, blood compatibility, stability, and better water permeability, which are among the most important and desirable properties for the healing of wounds. Adhesion, spread, and proliferation of the epidermal cells and fibroblast can also be influenced by the silk fibroin. But the mechanical strength of the silk fibroin at dry condition is very poor, which is its only drawback. To overcome this problem, naturally synthesized sodium alginate has been used in wound dressing. Sodium alginate not only increases the mechanical strength of the structure but also influences all advantageous properties in terms of wound-healing application. Finally, they observed that the angiogenesis and proliferation of the cells have been improved by using these blended films.

Panawes *et al.*, [25] studied the fruit hull of *Garcinia mangostana* (MT) extracts and alginate-based gauze as a wound dressing and evaluated its antibacterial characteristic. The concentration of the alginate in the gauze also affects wound characteristics in terms of swelling ratio, moist environment, and blood coagulation properties. Finally, they concluded that 0.5% AG and 50–55% MT are the suitable percentages to enhance the antibacterial properties of gauze-based wound dressing.

9.4.1.3 Zinc/Alginate- and Silver/Alginate-Based Wound Dressing

Zinc-incorporated alginate wound dressing has also been used tremendously in wound care application. As already mentioned in the above paragraph, the alginate itself has advantageous antimicrobial and proinflammatory reduction property. But incorporation of zinc in the alginate wound dressing significantly improves the antimicrobial and other activities. Zinc ions can also enhance the keratinocyte migration and endogenous growth factor in the wound dressing process [26].

Wiegand *et al.*, [27] investigated the functions of pure alginate and silver-based alginate in the cytotoxicity, antimicrobial activity, and binding capacity for pathophysiological factors in chronic wounds. They have developed three alginate-based wound dressings, namely, pure alginate, ionic silver-based alginate, and silver nanocrystals-based alginate. They finally observed that alginate binds a suitable amount of the elastase, reduces the proinflammatory cytokines, and also stops the free radical formation in the chronic wounds. Moreover, alginate also exposes the antimicrobial activity and biocompatibility. Addition of silver into the alginate improves the antibacterial activity and influences the antioxidant property of the alginate. Opasanon *et al.*, [28] also studied the wound-healing capability of the Askina Calgitrol Ag® and the 1% silver sulfadiazine (1% AgSD)-based wound dressing. The Askina Calgitrol Ag® wound dressing has been made from polyurethane foam layer, which can absorb the wound exudate, and an ionic silver alginate part that was used to act as an antimicrobial activator as well as to prevent any contaminants due to the external bacteria. The wound-healing property has been analyzed for both the wound dressings in terms of healing time, the number of dressing changes, pain score, and nursing time. The researchers have concluded the suitability of wound dressing by applying these in the partial thickness burn wound at Siriraj Hospital, Division of Trauma Surgery, Mahidol University, Thailand. The results from their experiment showed that the average pain score in the Askina Calgitrol Ag® dressing was lower as compared to 1% silver sulfadiazine dressing, i.e., 2.23 ± 1.87 versus 6.08 ± 2.33, respectively. Also, the number for the change of the wound dressing and healing time was found to be lower in the case of Askina Calgitrol Ag® dressing with respect to the other. One of the suitable properties of alginate is its controllable drug release action during wound dressing. Alginate gel consists of a nanoporous structure for which it can contain different small molecular weight drugs. The drugs can form various bonds with the alginate polymer so that a controlled release of the drug takes place during applications. Meng *et al.*, [29] studied silver sulfadiazine-incorporated chitosan/alginate polyelectrolyte-based composite membrane for wound dressing applications. Here, sulfadiazine acts as a drug that is released in a controlled manner during its application. The researchers concluded that the mechanical properties, water uptake capability, and biodegradability of the alginate-based membrane make it the best fit for wound dressing as well as drug delivery application. Young *et al.*, [30] developed a silver sulfadiazine-loaded absorbable alginate and gelatine-based sponge material by a suitable cross-linking method for using it in drug release and wound-healing applications. The porosity of the developed material was found to be increased with increasing gelatine percentage. The total absorption capacity of the sponge material has been improved, which is a very important factor in wound-healing treatment. As the absorption capacity of the sponge increases, it can easily absorb the wound exudate. Therefore, it is naturally expected that the wound would be cured quickly in less time. They used this sponge-type material in *in vivo* animal test using Wistar rats and finally, they observed that AgSD-loaded gelatine alginate-based sponge is more effective than a regular Vaseline gauze. In another study, Montaser *et al.*, [31] reported a wound-dressing material based on the silver/alginate/nicotinamide nanocomposite, wherein silver has been used to influence the antibacterial property and nicotinamide as a drug material. Alginate helps to synthesize silver nanoparticles as reducing and stabilizing agents. The size and shape of silver nanoparticles was found to be different by changing the concentration of the alginate. A nonwoven fabric was immersed in the composite solution of alginate/silver/nicotinamide.

Finally, this wound-dressing material has been used on the burn wound of a diabetic rat to evaluate the efficacy of the developed wound dressing.

9.4.1.4 Chitosan/Alginate- and Collagen/Alginate-Based Wound Dressing

Gils *et al.*, [32] developed a unique wound dressing based on collagen-alginate, and they have evaluated the functionality of this wound dressing in the postoperative management of chemical matricectomies. They observed that the wound-healing time required in the case of collagen-alginate dressing has been 24.4 days, whereas it took 35.8 days for control samples. Finally, the researchers suggested that collagen-alginate-based dressing is a good option for the postoperative management of chemical matricectomies in terms of less healing time, lower rate of infection, and improving the patient condition. In another study, Donaghue *et al.*, [33] also systematically studied the efficacy of collagen-alginate (Fibracol) and regular gauze-based wound dressing for diabetic foot ulcer treatment. They have applied these two wound dressings on 75 patients for treatment of foot ulcers. They observed that the wound reduction area is higher in the case of collagen-alginate-based dressing, i.e., 80.6%, whereas the same is 61.1% for local gauze dressing. Therefore, collagen-alginate-based dressing can be a good candidate for wound dressing application. Wang *et al.*, [34] fabricated a very narrow, translucent, and bendable polyelectrolyte composite-based sheet based on the chitosan-alginate biopolymers for potential application in wound dressing. These composite sheets have been developed by solution casting method. The researchers have explored the nontoxic characteristics of the fabricated sheets toward the mouse and human fibroblast cells. They also observed that the growth of the cells has not been disturbed by these composite membranes. Finally, it has been shown that a wound can be properly cured within 14 days in terms of a suitable epidermal structure, whereas the control sample has taken more time for the same.

9.4.1.5 Alginate Fiber-Based Wound Dressing

Alginate-based fibers can be used in a significant way in various biomedical applications due to their better flexibility, low cost, and higher surface area. Fibers made from different natural biopolymers like alginate, chitosan, etc. (mainly from the polysaccharides) have been used largely in wound care management because of their very advanced biocompatibility, nontoxicity, and effective bioactivity. Alginate has been used in different forms such as polymer matrix, polymer gel, and fiber (woven and nonwoven) for wound dressing application. Alginate fibers have been used to form a range of wound-dressing materials as early as the 1980s. Mainly, the alginate fibers or alginic acids have been produced by injecting an alginate (generally sodium alginate)-soluble solution through a spinneret in the acid (calcium salt)-containing solution of the coagulation bath, and then wound-dressing materials have been made from these produced alginate fibers [35–37]. In recent days, a surge of researchers have been working on alginate fiber-based wound dressing for wound-healing applications because of its suitable gel-forming as well as hydrophilic properties. When alginate fibers get into contact with the wound exudates, calcium ions from the fibers are exchanged with sodium ions of the wound exudates and form a gel. Furthermore, when water enters into the alginate fibers, the fibers become gel. Therefore, a suitable moist environment could be maintained by this alginate gel throughout the wound area [38]. The low cost, softness, nontoxicity, and simplicity in synthesis of the alginate fiber can also be important factors for using it in medical applications compared to other synthetic polymer-based fibers [39–41]. Alginate fibers can be used as a solution spun fiber or nanofiber in wound-dressing material. The main advantage of alginate fiber-based wound dressing is its higher moisture absorbance capability inside the fiber structure, so that a suitable moist environment could be maintained in the whole wound area [42]. Fan *et al.*, [43] have studied systematically the alginate-gelatin mixed-based solution spun fibers for wound care application. The fibers have been synthesized by a solution spinning method, where a mixed solution of the alginate and gelatin has been poured into $CaCl_2$ and ethanol containing coagulation bath. The fibers have been characterized by Fourier transform infrared (FTIR), X-ray diffractometer (XRD), scanning electron microscope (SEM), and thermogravimetric analyzer (TGA), and they also tested the moisture-retaining property, tensile strength, and breaking elongation of the above-mentioned blended fibers. Finally, they concluded that 30% gelatin into the alginate polymers has good moisture absorbance characteristics. Therefore, blend wound dressing can be more advantageous instead of the single alginate component. In another study, Fan *et al.*, [44] also investigated the properties of the alginate and N-succinyl-chitosan-based blend fibers, which is made by the solution spinning method. Finally, they concluded that the incorporation of the N-Succinyl-chitosan (SCS) has improved the moisture uptake property of the developed blend fibers compared to alginate as a single component. As moisture management is one of the most important functions in wound care, the blend fibers can be the best candidate compared to single fiber component.

The antimicrobial alginate fiber can be developed by incorporating different biomaterials such as neem, papaya, and their combination. In the past, researchers have already tried to develop antimicrobial alginate fiber-based wound-dressing material. Hussain *et al.,* [45] studied the antimicrobial calcium alginate fibers from neem and papaya leaves extract. As it is known that neem and papaya leaves extract has antimicrobial properties, their incorporation in the alginate-based biopolymer substantially enhances the antimicrobial function in the wound area of the patient. Herein, the authors mainly studied the tensile strength, liquid absorption, and antimicrobial properties of the developed wound-dressing materials and the results were compared with the control sample. They finally observed that the tensile strength of the neem-loaded sample has been increased compared to the papaya-loaded sample and the control sample. They also showed that the liquid absorption property has been decreased in the case of both samples, but the antibacterial effect has been increased for both samples. They also reported that 85% growth of the wound has been reduced in the neem- and papaya-loaded calcium alginate-based samples as compared to the control calcium-based wound dressing sample. Khajavi *et al.,* [46] developed eucalyptus essential oil (EEO)-incorporated alginate-based wet spun fibers and investigated their antibacterial properties in wound dressing treatment. Finally, they summarized that EEO loading enhances the antimicrobial effect of alginate-based fiber for its effective utilization in wound dressing, primarily because of the combined effects of the EEO and alginate. Knill *et al.,* [47] also developed unhydrolyzed and hydrolyzed chitosan modified alginate fiber-based wound-dressing materials. They studied the mechanical and antibacterial properties of the developed wound-dressing materials. The unhydrolyzed chitosan modified alginate fiber-based wound dressing has showed lower mechanical strength and antibacterial activity compared to the hydrolyzed chitosan modified alginate fiber-based wound-dressing material. In the case of unhydrolyzed chitosan modified alginate fiber, chitosan particles have not penetrated into the fiber so that a coating layer has been developed on the fiber surface. But, in the case of hydrolyzed chitosan particles, they have entered effectively inside the fiber. Therefore, hydrolyzed chitosan modified alginate fiber-based wound-dressing material has registered higher tenacity and antibacterial properties as compared to others. Sweeney *et al.,* [48] also studied the hydrolyzed chitosan modified alginate fiber-based wound-dressing material. The hydrolyzed chitosan has been used as the solution in the coagulation bath. They finally reported that the mechanical property (tenacity) and liquid absorption capacity of the developed material have been increased with increasing the chitosan percentage inside the alginate fibers. By the wet solution spinning technique, 4.5% to 5% chitosan has been incorporated into the alginate fibers. The amount of saline and deionized water absorbed by the developed fiber was >30 and >50 g/g, respectively.

The alginate in the form of nanofibrous web has drawn great interest in biomedical application, mainly in the area of wound care system. The nanofiber has a higher surface area compared to commercially available solution spun fiber, for which it could be the better option for wound dressing. As the surface area is high, the contact area of this nanofiber with the wound area would also be higher. Therefore, all properties of wound-dressing materials would increase in terms of wound-healing rate, moisture absorption capacity, and regeneration of the epithelial cells. In a study, Choi *et al.*, [49] developed a bioactive ingredient spirulina extract incorporated alginate-polycaprolactam nanofibrous-based wound-dressing materials. Spirulina is a photosynthetic filamentous cyanobacterium processing bioactive ingredient. It can be a good component for wound care application because of its suitable skin regeneration, anti-inflammatory, and antibacterial properties. polycaprolactone (PCL) has good biocompatible and biodegradable properties, for which it can be used in various biomedical applications. Mainly, they have developed pure PCL, spirulina/PCL, PCL/alginate, and spirulina/alginate/PCL nanofiber-based wound-dressing materials. As their main aim was to improve the absorption capacity and adherence properties of the wound-dressing material, alginate has been included in the mixture of PCL and spirulina extract. They have discussed the function of different nanofiber-based wound-dressing materials by using it in a rat model. Finally, it was concluded that the spirulina/alginate/PCL combination has better results than the other three materials in terms of water absorption capacity, water

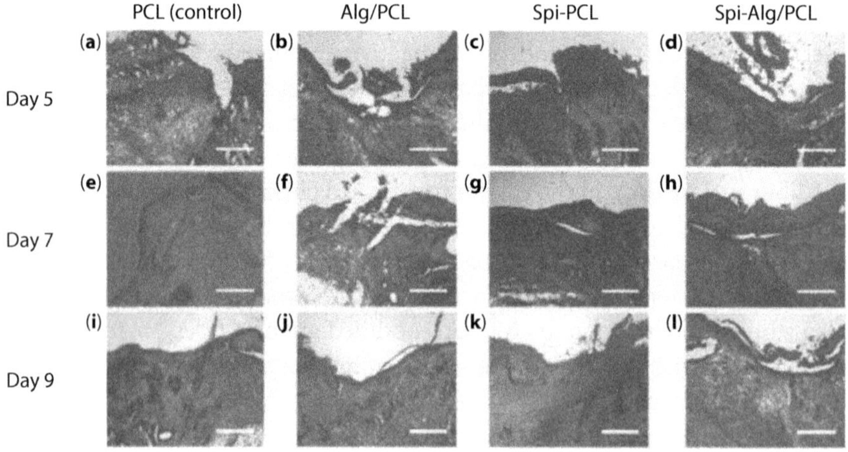

Figure 9.4 Cross-sectional images of regeneration of rat deep wound model stained by H&E. PCL (A, E, I), Alg/PCL (B, F, J), Spi-PCL (C, G, K), and Spi-Alg/PCL (D, H, L) nanofiber applied for 5, 7, and 9 days. 3.7% spirulina extract content of Spi-PCL (C, G, K) and SpiAlg/PCL (D, H, L) was used. Scale bars, 200 μm [49].

vapor rate, antibacterial properties, and wound-healing rate with a concentration of 4% only. When this developed wound-dressing material has been used in wound care, due to the higher water absorption capacity after 30 minutes of soaking in water, all the spirulina extract has been released from the wound dressing. Figure 9.4 confirmed the suitability of the spirulina/alginate/PCL nanofiber-based wound-dressing material compared to pure PCL, spirulina/PVA, and PCL/alginate-based materials.

9.4.1.6 Alginate Hydrogel-Based Wound Dressing

Hydrogels are three-dimensionally cross-linked structures made of hydrophilic polymers with high water contents [50]. They can be prepared from natural as well as synthetic polymers. Hydrogel is mainly categorized into two types, i.e., "physical hydrogel" and "chemical hydrogel." When a polyelectrolyte polymer interacts with multivalent ions-based opposite charges, then physical hydrogel will be formed. For instance, calcium alginate is one of the physical hydrogels. Chemical hydrogel will form when polymer network structure cross-linked covalently [51–53]. The physicochemical properties of the resultant hydrogels depend upon the cross-linking density, molecular weight, and chemical compositions of the polymer [54]. Various hydrophilic polymers from which a hydrogel can be synthesized are shown in Table 9.2. The polysaccharides-based polyanionic biopolymer alginate has been used tremendously in biomedical applications, mainly in wound care dressing, because of its different favorable properties, which are already mentioned in Section 9.3. Alginate can be used in different forms like direct polymer, fiber, and hydrogel. Among the different forms of alginate, hydrogel-based alginate has wide application in wound care dressing. Alginate hydrogel has been formed by cross-linking of alginate polymer with cross-linking agents like acrylic acid, acrylamide, methacrylamide, etc. It is known that alginate polymer consists of two types of acid block, i.e., alpha-L-guluronate (G block) and beta-D-mannuronate (M block). The G block is mainly cross-linked with divalent cations (e.g., $Ca+$) to form the alginate hydrogel material. Therefore, the length and molecular weight of the G block have great influence on the resultant hydrogel properties [55]. Alginate hydrogel-based wound-dressing material has higher biocompatibility, hydrophilicity, fast blood clotting rate, and high porosity. Due to the above-mentioned advantageous properties of alginate hydrogel-based wound dressing, it can be used highly in wound-healing application. Alginate hydrogel can also be used in cell transplantation in tissue engineering because of its mimicking characteristics with the extracellular membranes of living tissues [56]. Furthermore, in order to improve

Table 9.2 Different hydrophilic polymers for preparation of hydrogel [58].

Natural polymers and their derivatives	Synthetic polymers	Combinations of natural and synthetic polymers
Anionic polymers: HA, alginic acid, pectin, carrageenan, chondroitin sulfate, dextran sulfate. Cationic polymers: chitosan, polylysine Amphipathic polymers: collagen (and gelatin), carboxymethyl chitin, fibrin Neutral polymers: dextran, agarose, pullulan	Polyesters based: PEG-PLA-PEG, PEG-PLGA-PEG, PEG-PCL-PEG, PLA-PEG-PLA, PHB, P(PF-co-EG)±acrylate end groups, P(PEG/PBO terephthalate) Other polymers based: PEG-bis-(PLA-acrylate), PEG±CDs, PEG-g-P(AAm-co-Vamine), PAAm, P(NIPAAm-co-AAc), P(NIPAAm-co-EMA), PVAc/PVA, PNVP, P(MMA-co-HEMA), P(AN-co-allyl sulfonate), P(biscarboxy-phenoxy-phosphazene), P(GEMA-sulfate)	P(PEG-co-peptides), alginate-g-(PEO-PPO-PEO), P(PLGA-co-serine), collagen-acrylate, alginate-acrylate, P(HPMA-g-peptide), P(HEMA/Matrigel®), HA-g-NIPAAm

the overall properties of the alginate-based hydrogel, different ingredients can be mixed with it. For instance, Straccia *et al.*, [57] developed and characterized alginate hydrogel- and aloe vera gel-based wound-dressing materials. Finally, they observed that thermal stability, swelling capacity, and mechanical and chemical properties have been improved by using aloe vera gel with alginate hydrogel, and insolubility of the blended wound dressing has been increased.

In a study, Hoffman *et al.*, [59] developed an alginate hydrogel/nZnO particle-based nanocomposite wound dressing bandage material. The performance of the developed composite bandage in wound-healing application has been described including swelling ratio, blood clotting rate, antibacterial activity, and biodegradation rate. They observed that the porosity of the developed bandage has been 60% to 70% for which it can absorb more exudates from the wound area so that it can protect the wound from the risk of infection. The nano ZnO particles can improve the antibacterial properties of the bandage up to a considerable limit. But after a certain limit of the concentration of the nano ZnO particles, the composite structure becomes rigid. So, the concentration of the nanoparticles inside the alginate hydrogel is one of the most important factors in terms of maintaining the flexibility of the bandage. The ZnO nanoparticles also have influenced the biodegradability of the alginate hydrogel materials. Controlled biodegradation behavior of the bandage was observed, which could be an important factor to help in the faster healing of a wound. This alginate hydrogel and nano ZnO particles-based composite bandage can be used widely in highly bleeding wound, because of its very good hemostatic characteristics. Also, alginate hydrogel-based wound dressing could be a good candidate for chronic wounds. Higher pH values of chronic wounds have an effect in the proliferation process of the fibroblasts and hamper the synthesis of extracellular matrix. In addition, the lower level of oxygen at chronic wounds can trammel the growth and migration of cells. Chronic wound is a good source to develop bacterial colonization and biofilm [60–62]. All these above-mentioned problems with chronic wounds can be solved by using an acidic alginate hydrogel-based wound-dressing material. Acidic alginate hydrogel-based wound dressing can control the pH of the chronic wound. Generally, acidic alginate hydrogel-based wound dressing can be prepared by incorporating an acidic group into the alginate structure in the presence of hydrogel precursor-like poly(ethylene glycol) diacrylate (PEGDA). Koehler *et al.*, [63] studied pure PEGDA-, PEGDA/alginate-, and PEGDA/alginate/acrylic acid (AA)-based wound-dressing material. The different properties like mechanical properties, swelling ratio, and base-neutralizing capacity of these developed wound-dressing materials have been studied. It is known that the alginate as a single component has good biocompatibility and higher hydrophilicity, which help it to be used in various biomedical applications. The incorporation of acrylic acid into alginate can solve the problems in chronic wounds. The tensile stress is 38.0 ± 1.1 kPa for the pure PEGDA-based material, whereas 80.1 ± 13.8 kPa tensile stress was observed for PEGDA/alginate hydrogel-based wound material. But AA has some effect on the tensile stress due to the presence of a carboxylic group in it. The compressive strength has been decreased with increasing concentration of the AA into the wound material. They also observed that the swelling capacity of the developed wound-dressing materials has been varied with the concentration of the AA. The swelling capacity has been increased with the concentration of the AA, which can be observed from Figure 9.5. The reason behind increasing the swelling capacity is to increase the pore diameter by increasing the AA.

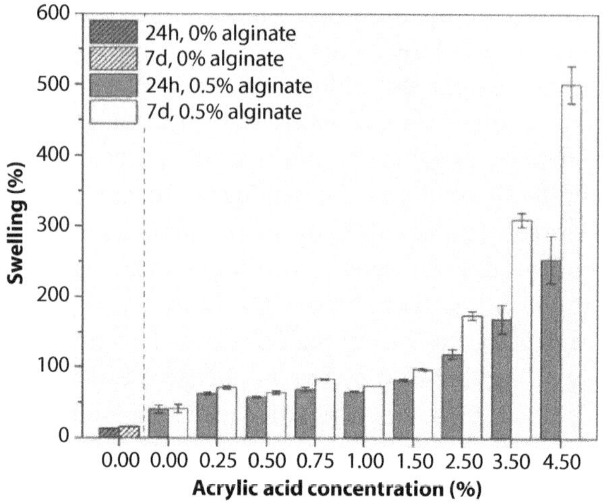

Figure 9.5 Swelling capacity of PEGDA/AA/alginate hydrogels with different monomer concentrations after 24 h and 7 d of incubation [63].

A moist environment is necessary to heal quickly the wound area. In this context, alginate hydrogel is one of the good candidates because of its higher hydrophilicity. This hydrophilicity can also be increased by incorporating different ingredients like chitosan and fucoidan into the alginate hydrogel structure. To maintain a better moist condition at the wound area, Murakami *et al.*, [64] developed an alginate hydrogel/chitosan/chitin/fucoidan-based composite sheet. This ACF-HS wound dressing has better beneficial properties than the only alginate treated and control wound dressing in terms of simplicity in application, removal, and better adherence properties. They basically used this developed wound dressing to treat a mitomycin c-treated impaired wound in rats. They observed that the ACF-HS-based wound is better than others (the only alginate treated and control wound) in terms of higher wound closure rate, which is depicted in Figure 9.6.

Balakrishnan *et al.*, [65] studied the *in situ* hydrogel-forming wound-dressing material based on alginate, gelatin, and borax in the absence of any external cross-linking agents. Hydrogel has been formed by cross-linking of the alginate with the gelatin in the presence of borax. In this composite hydrogel-based wound-dressing material, alginate acted as exudate absorptive part wherein gelatin acted as a good hemostatic controller and borax as an antiseptic agent. Therefore, a synergistic effect has been explored from this composite wound-dressing material. They tested the function of this developed wound dressing by using it in a rat model, and also histology

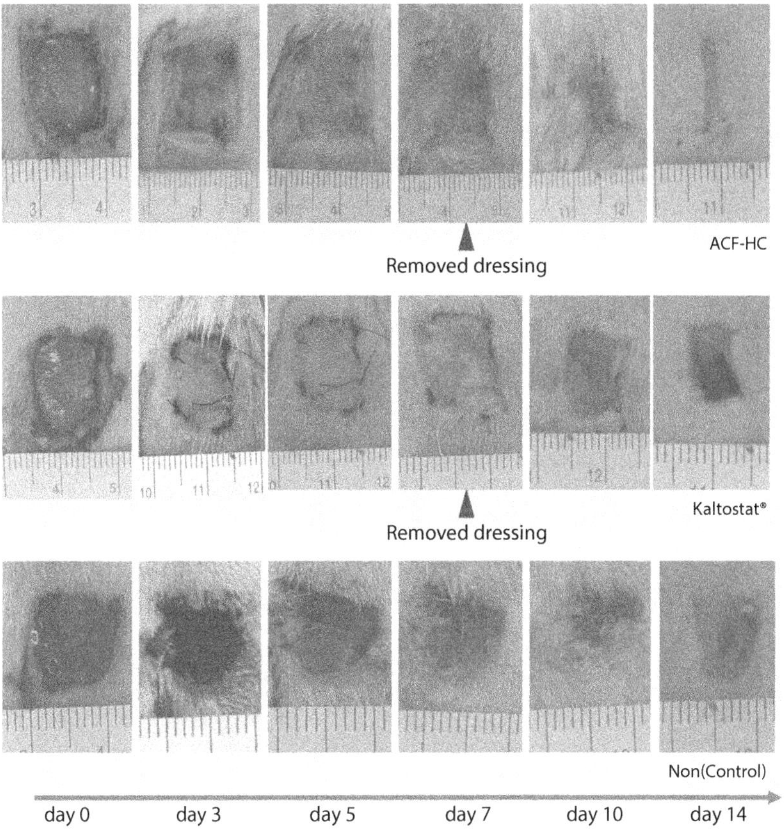

Figure 9.6 Photographic findings of wounds covered with ACF-HS or Kaltostat, and controls. Each wound on the indicated day is representative of eight wounds (four rats) covered with ACF-HS or Kaltostat, or not covered (control) [64].

has been studied. From this histological study shown in Figure 9.7, it can be cleared that the composite hydrogel-based wound dressing has better result than the control sample in terms of wound healing and epithelium cell generation. Elbadawy *et al.*, [66] fabricated a sodium alginate (SA)/ polyvinyl alcohol (PVA) hydrogel-based wound-dressing material. They also studied the different properties of the developed wound-dressing material such as gel-forming capacity, elasticity, swelling capacity, hydrolytic degradation rate, and antibacterial activity. The gel-forming capacity, elasticity, and elongation to break the wound-dressing material have been decreased, but other properties like swelling capacity, antibacterial activity, and hydrolytic degradability have been increased with the increase in the SA concentration in the PVA.

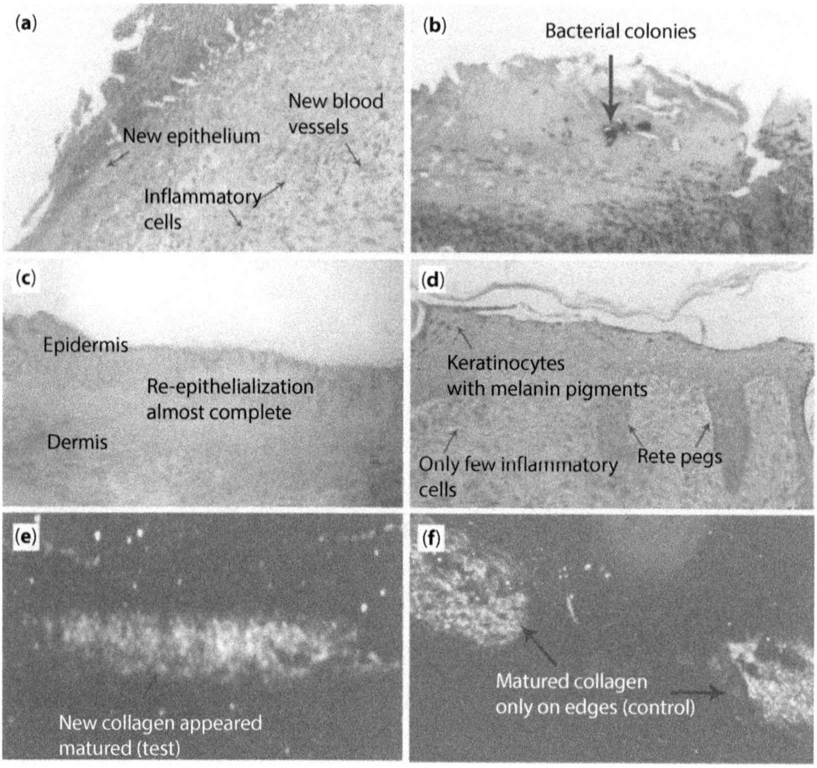

Figure 9.7 Histology of wound sections stained with hematoxylin and eosin. Epithelialization at test wound edges at 5 days (a, 150); bacterial colonies present in control wounds (b, 300); neat test wound section at 10 days (c, 60); test wound with Rete pegs at 10 days (d, 150); test wound (e, 15) and control wound (f, 15) under polarized light at 15 days [65].

9.5 Opportunities and Future Thrust

In recent years, researchers have shown great interest in naturally available biopolymers like alginate, chitosan, collagen, etc. in exploring them in different biomedical applications. Particularly, alginate could be a great solution over other commercially available synthetic or natural wound dressings since alginate has various attractive properties such as biocompatibility, slight gelation surroundings, and easy modifications to develop alginate by-products with diverse properties. Alginate wound dressing can provide a safe clinical use and high rate of healing in wound care application. It can also be used easily as an implanted part in different biomedical applications. Alginate hydrogel-based wound-dressing materials are also available in the market. Hydrogel-based materials have higher moisture uptake capacity, which can provide a suitable moist environment at around the wound area. However, alginate hydrogel has a limitation compared to another hydrogel in terms of its mechanical and physical properties. These present problems of alginate hydrogel can be overcome by using various cross-linking agents and using molecules with different molecular weights. A new type of alginate biopolymers with modified properties, unlike the differently available polymers, can be synthesized by changing the synthesis procedure and technology, which can be the future thrust in this area.

References

1. Percival, J.N., Classification of wounds and their management. *Surgery*, 20, 114–117, 2002.
2. Naradzay, F.X. and Alson, R., Burns, thermal. *Web MD*, 2005.
3. Harding, K.G., Morris, H.L., Patel, G.K., Science, medicine and the future: Healing chronic wounds. *Br. Med. J.*, 324, 160–163, 2002.
4. Boateng, J.S., Matthews, K.H., Stevens, H.N.E., Eccleston, G.M., Wound healing dressings and drug delivery systems: A review. *J. Pharm. Sci.*, 97, 2892–2923, 2008.
5. Draget, K.I., Moe, S.T., Skjåk-Bræk, G., Smidsrød, O., Alginates, in: *Hand book of Hydrocolloids*, A.M. Stephen and G.O. Phillips (Eds.), pp. 290–328, CRC Press and Sutherland, Boca Raton, 2016.
6. Williams, L., Alginates, in: *Biomaterials: Novel Materials from Biological Sources*, D. Byron (Ed.), New York, Stockton, 1991.
7. Franklin, M.J. and Ohman, D.E., Mutant analysis and cellular localization of the AlgI, AlgJ, and AlgF proteins required for O acetylation of alginate in *Pseudomonas aeruginosa*. *J. Bacteriol.*, 184, 3000–3007, 2002.
8. Franklin, M.J., Douthit, S.A., McClure, M.A., Evidence that the algI/algJ gene cassette, required for O acetylation of *Pseudomonas aeruginosa* alginate, evolved by lateral gene transfer. *J. Bacteriol.*, 186, 4759–4773, 2004.
9. Franklin, M.J. and Ohman, D.E., Identification of algF in the alginate biosynthetic gene cluster of *Pseudomonas aeruginosa* which is required for alginate acetylation. *J. Bacteriol.*, 175, 5057–5065, 1993.
10. Chapman, V.J. and Chapman, D.J., Algin and alginates, in: *Seaweeds and their Uses*, V.J. Chapman (Ed.), pp. 194–225, 1980.
11. Pawar, S.N. and Edgar, K.J., Alginate derivatization: A review of chemistry, properties and applications. *Biomaterials*, 33, 3279–3305, 2012.
12. Smidsrod, O. and Skjak-Braek, G., Microcapsules of alginate-chitosan—I: A quantitative study of the interaction between alginate and chitosan. *Biomaterials*, 19, 1815–1825, 1998.
13. Draget, K.I., Oestgaard, K., Smidsroed, O., Homogeneous alginate gels: A technical approach. *Carbohydr. Polym.*, 14, 159–178, 1990.

14. Martinsen, A., Skjaak-Braek, G., Smidsroed, O., Alginate as immobilization material: I. Correlation between chemical and physical properties of alginate gel beads. *Biotechnol. Bioeng.*, 33, 79–89, 1989.
15. Donati, I., Holtan, S., Mørch, Y.A., Borgogna, M., Dentini, M., Skjåk-Bræk, G., New hypothesis on the role of alternating sequences in calcium alginate gels. *Biomacromolecules*, 6, 1031–1040, 2005.
16. Lee, K.Y. and Mooneya, D.J., Alginate: Properties and biomedical applications. *Progr. Polym. Sci.*, 37, 106–126, 2012.
17. Chatfieid, S., A comparison of the efficacy of the alginate preparation, Gaviscon Advance, with placebo treatment of gastro-oesophageal reflux disease. *Curr. Med. Res. Opin.*, 15, 152–159, 1999.
18. Boateng, J.S., Matthews, K.H., Stevens, H.N.E., Eccleston, G.M., Wound healing dressings and drug delivery systems: A review. *J. Pharm. Sci.*, 97, 2892–923, 2008.
19. Queen, D., Orsted, H., Sanada, H., Sussman, G., A dressing history. *Int. Wound J.*, 1, 59–77, 2004.
20. Kennedy, J.F., Knill, C.J., Thorley, M., Natural polymers for healing wounds, in: *Recent Advances in Environmentally Compatible Polymers*, J.F. Kennedy, G.O. Phillips, P.A. Williams, H. Hatakeyama (Eds.), pp. 97–104, Cambridge, 2001.
21. Paul, W. and Sharma, P.C., Chitosan and alginate wound dressings: A short review. *Trends Biomater. Artif. Organs*, 18, 18–23, 2004.
22. Sayaq, J., Meaume, S., Bohbot, S., Healing properties of calcium alginate dressings. *J. Wound Care*, 5, 357, 1996.
23. Balakrishnan, B., Mohanty, M., Fernandez, A.C., Mohanan, P.V., Jayakrishnan, A., Evaluation of the effect of incorporation of dibutyryl cyclic adenosine monophosphate in an *in situ* forming hydrogel wound dressing based on oxidized alginate and gelatin. *Biomaterials*, 27, 1355–61, 2006.
24. Li, S., Li, L., Guo, C., Qin, H., Yu, X., A promising wound dressing material with excellent cytocompatibility and proangiogenesis action for wound healing: Strontium loaded Silk fibroin/Sodium alginate (SF/SA) blend films. *Int. J. Biol. Macromol.*, 104, 969–978, 2017.
25. Panawes, S., Ekabutr, P., Niamlang, P., Pavasant, P., Chuysinuan, P., Supapho, P., Antimicrobial mangosteen extract infused alginate-coated gauze wound dressing. *J. Drug Deliv. Sci. Technol.*, 41, 182–190, 2017.
26. Agren, M.S., Zinc in wound repair. *Arch. Dermatol.*, 135, 1273–1277, 1999.
27. Wiegand, C., Heinze, T., Hipler, U.C., Comparative *in vitro* study on cytotoxicity, antimicrobial activity, and binding capacity for pathophysiological factors in chronic wounds of alginate and silver-containing alginate. *Wound Repair Regener.*, 17, 511–532, 2009.
28. Opasanon, S., Muangman, P., Namviriyachote, N., Clinical effectiveness of alginate silver dressing in outpatient management of partial-thickness burns. *Int. Wound J.*, 7, 467–471, 2010.

29. Meng, X., Tian, F., Yang, J., He, C.N., Xing, N., Li, F., Chitosan and alginate polyelectrolyte complex membranes and their properties for wound dressing application. *J. Mater. Sci. Mater. Med.*, 21, 1751–1759, 2010.
30. Young, S.C., Sung, R.H., Young, M.L., Kang, W.S., Moon, H.P., Young, S.N., Study on gelatin-containing artificial skin: I. Preparation and characteristics of novel gelatin-alginate sponge. *Biomaterials*, 20, 409–417, 1999.
31. Montaser, A.S., Abdel-Mohsena, A.M., Ramadana, M.A., Sleemd, A.A., Sahffiee, N.M., Jancar, J., Hebeish, A., Preparation and characterization of alginate/silver/nicotinamide nanocomposites for treating diabetic wounds. *Int. J. Biol. Macromol.*, 92, 739–747, 2016.
32. Gils, C.C.V., Roeder, B., Chesler, S.M., Mason, S., Improved healing with a collagen-alginate dressing in the chemical matricectomy. *J. Am. Podiatr. Med. Assoc.*, 88, 452–456, 1998.
33. Donaghue, V.M., Chrzan, J.S., Rosenblum, B.I., Giurini, J.M., Habershaw, G.M., Veves, A., Evaluation of a collagen-alginate wound dressing in the management of diabetic foot ulcers. *Adv. Wound Care*, 11, 114–123, 1998.
34. Wang, L., Khor, E., Wee, A., Lim, L.Y.J., Chitosan-alginate PEC membrane as a wound dressing: Assessment of incisional wound healing. *Biomed. Mater. Res. Part B*, 63, 610–619, 2002.
35. Chen, X., Wells, G., Woods, D.M., Production of yarns and fabrics from alginate fibres for medical applications. *Proceedings of the 2nd International Conference*, 24th & 25th August, Bolton Institute, UK, pp. 20–29, 2001.
36. Miraftab, M., Qiao, Q., Kennedy, J.F., Anand, S.C., Collyer, G., Advanced materials for wound dressings: Biofunctional mixed carbohydrate polymers. *Proceedings of the 2nd International Conference*, 24th & 25th August, Bolton Institute, UK, pp. 164–172, 2001.
37. Miraftab, M., Qiao, Q., Kennedy, J.F., Groocock, M.R., Anand, S.C., Advanced wound care materials: Developing an alginate fibre containing branan ferulate. *J. Wound Care*, 11, 353–356, 2002.
38. Qin, Y., Gel swelling properties of alginate fibers. *J. Appl. Polym. Sci.*, 91, 1641–1645, 2004.
39. Kennedy, J.F., Knill, C.J., Thorley, M., Natural polymers for healing wounds. *Recent Advances in Environmentally Compatible Polymers Cellucon '99 Proceedings*, pp. 97–104, 2001.
40. Lloydb, L.L., Kennedya, J.F., Methacanona, P., Patersonb, M., Knilla, C.J., Carbohydrate polymers as wound management aids. *Carbohydr. Polym.*, 37, 315–322, 1998.
41. Park, S.A., Park, K.E., Kim, W., Preparation of sodium alginate/poly (ethylene oxide) blend nanofibers with lecithin. *Macromol. Res.*, 18, 891–896, 2010.
42. Qin, Y., The gel swelling properties of alginate fibers and their applications in wound management. *Polym. Adv. Technol.*, 19, 6–14, 2008.
43. Fan, L., Du, Y., Huang, R., Wang, Q., Wang, X., Zh, L., Preparation and characterization of alginate/gelatin blend fibers. *J. Appl. Polym. Sci.*, 96, 1225–1229, 2005.

44. Fan, L., Yu, L., Xu, Y., Yi, C., Cai, J., Li, M., Huang, J., The novel alginate/N-succinyl-chitosan antibacterial blend fibers. *J. Appl. Polymer. Sci.*, 116, 2151–2156, 2010.
45. Hussain, F., Khurshid, M.F., Masood, R., Ibrahim, W., Developing antimicrobial calcium alginate fibres from neem and papaya leaves extract. *J. Wound Care*, 26, 778–783, 2017.
46. Khajavi, R., Abbasipour, M., Barzi, M.G., Rashidi, A.R., Mohammad, K.R., Mirzababa, H.H., Eucalyptus essential oil-doped alginate fibers as a potent antibacterial wound dressing. *Adv. Powder Technol.*, 33, 1–7, 2014.
47. Knill, C.J., Kennedya, J.F., Mistrya, J., Miraftabb, M., Smartb, G., Groocockc, M.R., Williams, H.J., Alginate fibres modified with unhydrolysed and hydrolysed chitosans for wound dressings. *Carbohydr. Polym.*, 55, 65–76, 2004.
48. Sweeney, I.R., Miraftaba, M., Collyer, G., Absorbent alginate fibres modified with hydrolysed chitosan for wound care dressings—II. Pilot scale development. *Carbohydr. Polym.*, 102, 920–927, 2014.
49. Choi, J.I., Kim, M.S., Chung, G.Y., Shin, H.S., Spirulina extract-impregnated alginate-PCL nanofiber wound dressing for skin regeneration. *Biotechnol. Bioprocess Eng.*, 22, 679–685, 2017.
50. Sakiyama-Elbert, S.E. and Hubbell, J.A., Functional biomaterials: Design of novel biomaterials. *Ann. Rev. Mater. Res.*, 31, 183–201, 2001.
51. Wichterle, O. and Lim, D., Hydrophilic gels in biologic use. *Nature*, 185, 117–118, 1960.
52. Campoccia, D., Doherty, P., Radice, M., Brun, P., Abatangelo, G., Williams, D.F., Semisynthetic resorbable materials from hyaluronan esterification. *Biomaterials*, 19, 2101–2127, 1998.
53. Prestwich, G.D., Marecak, D.M., Marecak, J.F., Vercruysse, K.P., Ziebell, M.R., Controlled chemical modification of hyaluronic acid. *J. Controlled Release*, 53, 93–103, 1998.
54. Varghese, S. and Elisseeff, J.H., Hydrogels for musculoskeletal tissue engineering. *Adv. Polym. Sci.*, 203, 95–144, 2006.
55. Lee, K.Y. and Yuk, S.H., Polymeric protein delivery systems. *Progr. Polym. Sci.*, 32, 669–97, 2007.
56. George, M. and Abraham, T.E., Polyionic hydrocolloids for the intestinal delivery of protein drugs. *J. Control Release*, 114, 1–14, 2006.
57. Straccia, M.C., Ayala, G.G., Romano, I., Oliva, A., Laurienzo, P., Langer, R., Vacanti, J.P., Alginate hydrogels coated with chitosan for wound dressing tissue engineering. *Science*, 260, 920–926, 1993.
58. Pereira, R., Carvalho, A., Vaz, D.C., Gil, M.H., Mendes, A., Bártolo, P., Development of novel alginate based hydrogel films for wound healing applications. *Int. J. Biol. Macromol.*, 52, 221–230, 2013.
59. Hoffman, A.S., Hydrogels for biomedical applications. *Adv. Drug Deliv. Rev.*, 54, 3–12, 2002.

60. Mohandas, A., PT, S.K., Biswas, R., Kumar, V., Rangasamy, L., Kumar, J., Exploration of alginate hydrogel/nano zinc oxide composite bandages for infected wounds. *Int. J. Nanomedicine*, 10, 53–66, 2015.
61. Percival, S.L., McCarty, S., Hunt, J.A., Woods, E.J., The effects of pH on wound healing, biofilms, and antimicrobial efficacy. *Wound Repair Regen.*, 22, 174–186, 2014.
62. Schreml, S., Meier, R.J., Kirschbaum, M., Kong, S.C., Gehmert, S., Felthaus, O., Küchler, S., Sharpe, J.R., Wöltje, K., Weib, K.T., Albert, M., Seidl, U., Schröder, J., Morsczeck, C., Prantl, L., Duschl, C., Pedersen, S.F., Gosau, M., Berneburg, M., Wolfbeis, O.S., Landthaler, M., Babilas, P., Luminescent dual sensors reveal extracellular pH-gradients and hypoxia on chronic wounds that disrupt epidermal repair. *Theranostics*, 4, 721–735, 2014.
63. Koehler, J., Wallmeyer, L., Hedtrich, S., Goepferich, A.M., Brandl, F.P., pH-Modulating poly (ethylene glycol)/alginate hydrogel dressings for the treatment of chronic wounds. *Macromol. Biosci.*, 17, 1–11, 2017.
64. Murakami, K., Aoki, H., Nakamura, S., Nakamura, S., Takikawa, M., Hanzawa, M., Kishimoto, S., Hattori, H., Tanaka, Y., Kiyosawa, T., Sato, Y., Ishihara, M., Hydrogel blends of chitin/chitosan, fucoidan and alginate as healing-impaired wound dressings. *Biomaterials*, 31, 83–90, 2010.
65. Balakrishnana, B., Mohantyb, M., Umashankarc, P.R., Jayakrishnana, A., Evaluation of an *in situ* forming hydrogel wound dressing based on oxidized alginate and gelatin. *Biomaterials*, 26, 6335–6342, 2005.
66. Elbadawy, A.K., Kenawy, E.R.S., Tamer, T.M., Meligy, M.A.E., Eldin, M.S.M., Poly (vinyl alcohol)-alginate physically cross-linked hydrogel membranes for wound dressing applications: Characterization and bio-evaluation. *Arab. J. Chem.*, 8, 38–47, 2015.

10

Alginate-Based Biomaterials for Bio-Medical Applications

Reena Antil[1], Ritu Hooda[2], Minakshi Sharm[2] and Pushpa Dahiya[1]*

[1]Department of Botany, M.D. University, Rohtak, India
[2]Department of Zoology, M.D. University, Rohtak, India

Abstract

Alginate could be a natural novel bioactive compound isolated from seaweeds because of their high biocompatibility, nonimmunogenicity, nontoxicity, high versatility, and gelling properties used in medicine, pharmaceutical, and food industries. Alginate is nondegradable to the human alimentary canal because of the absence of alginase enzymes. There are different types of modified alginate according to their functional activity like calcium, sodium, ammonium, and potassium, and other modifications are propylene glycol alginate. These modifications form simply processable and assignable three-dimensional alginate, which could be simply transferred to human cell like hydrogels, foams, and microspheres. Alginate hydrogel features a pleasant potential to deliver with chemical synthesized drugs, proteins and cell delivery for tissue regeneration, cell transplantation, cell primarily based microparticles, proliferation, and producing contact lenses and wound dressing. Alginate conventionally plays a role in medical and pharmaceutical industries, most frequently used as oral dosage for the treatment of heartburn, esophagitis, obesity, and type-2 diabetic treatment, and to conjointly treat hypocholesterolemic activity. Alginate is conjointly used as an antioxidant, anti-inflammatory drug; its role is to reduce or stop the production of reactive oxygen species. The focus of this chapter is to elucidate the biological, pharmacological activity and medical application of alginates and to discuss the present use and future potentialities of alginates as a tool in drug formulation.

Keywords: Alginate, biomaterial, biomedical, pharmaceutical

**Corresponding author*: Pushpa.dahiya@hotmail.com

Shakeel Ahmed (ed.) Alginates, (179–204) © 2019 Scrivener Publishing LLC

10.1 Introduction

Alginate is an anionic polymer structural component of naturally obtained brown seaweed, algae, and some microorganisms. Alginate is a greatly useful and investigated polymer in the biomedical field because of its low toxic effects, biocompatibility, low price, and availability [1]. There are lots of gel formation by alginate. Alginate hydrogel is one of the best gel-forming alginates that is prepared by various cross-linking ways. The cross-linked hydrogel-based alginates show their similar structural properties to extracellular matrices of living tissues. Due to similarities in the structure of hydrogel-based alginates and living tissue matrices, there are huge applications of alginates like in wound healing, delivery of bioactive agents like proteins, and also as an aid in cell transplantation. Alginate-based hydrogel system can be used to exonerate macromolecular particles like protein to tiny microsized chemical drugs. The wound healing system of alginate is helpful to maintain a physiological microenvironment by diminishing microbial infection at the wound site. In the pharmaceutical, alginate gel is more applicable in a very invasive way to deliver drugs orally or through injection. Alginate gels are also auspicious for cell transplantation in tissue engineering. For patients who suffer from failures of organs or tissue, a recent advance technique helps to replace organ or tissue or can regenerate the tissues [2]. The regenerative tissues or cells can be delivered with the help of hydrogels formed by alginates [3].

10.2 Alginate: General Properties

Alginate is a structural forming polymer in seaweeds that provides mechanical strength by forming an intercellular alginate gel matrix analogous to pectin in higher plants. In the extracellular matrix, alginates exist with mixed cations found in ocean water like Mg^{2+}, Ca^{2+}, Sr^{2+}, Ba^{2+}, and Na^{+}. Bacterial alginates produced from *Azotobacter* and *Pseudomonas* are functionally different from plant alginates because of their more determined chemical and physical properties. *Azotobacter vinelandii* is gram-negative soil microorganism. In unsuitable environmental conditions, *A. vinelandii* differentiates itself into desiccated cyst. Alginate is consistently extracted from brown algae (Phaeophyceae), as well as *Laminaria hyperborea*, *Laminaria digitata*, *Laminaria japonica*, *Ascophyllum nodosum*, and *Macrocystis pyrifera* [4]. The extraction was done by treatment with NaOH [5]. The precipitation of alginate is filtered with each of two chloride or sodium, and these alginate salts are transformed from alginate into alginic acid by reacting with dilute HCl; after purification, water-soluble powdered sodium alginates are formed [6]. The pathway that forms the alginate has the following steps: (i) synthesis of precursor substrate, (ii) polymerization and cytoplasmic membrane transfer, (iii) periplasmic transfer and modification, and (iv) export through the outer membrane [7].

10.2.1 Chemical Properties, Structure, and Characterization

Alginates are indeed linear block copolymers. The ratio of guluronate to mannuronate in alginates differs; it depends on the sources from where alginates are extracted [8]. Alginates are composed of alternating M and G or consecutive G residues (GGGGGG), and M residues (MMMMMM) that are linked by block (1,4)-linkage of -d-mannuronate (M) and -l-guluronate (G) residues (Figure 10.1). The content of M and G as well as the length of each block depends on the sources of alginates. There are more than 200 different alginates that are currently being manufactured [9]. *L. hyperborea* are large brown algae having 60% of G-block (α-L-guluronic acid) content [10], whereas alginates from *A. nodosum* and *L. japonica* have a low content of α-L-guluronic acid blocks. G-blocks of alginate are supposed to play a role in intermolecular cross-linking with divalent cations like Ca^{2+} to form hydrogels (Figure 10.2); thus, the formation of hydrogel depends on physical properties of alginates and the ratio of M/G, length of G-block, and molecular weight [10].

The characterization of alternative polysaccharides of alginates is consolidated. Circular dichroism spectroscopy has been used to match the linear spectra of the alginate to model samples of well-characterized homopolymeric blocks [13]. NMR spectrum analysis has confirmed the monomer composition, along with the frequencies of the four doable diad structures FGG, FMG, FMM, and FGM (G = α-L-guluronic acid; M = β-D-mannuronic acid) [14]. Characterization of alginate samples by gel permeation chromatography indicates a polydispersed size distribution [15]. In the case of *A. vinelandii* and *Pseudomonas aeruginosa*, there is variation in average molecular weights from 80 to 290 kDa that was confirmed by light scattering [16]. The molecular weights

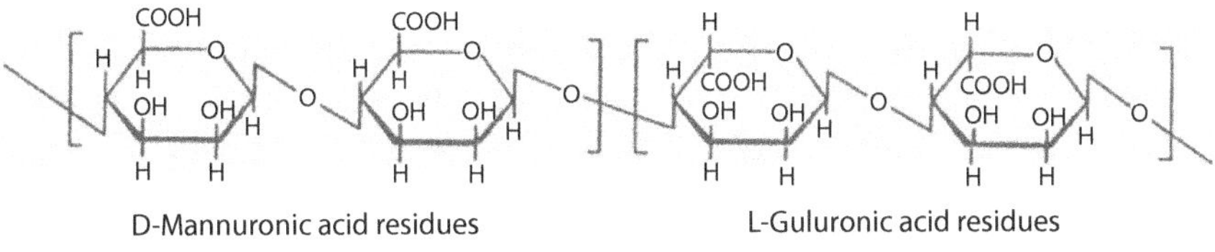

Figure 10.1 The (1-4)-linkage of alternate M and G residues of alginate [11].

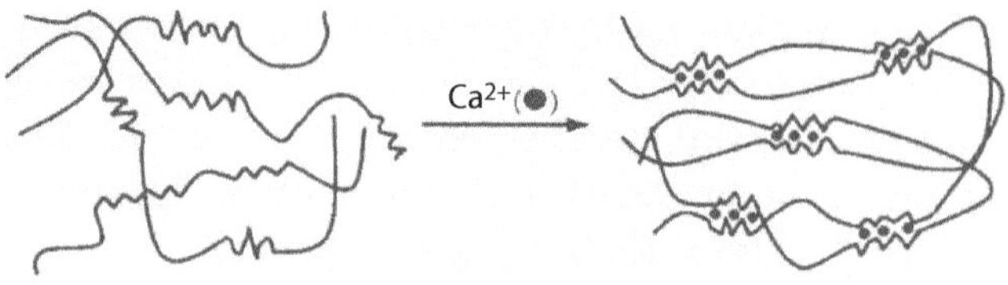

Figure 10.2 The intermolecular cross-linkage of G residues of alginate with Ca2+ ions [12].

of commercially available sodium alginates vary between 32,000 and 40,000 g/mol, whereas intrinsic viscosity (mL/g) and Mv are the viscosity-average molecular weight (g/mol) [18]. The rate of drug delivery, rate of releasing bioactive agents, and encapsulated cell function all depend on adherence of alginate gels. Increased length of G-block and molecular weight of alginates provide mechanical strength to alginate hydrogels. Mechanical strength also depends on sources from where alginates are obtained like *Azotobacter*, which encompasses a high concentration of G-blocks, and the hydrogels that were made by *Azotobacter* alginates have a comparatively high stiffness [17].

The viscosity of alginate solutions depends on the molecular weight of the material and pH. As decrease in pH 3–3.5, there is protonation of carboxylate group from the hydrogen bond in the backbone of the alginate. Increased-molecular-weight alginates improve the quality of alginates-based gel system. Nonetheless, high-molecular-weight alginates are highly viscous, which is usually unwanted in the process of commercially available alginates [19]. That is why the high-viscosity-based alginates are not preferable for the delivery of proteins and cells into the body because of introduction of high shear forces that directly affect the delivered proteins and cells [20]. Highly elastic modulus of alginate gels is generally increased considerably, with the combination of low viscosity and high and low molecular weight alginate polymers [21].

10.3 Extraction and Preparation

For the preparation of alginates, seaweeds, especially algae, are harvested and air-dried before they are further processed. The algal material is treated with dilute mineral acid for the degradation of neutral homopolysaccharides like laminarin and fucoidin that are associated with alginates. Concurrently, there is an exchange of cations for H^+ ions and conversion of insoluble alginates to soluble alginates by addition of sodium carbonate at a pH scale below 10 (Figure 10.3).

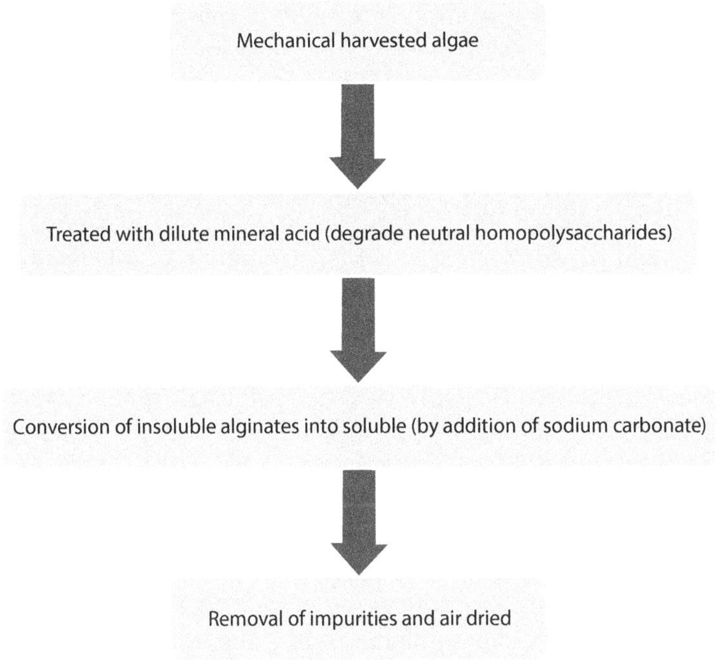

Figure 10.3 Preparation and extraction of alginate from natural resources.

Once alginate is extracted, the refining of alginate is a must because of varieties of impurities present in extracted alginates from the natural sources and additionally for the conversion of either salt or acid alginates [2]. These impurities may be heavy metals, endotoxin, proteins, other carbohydrates, and polyphenols [22]. The low level of impurities in the processed alginates does not cause any mess in the food and beverage industry, but for pharmaceutical applications, there is notable removal of impurities from processed alginates.

10.3.1 Gelation and Cross-Linking of Alginate

Gelation and cross-linking are prepared by exchange of ions in a solution like Na^{2+} alginate ions from α-L-guluronic acids with divalent cross-link solutions such as Ca^{2+}, Sr^{2+}, or Ba^{2+}. There is drawback of monovalent cations and Mg^{2+} ions in the gelation that affects the elasticity of alginate during gelation [2], although Ba^{2+} and Sr^{2+} ions produce active alginate gels [23, 1]. Alternative divalent cations such as Pb^{2+}, Cu^{2+}, Cd^{2+}, Co^{2+}, Ni^{2+}, Zn^{2+}, and Mn^{2+} also formed cross-linked alginate gels, but due to their toxicity, there is limited use of these ions in the formation of gel [24].

The gel of alginate is formed by dimerization of chain that is further cross-linked to other chains [25].

10.3.2 Ionic Cross-Linking

Alginate is a copolymer that has distinct affinities against the distinct divalent ions. The stability, permeability, and strength of alginate gel depend on the cation that will be used in the formation of gel [26]. The exchange monovalent cations from the surrounding environment and the divalent cations affect the stability of hydrogel by weakening the mechanical properties [4]. The concentration, source, degree and type of cross-linking, and molecular weight limit the delivery system of drugs, protein agents, and cells [27]. Besides these factors, the physiological conditions, whether they are *in vitro* or *in vivo*, affect the stability. One of the divalent cation Ca^{2+}-based hydrogel alginates is a very less stable cross-linking hydrogel. If 0.9 weight % sodium chloride is added in the solution, there is an exchange of calcium by non-gelling monovalent sodium ions as well as by chelators such as citrate or phosphate present in the solution [28]. The monovalent sodium ions act as de-cross-linking agents. The cell culture medium *in vitro* condition has sufficiently high calcium concentration that counterbalances such kind of effects, but this hydrogel nevertheless remains stable for weeks [29].

10.3.3 External Gelation

Calcium chloride ($CaCl_2$) is the most common and frequently used soluble salt, which acts as ionic cross-linker agents in external gelation. External gelation is a very transparent process that creates a nontoxic environment for cell entrapment [4]. However, although gelation occurs almost instantaneously, this process frequently results in unbalanced cross-linking density and a polymer concentration gradient within the formed hydrogel [30]. Gel beads are formed by exuding a sodium alginate solution into an aqueous solution of calcium ions. Generally, these microbeads are produced by coaxial airflow that controls the size of the droplets by blowing them from a needle tip into a $CaCl_2$ bath. In tissue regeneration, cell delivery system alginate beads have been widely used [31–33].

The size of microbeads is dependent on the flow of air and the diameter of the solution. The electrostatic bead generation technique is used to attain smaller size beads up to 150 ml [34]. For the production of spherical hydrogels, a novel method was used that utilizes super hydrophobic substrates. Without any precipitation, the spherical hydrogels entangle the rat MSCs that were isolated from bone marrow and fibronectin [35]. Microdrops of low viscosity alginate were applied on the super hydrophobic surface, and then $CaCl_2$ was added on top of the hydrophobic surface and allowed for cross-linking. Over the conventional technique, the super hydrophobic surface method is advantageous in tissue engineering in reducing mechanical forces and aggregation of particles, and has also shown excellent permissibility for nutrients and oxygen, although super hydrophobic surface-based alginate gels administered optimum survivability of cell during entrapment for several days [35].

10.3.4 Internal Gelation

Internal gelation process is a process not much explored like external gelation, but in the future, there will be chances of increasing the use of this process because of direct site injection. Internal gelation is an *in situ* hydrogel formation process. By internal gelation, the alginate hydrogels are formed at the site of interest by using polymeric solution with a combination of cells [36–40].

The most acceptable approach in the internal gelation is gradual gelation with low solubility using divalent cation salts that slow down gelation rate with time. Most commonly used divalent cations are calcium carbonate ($CaCO_3$) and calcium sulfate ($CaSO_4$). Both of these divalent cations have low solubility in pure water at neutral pH, but may be soluble under acidic conditions. This low solubility allows uniform distribution in the alginate solution [30, 41]. Later on, at acidic pH, free calcium ions can be released with glucone-d-lactone (GDL). To attain a neutral pH, the molar ratio of $CaCO_3$, the $CaCO_3$/GDL molar ratio, can be set [30, 36]. A photoacid generator (PAG) is an alternative to GDL that also helps to release out free Ca^{2+} and H^+ [42]. The use of Ca^{2+} chelators also advocates photoactivated internal gelation, which was mixed with alginate solutions and exposed to light that decreases the affinity to Ca^{2+} in an irreversible manner [43]. Internal gelation method is most applicable in improvements in mechanical properties and homogeneity.

10.3.5 Covalent Cross-Linking

Alginate hydrogels are produced through a various methods; out of these methods, covalent cross-linking provides a more mechanically stable and stronger gel than other linking methods. Once a material is covalently linked, it no longer meets the injectability criteria [44, 45]. PEG-diamine molecules with different molecular weights are used as a covalent cross-linker. The elasticity of gel depends on the molecular weight that can be changed according to the molecular weight of the PEG molecules. PEG hydrogels can be formed by either adipic acid dihydrazide (AAD) as a bifunctional cross-linking molecule or poly(acrylamide-co-hydrazide) as a multifunctional cross-linking molecule. In covalent cross-linking, multifunctional cross-linking provides a wide range, control over degradation rates, and mechanical stiffness for stronger hydrogel [46]. PAG hydrogels were formed with either poly(acrylamide-co-hydrazide) as a multifunctional cross-linking molecule or adipic acid dihydrazide (AAD) as a bifunctional cross-linking molecule. This multicross-linking strategy led to the formation of stronger hydrogels.

10.3.6 Large Bead Preparation

For the preparation of a larger diameter bead, there is a need for a larger size needle to be used in the formation of beads and high viscosity solution. Viscosity also affects the shape of beads. Beads of size greater than 1.0 mm were prepared by either using a needle or a pipette [27, 47–51]. Sodium alginate is a good solution for the formation of larger beads, as with the increase in the viscosity of solution, formation of spherical beads occurs. There is a drawback of sodium alginates due to the presence of soluble proteins in sodium alginate solution that disturb the divalent cross-linking [52]. The formed beads are totally cured by cross-linking by divalent ions before rinsing with distilled water. The finally formed external alginate is coated with poly-L-lysine. Fourier transform infrared spectroscopy has shown that high β-D-mannuronic acid content in alginate beads is strongly coated with poly-L-lysine [53] and stored in 0.9% NaCl solution.

10.3.7 Microbead Preparation

Atomization, emulsification, and coacervation are the three different methods for preparation of small size beads less than 0.2 mm in diameter. The more often used technique is atomization or spraying method [54–59]. Formation of microbead procedure is as follows. Solutions containing the alginate and protein, as described above in the preparation of large beads, are well mixed and loaded into a syringe mounted on a syringe pump. The mixture of alginate and protein solution that was already mentioned in the large beads method is forwarded through an atomization device with a smaller diameter orifice at the tip. The size of beads depends on the used pressure of nitrogen gas, rate of flow through syringe, and distance between orifice and surface. Microbeads are finally coated with poly-L-lysine. Microbeads formation is simpler than larger beads, which comprise only aqueous solutions [60].

10.4 Alginate Hydrogels

Alginates are unbranched polysaccharides consisting of 1,4-linked b-D mannuronic acid (M) and a-L-guluronic acid (G) units. Depending on the alginate source, alginates are covalently linked in sequence and in different manner with their polymer chain [46]. The function of alginate depends on the sequence, range of molecular weight (typically 101–103 kDa), and monomer composition such as MG blocks (MGMGMGM), which form the most flexible chains, and G blocks (GGGGGGG), which form stiff chains [1, 47]. At present, alginates are used in certain clinical applications such as in heartburn treatment, acid reflux, appetite suppressant, weight control, and type I diabetes treatment, and are also used in cardiac remodeling [48–50].

Alginate is considered to be nonimmunogenic and has shown great potential as a cell delivery vehicle [46, 50–53]. The formation of hydrogels (with water) *in situ* with gelling process can be accomplished in physiological condition by using nontoxic solvents. These hydrogels maintain softness in gel that their physical property is very similar to native tissues [54]. The major drawback of alginate hydrogels is that they are noninteractive to delivered cell and show slow biodegradation [55]. Despite this fact, the chain of alginate cannot cleave by mammalian enzymes, but alginate can be degradable in *in vivo* condition by alginate partial oxidation applying sodium periodate, altering the chain confirmation by cleavage in carbon–carbon bond in urinate residue [55]. The degradation rate of oxidized alginate can be regulated by adjustment of molecular weight without disruption in flexibility and gel formation ability [35]. The oxidized binary hydrogels improved the formation of bone tissue compared to nonmodified alginate, since a faster degradation occurs that facilitates the formation of new bone tissues [35]. There is a vast application of oxidized hydrogels such as helping in the formation of bone tissue by addition of regenerative cells; additionally, alginate hydrogels help in easy incorporation of biochemical substances to help engineer specific cell responses. Sometimes, cell has no specific receptor for attachment to the alginate beads; in such condition, several methods to promote cell attachment to alginate matrices have been developed. One of these methods is coupling of ECM proteins that are not fully integrated proteins because the entire protein leads to nonspecific interaction that may enhance immune response to those specific cells. Examples of ECM proteins are laminin, collagen, and fibronectin that couple with alginate [56–59]).

The use of small immobilized peptides is popular in the attachment of cell to alginate, particularly with RGD (arginine–glycine–aspartic acid) sequence of first peptides, and promotes cell adhesion on a biomaterial. This tripeptide motif sequence has been identified in ECM proteins such as fibronectin, collagen, laminin, osteopontin, and vitronectin [57]. Several other combined modifications have been investigated to improve alginate properties such as ECM-like 3-D cellular microenvironment for alginate hydrogel. Other than the RGD sequence for grafting, there are several sequences used for the attachment such as protease-labile cross-linking peptide (proline–valine–glycine–leucine–isoleucine–glycine, PVGLIG) that is cleavable by metalloproteinases (MMPs) produced by cells [16, 36], but some sensitive hydrogels are remodeled by cell-driven proteolytic mechanisms, which increase evasion/invasion. Despite these peptide linkages, alginate can functionalize with cell-signaling moieties such as galactose to improve hepatocyte cell recognition [61–63]. Several other alginate modifications have been expected for the improvement of cell delivery and its behavior [64].

10.5 Photocross-Linking

For better regulation of mechanical properties, swelling ratios, and degradation rates, photocross-linking method is introduced in the formation of alginate gel. The photocross-linking method is most popularly used in tissue engineering [44, 46, 65, 66]. Photocross-linked alginate delivers the cell rapidly by less interference *in situ* following a brief exposure to ultraviolet (UV) light [69]. This process has been done in mild conditions. For photocross-linking, alginate was modified with 2-aminoethyl methacrylate that is directly photocross-linked using UV light with a photoinitiator [44]. In a recent study, photocross-linkable alginate with cell adhesive properties is shown to be of great interest in tissue engineering applications [44, 67, 68, 70, 71]. Photocross-linking method is a newly formed method and learning their effectiveness requires further study.

10.6 Shape-Memory Alginate Scaffolds

According to the desired shape and size, a novel type of method has been developed that is based on covalent cross-linking from a macroporous structure. *In vitro* condition shape-memory alginate modulates the shape according to delivery and targeting of biomaterial with alginate and further reformed its original shape by rehydration [45, 72]. Shape-memory scaffolds method demonstrates greater shape by maintaining a range of physical and mechanical properties. Shape-memory scaffold alginate hydrogel is covalently cross-linked with AAD and carbodiimide chemistry using 1-ethyl-(dimethyl aminopropyl) carbodiimide, 1-hydroxybenzotriazole [45, 73]. Lee *et al.*, [45] were able to produce macroporous alginate hydrogel scaffolds with defined geometry offensively delivered to the dorsal subcutaneous space of a mouse with minimal scaffold and rehydrated by PBS. Shape-memory scaffolds method has also been used for skeletal muscle cell survival, proliferation, and migration. The results confirm the potential of these shape-memory alginate scaffolds as cell delivery systems for tissue regeneration, although this strategy needs further exploration.

10.7 Biodegradation of Alginate

Due to biochemical properties and lack of enzymes in mammalian system, alginate is nondegradable. Despite the covalently cross-linked hydrogel, ionically cross-linked alginate gels can be easily degradable in mammals by direct exchange of monovalent ions to the divalent ions into the surrounding media [73]. The most acceptable method for the degradation of alginates in physiological conditions is partial oxidation of alginate chains. In an aqueous medium, alginate is oxidized with sodium ions that directly help in hydrolysis. The use of higher AAD concentration in shape-memory scaffolds for the formation of gels decreases the degradation rate of alginate. The degradation rate and mechanical properties are two critical factors in new tissue formation in tissue engineering and delivery of cells and tissue, but these two are decoupled by adjusting the molecular weight distribution of alginate. A different chemical has high content of single-end AAD molecules that degrade easily [74]. These studies clearly indicate that soft gels degrade slowly over time, unlike conventional gels. Partially oxidized alginate can be formed by binary alginate gels with low and high molecular weights by either ionic or covalent cross-linking. Compared with high MW alginate gels, increasing the fraction of low MW alginate maintains the mechanical stiffness up to 0.50, but it leads to faster degradation, irrespective of the cross-linking method [75, 76]. These various approaches may be useful alone or in combination in manipulating the physical properties of various hydrogels in the development of drug delivery and cell transplantation vehicles [77].

10.8 Biomedical Application of Alginates

There is immense requirement of alginate-based biomaterials in the field of regeneration and drug delivery. Especially, stem cells have more important function in the field of regenerative remedy [78, 79], and the mixture and relations involving stem cells and alginate-based materials have been distinctively emphasized. Both cytotoxicity assay (*in vitro*) and implantation (*in vitro*) have shown that microcapsules and scaffolds (alginate-based) revealed minimal or negligible cytotoxicity and are histocompatible [80–82]. For increasing hematoma-like crack repair, these *in vitro* consequences recommended tunable interactions between the biocomposites and the multiple platelet releasate-derived bioagents. Furthermore, for *in situ* curing of implant systems, modestly invasive liberation was established in rat tail vertebrae via vaccination by means of microcomputed tomography. These consequences confirmed that alginate-based scaffolds were capable to mortify, permitted vascularization, and elicited little inflammatory reactions following transplantation. Thus, alginate-based scaffolds can give suitable characteristics like potential cell and medicine carriers for tissue regeneration. Important functions of alginate in pharmaceutics include the following: stabilizing agents and they help in gel making and thickening, because alginate can play an important function in restricted discharge drug products. Although oral dose forms are presently the most common use of alginate in pharmaceutical applications, the use of alginate hydrogels like depots for tissue localized drug delivery is rising. At this juncture, we briefly explain fresh development in biomedical function of alginate and/or its derivatives.

10.8.1 Controlled Chemical and Protein Drug Delivery

For the delivery of many small molecular weight drugs, alginate gels have been checked. The bond (primary or secondary) between the drug and the alginate was used for the kinetics regulation of drug release. Alginate gels usually have pore sizes in the nanometer range (~5 nm) [83], which leads to quick diffusion of minute molecules via the gel; for instance, the release of flurbiprofen from ionically cross-linked alginate gels (partially oxidized) is approximately complete in 1.5 h. On the other hand, assimilation into beads created using alginate gels (partially oxidized) in the occurrence of both adipic acid (dihydrazide) and calcium ions (mixture of ionic and covalent cross-linking) leads to an extended release because of the amplified quantity of cross-links and consequently decreased inflammation [84]. By using partially oxidized alginate gels, the release of antineoplastic agents (controlled and localized) has also been gained. For concurrent or in order delivery, several drugs can be loaded into alginate-based gels because the chemical makeup and mode of incorporation of the drug will significantly change the release kinetics. For example, a noninteractive drug with alginate, methotrexate, was quickly released by diffusion, while covalently linked to the alginate; doxorubicin was released by cross-linker chemical hydrolysis. The drug linked ionically to alginate, mitoxantrone, was barely released following gel dissociation [85]. To alter the discharge of hydrophobic drugs, amphiphilic gel beads have also been formed. For controlled release of theophylline, which is a model drug of deprived water solubility, grafted alginate and poly(ε-caprolactone) (PCL) were cross-linked through calcium ions. The length of the PCL chains regulates the inflammation activities of the gel beads and restricts the discharge of theophylline. The drug discharge for alginate-g-PCL/Ca^{2+} beads was completed in 2 h, but for alginate/Ca^{2+} beads, it was 1 h [86]. The persistent release of theophylline was also done by carbon nanotube (CNT) integrated alginate microspheres; CNT increases the mechanical stability of gels, without affecting the structure and morphology of the microspheres, and no major cytotoxicity was seen, signifying important functions, e.g., as a release transporter to the colon and intestine [87]. The combination of alginate with chitosan makes ionic complexes broadly used in lots of drug delivery applications. Chitosan has a repeating composition of (1, 4) connected glucosamine, by a clear pK of 6.5; it is a chitin derivative and the second most abundant innate polymer on earth [88]. Chitosan has been commonly used due to its complimentary properties and biocompatibility in the field of cosmetics, foodstuff, and pharmaceutical and biomedical applications [89]. A multifunctioning system of alginate and chitosan having triamcinolone was formed by an ionotropic/coacervation gelation process for drug delivery for colon. An advanced inflammation degree and quicker drug release were analyzed from the particulate systems in an influence enteric situation (pH 7.5), in comparison to a simulated gastric situation (pH 1.2) [90]. Chitosan beads with magnetic alginate were prepared with albendazole (ABZ) for inactive targeting by means of physical arrest process (e.g., pH, magnetic field) to the gastrointestinal region. These beads give distinctive pH-dependent inflammation behaviors and a nonstop discharge of ABZ [91]. Chitosan-linked alginate microparticles having all-trans retinoic acid (ATRA) improve localization in dermal beneath the epidermis beneath the epidermis and gained constant discharge of ATRA into the skin [92]. Chitosan-linked alginate beads also entrap metronidazole by a gelation (ionotropic) technique; these beads were useful in suppression of *Helicobacter pylori* once orally given to mice [93, 94]. Alginate is an outstanding aspirant for deliverance of protein drugs because proteins can be integrated into alginate-based formulations beneath comparatively soft conditions to reduce their denaturation, while until their release, the gels can defend them from degradation. A range of methods have been analyzed to manage the speed of protein discharge from alginate gels. Overall, because of the hydrophilic nature and the intrinsic porosity of the gels, the discharge speed of proteins from alginate gels is fast. However, fundamental fibroblast growth factor (bFGF) or heparin binding growth factors like vascular endothelial growth factor (VEGF) have the same, reversible binding to alginate hydrogels, allowing a constant and localized discharge [95, 96]. The discharge in this situation can be controlled by changing the degradation speed of the gels (e.g., utilization of partially oxidized alginate), so as to build protein discharge at minimum partially reliant on the degradation response [96]. Lots of efforts have been done mainly for factors that are not heparin binding, to more regulate the discharge of angiogenic molecules from alginate gels. Lysozyme and chymotrypsin, the high pI proteins, emerge to bodily cross-link with sodium alginate, efficiently encapsulating ionically cross-linked alginate as microspheres allowing for more constant release [97]. A poly((2-dimethylamino) ethyl methacrylate) end with amino group reacted with oxidized alginate devoid of gel beads and catalyst and was formed by reducing the alginate derivative aqueous solution into a $CaCl_2$ aqueous solution to make particles for oral release of proteins [98]. The formation of acetal-linked polymer network (tetra-functional) for stimuli-responsive gels of alginate with flexible pore sizes acted as a building block. The gels sheltered proteins (acid-labile) from denaturation, like insulin in gastric surroundings (pH 1.2); when discharged, the loaded protein in neutral surroundings is at close to zero-order kinetics [10]. The quick release of many proteins through alginate gels and small efficiency of encapsulation can also be targeted by different encapsulation techniques or cross-linking and/or by increasing hydrogel interactions with proteins [99]. For instance, alginates loaded with insulin microspheres were formed by mixing anionic polymers among alginates, followed by coating of chitosan so as to defend insulin from gastric pH and to get their continued release at intestinal pH [100]. Alginate microspheres layered with silk fibroin (*Bombyx mori*) by means of deposition layer-by-layer techniques give mechanically secure missiles in addition to diffusion obstruction to the encapsulated proteins [101]. Microspheres loaded with hydrogels were formed by suspension of poly(d,l-lactide-co-glycolide) (PLGA) microspheres encapsulated in alginate prior to ionic cross-linking. Alginate gels discharging proteins are extensively used in regeneration and tissue engineering, as discussed in the following sections on blood vessel, bone, and muscle regeneration.

10.8.2 Wound/Injury Dressings

Usually, injury dressings (e.g., gauze) keep the injury dry by facilitating the fading of injury exudates and at the same time protecting the entry of pathogen into the injury; they act mostly as a barrier. Curing chronic and acute wound is an urgent need in a lot of facets of drugs, and alginate-based wound dressings put forward numerous beneficial features [102]. Current dressings (e.g., dressings by alginate) give wet injury surroundings and allow injury healing [103]. Alginate dressings are usually formed by ionic cross-linking of calcium ions with an alginate aqueous solution to make a gel, and then processing them to make freeze-dried spongy sheets (i.e., foam) and stringy nonwoven dressings. Alginate dressings soak up the injury fluid with the dry form to re-gel and to maintain a moist physiological microsurrounding and reduce microbial infection at the injured area; the gel afterward can provide water to a dry injury. These roles can also endorse granulated tissue development, fast epithelialization, and curing. A variety of alginate dressings like Sorbsan (UDL Laboratories), AlgiSite (Smith & Nephew), Algicell (Derma Sciences), Comfeel Plus (Coloplast), Tegagen (3M Healthcare), and Kaltostat (ConvaTec) are commercially offered. A diversity of extra functions of bioactive alginate-based wound dressings have also been studied to date. The constant discharge of a controller of human keratinocyte production, cyclic adenosine monophosphate dibutyryl, partly oxidized gels of alginate that triggers injury healing and leads to the end of re-epithelialization of injury in 10 days in a rat model [104]. Alginate gels secreting stromal factor-1 (cell-derived) were also helpful in triggering injury end rates and minimizing scar development in pigs by severe surgical injury [105]. Alginate dressings integrated with silver-improved antibacterial properties and enhanced the binding efficiency for proinflammatory cytokines (e.g., IL-8), matrix metalloproteases-2 (MMP-2), and elastase. The adding up of silver into dressings of alginate also improved the antioxidant ability [106]. The adding up of zinc ions into alginate fibers has also been anticipated for injury dressings because zinc ions might produce immunomodulatory and antibacterial effects, in addition to improved keratinocyte passage and improved stages of endogenous growth factors [107]. The mixture of chitosan, alginate, and fucoidan gels has been found to give wet healing surroundings in rats, with no difficulty of application and elimination [108].

10.8.3 Cell Culture

The applications of alginate gels are increasing like a model system in biomedical studies for mammalian cell culture. These are gladly modified to supply as both 2-D and 3-D culture systems. The mammalian cell has no receptor for alginate. The mammalian cells and alginate gels were united with the small protein adsorption, allowing these supplies to be provided in several ways like a perfect blank slate, on which extremely precise and quantitative processes for cell adhesion can be integrated (e.g., combining artificial peptides particularly for cellular adhesion receptors). Additional essential findings exposed that *in vitro* observation can be gladly converted *in vivo*, because of the biocompatibility and simple penetration of alginate into the body. Alginate gels modified with RGD have also been mainly and often utilized *in vitro* cell culture. The RGD part in alginate gels facilitates one to regulate the phenotype of myoblasts [67], chondrocytes [109], osteoblasts [110], and ovarian follicle [111], in addition to bone marrow stromal cells [112, 113]. For instance, RGD peptides chemically conjugate to the alginate spine and there was severely improved linkage and production of cultured myoblasts compared to nonconjugated alginate gels (Figure 10.4). Additionally, the amount of cells adherent and the growth speed were robustly connected to the mass RGD density in the gels. A key factor in the controlling of cellular reaction is the extent of the spacer arm connecting the RGD peptide and the alginate sequence. The linkage and growth of main human fibroblasts cultivated on alginate gels customized with a peptide with the chain of (glycine)*n*–arginine–glycine–aspartic acid–serine–proline (G*n*RGDSP) were severely affected by the spacer arm extent, in spite of the entirely similar concentration of the peptides in the gels (Figure 10.5) [114].

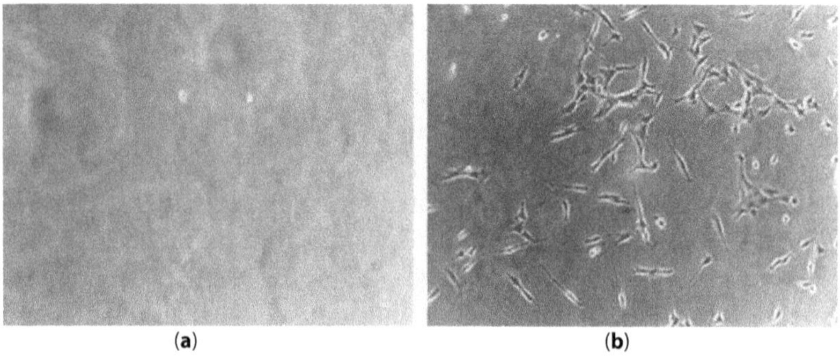

(a) (b)

Figure 10.4 Images of optical microscopic C_2C_{12} myoblasts linked to the shell of (a) nonpeptide-modified alginate gels and (b) alginate gels modified with RGD [114].

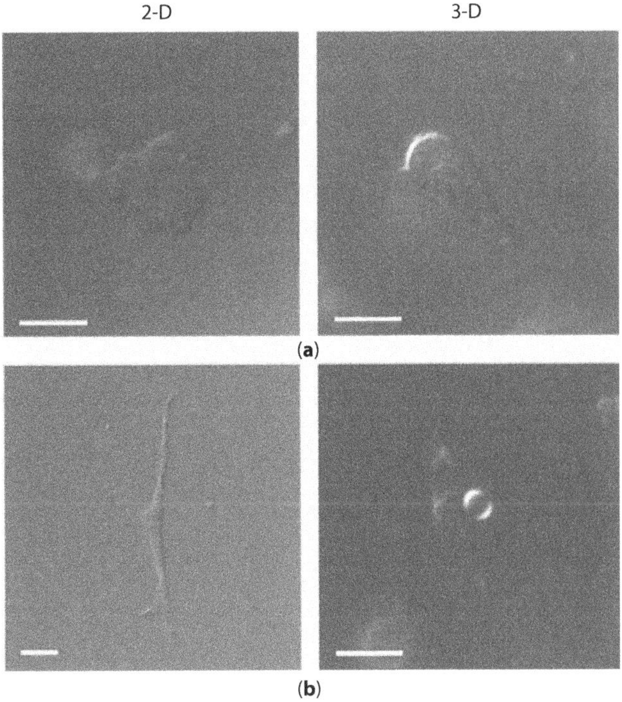

Figure 10.5 Images of confocal microscopic main human fibroblasts cultivated on alginate gels (2-D) modified with either (a) RGDSP or (b) G_{12}RGDSP, and cells encapsulated within the same two types of gels (3-D) [115].

For suitable binding to cellular receptors, a minimum of four glycine units as a spacer arm is permissible, but there was no additional progress in growth and cell adhesion using more than 12 glycine units [116].

10.8.4 Tissue Regeneration

Alginate gels act as a carrier to transport proteins, which can regulate the restoration or production of a variety of organs and tissues in the body, and have been broadly used over numerous decades. Alginate gels have been used in a range of applications like gelling approaches, cell adhesion, and breakdown behavior. Due to the hole size of a regenerative agent (~5 nm), there are restrictions to the size that can be discharged via diffusion from alginate hydrogels. The majority of proteins can spread out from alginate gels, yet in the dearth of gel deprivation [96]. Too big molecules that have major diffusion-force discharge can yet be transferred if the gel breaks down. For instance, antibody and coiled DNA of plasmid (range ~100 nm)

[118] are possibly released from breaking down alginate gels [119] by similar mechanism. When alginate hydrogels break down, cells should migrate out or be released. There are a number of studies of cell relocation in different alginate gels (nanoporous) in which relocation was found; however, it was not as quantitative as observed [120]. The amount of cells relocating externally as a role of the RGD appearance in alginate gels (macroporous) and porosity has been quantified [117, 121, 122].

References

1. Gombotz, W.R. and Wee, S.F., Protein release from alginate matrices. *Adv. Drug Deliv. Rev.*, 31, 267–85, 1998.
2. Langer, R. and Vacanti, J.P., Tissue engineering. *Science*, 260, 920–6, 1993.
3. Lee, K.Y. and Mooney, D.J., Hydrogels for tissue engineering. *Chem. Rev.*, 101, 1869–79, 2001.
4. Smidsrod, O. and Skjak-Bræk, G., Alginate as immobilization matrix for cells. *Trend. Biotechnol.*, 8, 71–8, 1990.
5. D.E. Clark and H.C. Green, Alginic acid and process of making same. *US. Patent,* 2036922, 1936.
6. Rinaudo, M., Main properties and current applications of some polysaccharides as biomaterials. *Polym. Int.*, 57, 397–430, 2008.
7. Remminghorst, U. and Rehm, B.H.A., Bacterial alginates: From biosynthesis to applications. *Biotechnol. Lett.*, 28, 1701–12, 2006.
8. Haug, A., Fractionation of alginic acid. *Acta. Chem. Scand.*, 13, 601–3, 1959.
9. Tonnesen, H.H. and Karlsen, J., Alginate in drug delivery systems. *Drug. Dev. Ind. Pharm.*, 28, 621–30, 2002.
10. George, M. and Abraham, T.E., Polyionic hydrocolloids for the intestinal delivery of protein drugs. *J. Control. Release*, 114, 1–14, 2006.
11. MacGregor, E.A. and Greenwood, C.T., *Polymers in Nature in Polysaccharides*, pp. 240–329, Wiley, New York, 1980.
12. Lee, K.Y. and Yuk, S.H., Polymeric protein delivery systems. *Progr. Poly. Sci.*, 32, 669–97, 2007.
13. Morris, E.R., Rees, D.A., Thom, D., Characterization of alginate composition and block structure by circular dichroism. *Carbohydr. Res.*, 81, 305–314, 1980.
14. Grasdalen, H., High field 1H-NMR spectroscopy of alginate sequential structure and linkage conformations. *Carbohydr. Res.*, 118, 255–260, 1983.
15. Berth, G., Methodical aspects of characterization of alginate and pectate by light scattering and viscometry coupled with GCP. *Carbohydr. Polym.*, 19, 1–9, 1992.
16. Martinsen, A., Skjak-Braek, G., Smidsrod, O., Zanetti, F., Paoletti, S., Comparison of different methods for determination of molecular weight and

molecular weight distribution of alginates. *Carbohydr. Polym.*, 15, 171–193, 1991.
17. Hay, I.D., Rehman, Z.U., Ghafoor, A., Rehm, B.H.A., Bacterial biosynthesis of alginates. *J. Chem. Technol. Biotechnol.*, 85, 752–9, 2010.
18. Rinaudo, M., On the abnormal exponents a_ and aD in Mark–Houwink type equations for wormlike chain polysaccharides. *Polym. Bull.*, 27, 585–9, 1992.
19. LeRoux, M.A., Guilak, F., Setton, L.A., Compressive and shear properties of alginate gel: Effects of sodium ions and alginate concentration. *J. Biomed. Mater. Res.*, 47, 46–53, 1999.
20. Kong, H.J., Smith, M.K., Mooney, D.J., Designing alginate hydrogels to maintain viability of immobilized cells. *Biomaterials*, 24, 4023–9, 2003.
21. Kong, H.J., Lee, K.Y., Mooney, D.J., Decoupling the dependence of rheological mechanical properties of hydrogels from solids concentration. *Polymer*, 43, 6239–46, 2002.
22. Smidsrod, O., The relative extension of alginates having different chemical composition. *Carbohydr. Res.*, 27, 107–118, 1973.
23. Clark, A.H. and Ross-Murphy, S.B., Structural and mechanical properties of biopolymer gels. *Adv. Polym. Sci.*, 83, 57–192, 1987.
24. Rees, D.A., Polysaccharide shapes and their interactions—Some recent advances. *Pure. Appl. Chem.*, 53, 1–14, 1987.
25. Rees, D.A. and Welsh, E.J., Secondary and tertiary structure of polysaccharides in solution and gels. *Angew. Chem. Int. Ed.*, 16, 214–224, 1977.
26. Morch, Y.A., Donati, I., Strand, B.L., Effect of Ca^{2+}, Ba^{2+}, and Sr^{2+} on alginate microbeads. *Biomacromolecules*, 7, 1471–80, 2006.
27. Martinsen., A., Skjåk-Bræk, G., Smidsrød, O., Alginate as immobilization material Correlation between chemical and physical properties of alginate gel beads. *Biotechnol. Bioeng.*, 33, 79–89, 1989.
28. Shoichet, M.S., Li, R.H., White, M.L., Winn, S.R., Stability of hydrogels used in cell encapsulation: An *in vitro* comparison of alginate and agarose. *Biotechnol. Bioeng*, 50, 374–81, 1996.
29. Kuhtreiber, W.M., Lanza, R.P., Chick, W.L., Chang, T.M.S., *Cell Encapsulation Technology and Therapeutics (Tissue Engineering)*, Birkhauser, Boston, MA, 1999.
30. Kuo, C.K. and Ma, P.X., Ionically cross-linked alginate hydrogels as scaffolds for tissue engineering. Part 1. Structure, gelation rate and mechanical properties. *Biomaterials*, 22, 511–21, 2001.
31. Bidarra, S.J., Barrias, C.C., Barbosa, M.A., Soares, R., Granja, P.L., Immobilization of human mesenchymal stem cells within RGD-grafted alginate microspheres and assessment of their angiogenic potential. *Biomacromolecules*, 11, 1956–64, 2010.
32. Grellier, M., Granja, P.L., Fricain, J.C., Bidarra, S.J., Renard, M., Bareille, R. *et al.*, The effect of the co-immobilization of human osteoprogenitors and endothelial cells within alginate microspheres on mineralization in a bone defect. *Biomaterials*, 30, 3271–8, 2009.

33. Evangelista, M.B., Hsiong, S.X., Fernandes, R., Sampaio, P., Kong, H.J., Barrias, C.C. *et al.*, Upregulation of bone cell differentiation through immobilization within a synthetic extracellular matrix. *Biomaterials*, 28, 3644–55, 2007.
34. Klokk, T.I. and Melvik, J.E., Controlling the size of alginate gel beads by use of a high electrostatic potential. *J. Microencapsul.*, 19, 415–24, 2002.
35. Lima, A.C., Batista, P., Valente, T.A.M., Silva, A.S., Correia, I.J., Mano, J.F., Novel methodology based on biomimetic superhydrophobic substrates to immobilize cells and proteins in hydrogel spheres for applications in bone regeneration. *Tissue Eng A.*, 19, 1175–87, 2013.
36. Fonseca, K., Bidarra, S.J., Oliveira, M.J., Granja, P.L., Barrias, C.C., Molecularly-designed alginate hydrogels susceptible to local proteolysis as 3D cellular microenvironments. *Acta Biomater.*, 7, 1674–82, 2011.
37. Lee, K.Y. and Mooney, D.J., Alginate: Properties and biomedical applications. *Prog. Polym. Sci.*, 37, 106–26, 2012.
38. Chang, S.C.N., Chung, H.Y., Tai, C.L., Chen, P.K.T., Lin, T.M., Jeng, L.B., Repair of large cranial defects by hBMP-2 expressing bone marrow stromal cells: Comparison between alginate and collagen type I systems. *J. Biomed. Mater. Res. A.* 94A, 433–41, 2012.
39. Kang, S.W., Cha, B.H., Park, H., Park, K.S., Lee, K.Y., Lee, S.H., The effect of conjugating RGD into 3D alginate hydrogels on adipogenic differentiation of human adipose-derived stromal cells. *Macromol. Biosci.*, 11, 673–9, 2011.
40. Oliveira, S.M., Barrias, C.C., Almeida, I.F., Costa, P.C., Ferreira, M.R.P., Bahia, M.F. *et al.*, Injectability of a bone filler system based on hydroxyapatite microspheres and a vehicle with *in situ* gel-forming ability. *J. Biomed. Mater. Res. B. Appl. Biomater.*, 87B, 49–58, 2008.
41. Javvaji, V., Baradwaj, A.G., Payne, G.F., Raghavan, S.R., Light-activated ionic gelation of common biopolymers. *Langmuir*, 27, 12591–6, 2011.
42. Cui, J., Wang, M., Zheng, Y., Rodriguez, M.G.M., Del, C.A., Light-triggered cross-linking of alginates with caged Ca^{2+}. *Biomacromolecules*, 14, 5, 1251–6, 2013.
43. Jeon, O., Bouhadir, K.H., Mansour, J.M., Alsberg, E., Photocross-linked alginate hydrogels with tunable biodegradation rates and mechanical properties. *Biomaterials*, 30, 2724–34, 2009.
44. Thornton, A.J., Alsberg, E., Hill, E.E., Mooney, D.J., Shape retaining injectable hydrogels for minimally invasive bulking. *J. Urol.*, 172, 763–8, 2004.
45. Lee, K.Y., Bouhadir, K.H., Mooney, D.J., Controlled degradation of hydrogels using multi-functional cross-linking molecules. *Biomaterials*, 25, 2461–6, 2004.
46. Wee, S.F. and Gombotz, W.R., Controlled release of recombinant human tumor necrosis factor receptor from alginate beads. *Proc. Int. Symp. Control. Release Bioact. Mater.*, 21, 730–731, 1994.

47. Austin, L., Bower, J.J., Muldoon, C., The controlled release of leukaemia inhibitory factor (LIF) from alginate gels. *Proc. Int. Symp. Control. Release Bioact. Mater.*, 23, 739–740, 1996.
48. Elqin, Y.M., Encapsulation of urease enzyme in xanthan–alginate spheres. *Biomaterials*, 18, 1157–1161, 1995.
49. Gray, C.J. and Dowsett, J., Retention of insulin in alginate gel beads. *Biotechnol. Bioeng.*, 31, 607–612, 1988.
50. Bodemeier, R., Chen, H., Paeratakul, O., A novel approach to the oral delivery of micro- or nanoparticles. *Pharm. Res.*, 6, 413–417, 1989.
51. Badwan, A.A., Abumalooh, A., Sallam, E., Abukalaf, A., Jawan, O., A sustained release drug delivery system using calcium alginate beads. *Drug. Dev. Ind. Pharm.*, 11, 239–256, 1985.
52. Dupuy, B., Arien, A., Minnot, A.P., FT-IR of membranes made with alginate polylysine complexes—Variations with the mannuronic or guluronic content of the polysaccharides. *Artif. Cells Blood Subs. Immob. Biotechnol.*, 22, 71–82, 1994.
53. Ke, C., Dixit, V., Shaw, W., Gitnick, G., *In vitro* slow release profile of endothelial cell growth factor immobilized within calcium alginate microbeads. *Artif. Cells Blood Subs. Immob. Biotechnol.*, 23, 143–151, 1995.
54. Wee, S.F., Gombotz, W.R., Fanslow, W., Evaluation of alginate microbeads for intranasal delivery of ovalbumin. *Proc. Int. Symp. Control. Release Bioact. Mater.*, 22, 566–567, 1995.
55. Bowersock, T.L., HogenEsch, H., Suckow, M., Porter, R.E., Jackson, R., Park, K., Oral vaccination with alginate microsphere systems. *J. Control. Release*, 39, 209–220, 1996.
56. Matsumoto, S., Kobayashi, H., Takashima, Y., Production of mono dispersed capsules. *J. Microencapsul.*, 3, 25–31, 1986.
57. Kwok, K.K., Groves, M.J., Burgess, D.J., Sterile microencapsulation of BCG in alginate- poly-L-lysine by an air spraying technique. *Proc. Int. Symp. Control. Release Bioact. Mater.*, 16, 170–171, 1989.
58. Kwok, K.K., Groves, M.J., Burgess, D.J., Production of 5–15 μm diameter alginate–polylysine microcapsules by an air atomization technique. *Pharm. Res.*, 8, 341–344, 1991.
59. Abraham, S.M., Vieth, R.F., Burgess, D.J., Novel technology for the preparation of sterile alginate–poly-L-lysine microcapsules in a bioreactor. *Pharm. Dev. Technol.*, 1, 63–68, 1996.
60. Donati, I., Coslovi, A., Gamini, A., Vetere, A., Campa, C., Paoletti, S., Galactose substituted alginate. 2. Conformational aspects. *Biomacromolecules*, 5, 186–96, 2004.
61. Donati, I., Vetere, A., Gamini, A., Coslovi, A., Campa, C., Paoletti, S., Galactose substituted alginate: Preliminary characterization and study of gelling properties. *Biomacromolecules*, 4, 624–31, 2003.
62. Leone, G., Torricelli, P., Chiumiento, A., Facchini, A., Barbucci, R., Amidic alginate hydrogel for nucleus pulposus replacement. *J. Biomed. Mater. Res. A*, 84A, 391–401, 2008.

63. Eiselt, P., Lee, K.Y., Mooney, D.J., Rigidity of two-component hydrogels prepared from alginate and poly(ethylene glycol)-diamines. *Macromolecules*, 32, 5561–6, 1999.
64. Jeon, O., Alt, D.S., Ahmed, S.M., Alsberg, E., The effect of oxidation on the degradation of photocross-linkable alginate hydrogels. *Biomaterials*, 33, 3503–14, 2012.
65. Rouillard, A.D., Berglund, C.M., Lee, J.Y., Polacheck, W.J., Tsui, Y., Bonassar, L.J. *et al.*, Methods for photocross-linking alginate hydrogel scaffolds with high cell viability. *Tissue Eng. C. Methods*, 17, 173–9, 2011.
66. Jeon, O. and Alsberg, E., Photofunctionalization of alginate hydrogels to promote adhesion and proliferation of human mesenchymal stem cells. *Tissue Eng. A*, 19, 1424–32, 2013.
67. Jeon, O., Powell, C., Ahmed, S.M., Alsberg, E., Biodegradable, photo-cross-linked alginate hydrogels with independently tailorable physical properties and cell adhesivity. *Tissue Eng. A*, 16, 2915–25, 2010.
68. Smeds, K.A. and Grinstaff, M.W., Photocross-linkable polysaccharides for *in situ* hydrogel formation. *J. Biomed. Mater. Res.*, 54, 115–21, 2001.
69. Chou, A.I., Akintoye, S.O., Nicoll, S.B., Photocross-linked alginate hydrogels support enhanced matrix accumulation by nucleus pulposus cells *in vivo*. *Osteoarthritis Cartilage*, 17, 1377–84, 2009.
70. Chandler, E.M., Berglund, C.M., Lee, J.S., Polacheck, W.J., Gleghorn, J.P., Kirby, B.J. *et al.*, Stiffness of photo cross-linked RGD–alginate gels regulates adipose progenitor cell behavior. *Biotechnol Bioeng.*, 108, 1683–92, 2011.
71. Wang, L., Shansky, J., Borselli, C., Mooney, D., Vandenburgh, H., Design and fabrication of a biodegradable, covalently cross-linked shape-memory alginate scaffold for cell and growth factor delivery. *Tissue Eng. A*, 18, 2000–7, 2012.
72. Lee, K.Y., Rowley, J.A., Eiselt, P., Moy, E.M., Bouhadir, K.H., Mooney, D.J., Controlling mechanical and swelling properties of alginate hydrogels independently by cross-linker type and cross-linking density. *Macromolecules*, 33, 4291–4, 2000.
73. Al-Shamkhani, A. and Duncan, R., Radioiodination of alginate via covalently-bound tyrosinamide allows monitoring of its fate *in vivo*. *J. Bioact. Compat. Polym.*, 10, 4–13, 1995.
74. Lee, K.Y., Bouhadir, K.H., Mooney, D.J., Degradation behavior of covalently cross-linked poly (aldehyde guluronate) hydrogels. *Macromolecules*, 33, 97–101, 2000.
75. Kong, H.J., Kaigler, D., Kim, K., Mooney, D.J., Controlling rigidity and degradation of alginate hydrogels via molecular weight distribution. *Biomacromolecules*, 5, 1720–7, 2004.
76. Kong, H.J., Alsberg, E., Kaigler, D., Lee, K.Y., Mooney, D.J., Controlling degradation of hydrogels via the size of cross-linked junctions. *Adv. Mater.*, 16, 1917–21, 2004.

77. Tan, H. and Marra, K.G., Injectable, biodegradable hydrogels for tissue engineering applications. *Materials*, 3, 1746–1767, 2010.
78. Tememoff, J.S. and Mikos, A.G., Injectable biodegradable materials for orthopedic tissue engineering. *Biomaterials*, 21, 2405–2412, 2000.
79. McCanless, J.D., Jennings, L.K., Bumgardner, J.D., Cole, J.A., Haggard, W.O., Hematoma-inspired alginate platelet releasate $CaPO_4$ composite: Initiation of the inflammatory-mediated response associated with fracture repair *in vitro* and *ex vivo* injection delivery. *J. Mater. Sci. Mater. Med.*, 23, 1971–1981, 2012.
80. De Vos, P., Spasojevic, M., de Haan, B.J., Faas, M.M., The association between *in vitro* physicochemical changes and inflammatory responses against alginate based microcapsules. *Biomaterials*, 33, 5552–5559, 2012.
81. Vanacker, J., Luyckx, V., Dolmans, M.M., Des Rieux, A., Jaeger, J., van Langendonckt, A., Donnez, J., Amorim, C.A., Transplantation of an alginate-matrigel matrix containing isolated ovarian cells: First step in developing a biodegradable scaffold to transplant isolated preantral follicles and ovarian cells. *Biomaterials*, 33, 6079–6085, 2012.
82. Boontheekul, T., Kong, H.J., Mooney, D.J., Controlling alginate gel degradation utilizing partial oxidation and bimodal molecular weight distribution. *Biomaterials*, 26, 2455–65, 2005.
83. Maiti, S., Singha, K., Ray, S., Dey, P., Sa, B., Adipic acid dihydrazide treated partially oxidized alginate beads for sustained oral delivery of flurbiprofen. *Pharm. Develop. Technol.*, 14, 461–70, 2009.
84. Bouhadir, K.H., Alsberg, E., Mooney, D.J., Hydrogels for combination delivery of antineoplastic agents. *Biomaterials*, 22, 2625–33, 2001.
85. Colinet, I., Dulong, V., Mocanu, G., Picton, L., Le Cerf, D., New amphiphilic and pH-sensitive hydrogel for controlled release of a model poorly water-soluble drug. *Eur. J. Pharm. Biopharm.*, 73, 345–50, 2009.
86. Zhang, X.L., Hui, Z.Y., Wan, D.X., Huang, H.T., Huang, J., Yuan, H., Yu, J.H., Alginate microsphere filled with carbon nanotube as drug carrier. *Int. J. Biol. Macromol.*, 47, 389–95, 2010.
87. Sandford, P.A. and Steinnes, A., Biomedical application of high-purity chitosan, in: *Water Soluble Polymers: Synthesis, Solution Properties, and Applications*, vol. 467, S.W. Shalaby, C.L. McCormick, G.B. Butler (Eds.), pp. 430–45, American Chemical Society, Washington, DC, 1991.
88. Rinaudo, M., Chitin and chitosan: Properties and applications. *Progr. Polym. Sci.*, 31, 603–32, 2006.
89. Lucinda-Silva, R.M., Salgado, H.R.N., Evangelista, R.C., Alginate–chitosan systems: *In vitro* controlled release of triamcinolone and *in vivo* gastrointestinal transit. *Carbohydr. Polym.*, 81, 260–8, 2010.
90. Wang, F.Q., Li, P., Zhang, J.P., Wang, A.Q., Wei, Q., A novel pH-sensitive magnetic alginate74 chitosan beads for albendazole delivery. *Drug. Dev. Ind. Pharm.*, 36, 867–77, 2010.

91. Lira, A.A.M., Rossetti, F.C., Nanclares, D.M.A., Neto, A.F., Bentley, M.V.L.B., Marchetti, J.M., Preparation and characterization of chitosan-treated alginate microparticles incorporating all-trans retinoic acid. *J. Microencapsul.*, 26, 243–50, 2009.
92. Ishak, R.A.H., Awad, G.A.S., Mortada, N.D., Nour, S.A.K., Preparation, *in vitro* and *in vivo* evaluation of stomach-specific metronidazole-loaded alginate beads as local anti-Helicobacter pylori therapy. *J. Control. Release*, 119, 207–14, 2009.
93. Chang, C.H., Lin, Y.H., Yeh, C.L., Chen, Y.C., Chiou, S.F., Hsu, Y.M., Chen, Y.S., Wang, C.C., Nanoparticles incorporated in pH-sensitive hydrogels as amoxicillin delivery for eradication of *Helicobacter pylori*.. *Biomacromolecules*, 11, 133–42, 2010.
94. Lee, K.Y., Peters, M.C., Mooney, D.J., Comparison of vascular endothelial growth factor and basic fibroblast growth factor on angiogenesis in SCID mice. *J. Control, Release*, 87, 49–56, 2003.
95. Silva, E.A. and Mooney, D.J., Effects of VEGF temporal and spatial presentation on angiogenesis. *Biomaterials*, 31, 1235–41, 2010.
96. Wells, L.A. and Sheardown, H., Extended release of high pI proteins from alginate microspheres via a novel encapsulation technique. *Eur. J. Pharm. Biopharm.*, 65, 329–35, 2007.
97. Gao, C.M., Liu, M.Z., Chen, S.L., Jin, S.P., Chen, J., Preparation of oxidized sodium alginate-graft poly((2-dimethylamino) ethyl methacrylate) gel beads and *in vitro* controlled release behavior of BSA. *Int. J. Pharm.*, 371, 16–24, 2009.
98. Chan, A.W. and Neufeld, R.J., Tuneable semi-synthetic network alginate for absorptive encapsulation and controlled release of protein therapeutics. *Biomaterials*, 31, 9040–7, 2010.
99. Silva, C.M., Ribeiro, A.J., Ferreira, D., Veiga, F., Insulin encapsulation in reinforced alginate microspheres prepared by internal gelation. *Eur. J. Pharm. Sci.*, 29, 148–59, 2006.
100. Wang, X., Wenk, E., Hu, X., Castro, G.R., Meinel, L., Wang, X., Li, C., Merkle, H., Kaplan, D.L., Silk coatings on PLGA and alginate microspheres for protein delivery. *Biomaterials*, 28, 4161–9, 2007.
101. Boateng, J.S., Matthews, K.H., Stevens, H.N.E., Eccleston, G.M., Wound healing dressings and drug delivery systems: A review. *J. Pharm. Sci.*, 97, 2892–923, 2008.
102. Queen, D., Orsted, H., Sanada, H., Sussman, G., A dressing history. *Int. Wound J.*, 1, 59–77, 2004.
103. Balakrishnan, B., Mohanty, M., Fernandez, A.C., Mohanan, P.V., Jayakrishnan, A., Evaluation of the effect of incorporation of dibutyryl cyclic adenosine monophosphate in an *in situ* forming hydrogel wound dressing based on oxidized alginate and gelatin. *Biomaterials*, 27, 1355–61, 2006.
104. Rabbany, S.Y., Pastore, J., Yamamoto, M., Miller, T., Rafii, S., Aras, R., Penn, M., Continuous delivery of stromal cell-derived factor-1 from alginate scaffolds accelerates wound healing. *Cell Transplant*, 19, 399–408, 2010.

105. Wiegand, C., Heinze, T., Hipler, U.C., Comparative *in vitro* study on cytotoxicity, antimicrobial activity, and binding capacity for pathophysiological factors in chronic wounds of alginate and silver-containing alginate. *Wound Repair Regener.*, 17, 511–21, 2009.
106. Agren, M.S., Zinc in wound repair. *Arch. Dermatol.*, 135, 1273–4, 1999.
107. Murakami, K., Aoki, H., Nakamura, S., Nakamura, S., Takikawa, M., Hanzawa, M., Kishimoto, S., Hattori, H., Tanaka, Y., Kiyosawa, T., Sato, Y., Ishihara, M., Hydrogel blends of chitin/chitosan, fucoidan and alginate as healing-impaired wound dressings. *Biomaterials*, 31, 83–90, 2010.
108. Rowley, J.A., Sun, Z.X., Goldman, D., Mooney, D.J., Biomaterials to spatially regulate cell fate. *Adv. Mater.*, 14, 886–9, 2002.
109. Degala, S., Zipfel, W.R., Bonassar, L.J., Chondrocyte calcium signaling in response to fluid flow is regulated by matrix adhesion in 3-D alginate scaffolds. *Arch. Biochem. Biophys.*, 505, 112–7, 2011.
110. Kreeger, P.K., Deck, J.W., Woodruff, T.K., Shea, L.D., The *in vitro* regulation of ovarian follicle development using alginate-extracellular matrix gels. *Biomaterials*, 27, 714–23, 2006.
111. Hsiong, S.X., Carampin, P., Kong, H.J., Lee, K.Y., Mooney, D.J., Differentiation stage alters matrix control of stem cells. *J. Biomed. Mater. Res. Part A*, 85, 145–56, 2008.
112. Wang, L., Shelton, R.M., Cooper, P.R., Lawson, M., Triffitt, J.T., Barralet, J.E., Evaluation of sodium alginate for bone marrow cell tissue engineering. *Biomaterials*, 24, 3475–81, 2003.
113. Rowley, J.A., Madlambayan, G., Mooney, D.J., Alginate hydrogels as synthetic extracellular matrix materials. *Biomaterials*, 20, 45–53, 1999.
114. Lee, J.W., Park, Y.J., Lee, S.J., Lee, S.K., Lee, K.Y., The effect of spacer arm length of an adhesion ligand coupled to an alginate gel on the control of fibroblast phenotype. *Biomaterials*, 31, 5545–51, 2010.
115. Comisar, W.A., Hsiong, S.X., Kong, H.J., Mooney, D.J., Linderman, J.J., Multi-scale modeling to predict ligand presentation within RGD nanopatterned hydrogels. *Biomaterials*, 227, 2322–9, 2006.
116. Lee, K.Y., Alsberg, E., Hsiong, S., Comisar, W., Linderman, J., Ziff, R., Mooney, D., Nanoscale adhesion ligand organization regulates osteoblast proliferation and differentiation. *Nano Lett.*, 4, 1501–6, 2004.
117. Comisar, W.A., Kazmers, N.H., Mooney, D.J., Linderman, J.J., Engineering, R.G.D., nanopatterned hydrogels to control preosteoblast behavior: A combined computational and experimental approach. *Biomaterials*, 28, 4409–17, 2007.
118. Gu, F., Amsden, B., Neufeld, R., Sustained delivery of vascular endothelial growth factor with alginate beads. *J. Control. Release*, 96, 463–72, 2004.
119. Jay, S.M. and Saltzman, W.M., Controlled delivery of VEGF via modulation of alginate microparticle ionic cross-linking. *J. Control. Release*, 134, 26–34, 2009.
120. Sun, Q.H., Silva, E.A., Wang, A.X., Fritton, J.C., Mooney, D.J., Schaffler, M.B., Grossman, P.M., Rajagopalan, S., Sustained release of multiple growth factors

from injectable polymeric system as a novel therapeutic approach towards angiogenesis. *Pharm. Res.*, 27, 264–71, 2010.
121. Hao, X.J., Silva, E.A., Mansson-Broberg, A., Grinnemo, K.H., Siddiqui, A.J., Dellgren, G., Wardell, E., Brodin, L.A., Mooney, D.J., Sylven, C., Angiogenic effects of sequential release of VEGFA(165) and PDGF-BB with alginate hydrogels after myocardial infarction. *Cardiovasc. Res.*, 75, 178–85, 2007.
122. Ruvinov, E., Leor, J., Cohen, S., The promotion of myocardial repair by the sequential delivery of IGF-1 and HGF from an injectable alginate biomaterial in a model of acute myocardial infarction. *Biomaterials*, 32, 565–78, 2011.

Part 3

ALGINATES IN FOOD INDUSTRY

11

Alginates for Food Packaging Applications

Radhika Theagarajan, Sayantani Dutta, J.A. Moses and C. Anandharamakrishnan*

Computational Modeling and Nano Scale Processing Unit, Indian Institute of Food Processing Technology (IIFPT), Ministry of Food Processing Industries, Government of India, Thanjavur, India

Abstract

Food packaging plays a key role in protecting the quality and safety of food in the food supply chain. Packaging helps to protect food from physical, chemical, biological, and environmental contaminants. The choice of packaging material and design plays a vital role in product's shelf life also. In recent years, biopolymers are being used as edible food packaging material aiming to address issues of petroleum-based packaging materials and their recalcitrant nature. Several biopolymers, namely chitosan, alginate, carrageenan, pectin, cellulose, gellan, etc. are used for food packaging applications. Among these, alginates are most favored, that are natural hydrophilic polysaccharide biopolymers extracted from marine brown algae (*Phaeophyceae* sp.). They have good film-forming property and help to retain moisture, reduce shrinkage, and improve the sensory characteristics of food products. Alginate coatings can reduce microbial counts, and retard oxidative off-flavors; it can enhance sensory acceptance, and minimize cooking losses. This chapter presents the application of alginates in food packaging, highlighting recent advancements in research and commercialization.

Keywords: Alginates, biopolymer, food packaging, biodegradable, edible coating, eco-friendly

11.1 Introduction

Today, packaging is one of the most important aspects in ensuring food's acceptance, safety, portability, storability, and shelf stability, besides gaining

**Corresponding author*: anandharamakrishnan@iifpt.edu.in

Shakeel Ahmed (ed.) Alginates, (207–232) © 2019 Scrivener Publishing LLC

consumer attention. A myriad number of foods and their types require flexible yet conducive packaging and packaging aids. Usage of petroleum products in packaging is of course conducive and has been in practice in the past, but it is not without the concerns of material itself that are unfriendly to the environment. The focus has now shifted to natural alternatives that would offer similar protection, and also be environmentally benign. The development of biodegradable packaging has been one of the focused themes in the area of food packaging. Biopolymers are a renewable source of food packaging. They are produced from natural sources, such as starch and polysaccharide (mainly from pectin, cellulose, alginates, carrageenan, and chitosan) [1–4].

Biopolymers have different material properties compared with usual plastics; for example, they exhibit low tensile strength. These biopolymers have some unique properties such as biodegradability, edibility, antimicrobial activity, sustainability, biocompatibility, and eco-friendly nature. In addition, the quality of the food packed is very well maintained [5–7]. Even recent trends highlight more consumer preference for biopolymers [8].

There are two major classifications of biopolymers: polylactic acid and starch-based polymers. Only these two completely biodegradable biopolymers are used in commercial market simultaneously with other nonbiodegradable polymers [8]. Nowadays, awareness about biopolymer-based food packages has increased around the world. Therefore, biopolymer food packaging technologies are gaining huge potential to guarantee the quality and safety of food materials packed in them.

Alginates are one of the efficient polysaccharides that are used as edible coating for food products. Alginate packaging is known to possess specific characteristics such as good tensile strength, flexibility, tear resistance, rigidity, mechanical resistance, tastelessness, oil resistance, glossiness, and odorlessness. However, due to their porous structure, they exhibit high permeability toward water and oxygen [9]. This chapter aims to explain about alginate packaging in terms of its characteristics and properties, and also from the customer point of view.

11.2 Biopolymer in Food Industry

Biopolymers are used as an alternative food packaging for their eco-friendly nature. Cha and Chinnan [10] have classified biopolymers in categories such as (a) polymers extracted from natural materials, like protein, cellulose, starch, and sea prokaryotes; (b) polymers chemically synthesized from biobased monomer; and (c) polymers produced by microbes, such as hydroxyl valerate and hydroxyl butyrate (Figure 11.1).

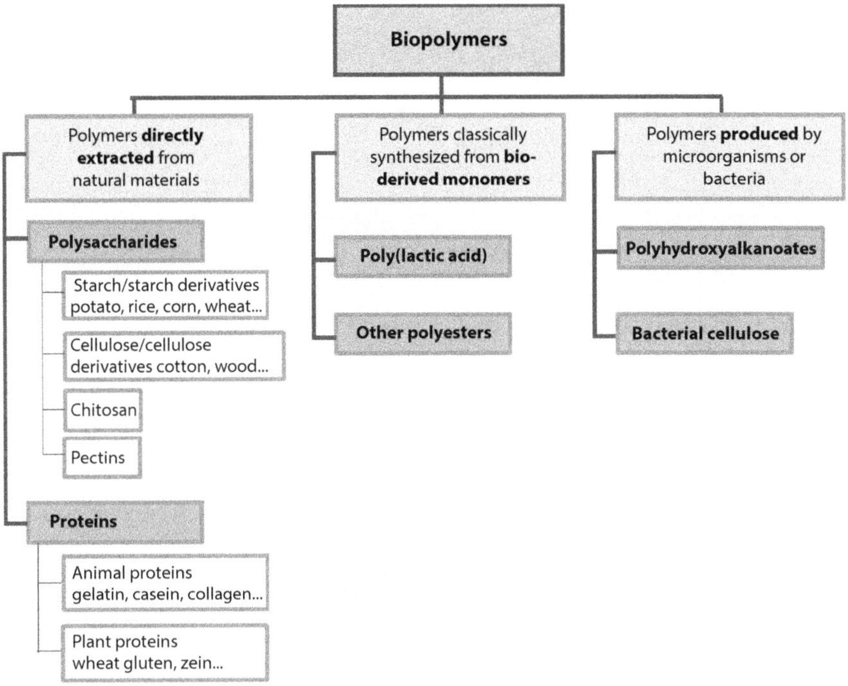

Figure 11.1 Schematic overview of different biopolymers [15].

Edible films and coatings for food products are emerging technologies that are gaining demand in this era of utilization of nonbiodegradable food packaging [9]. They are superior in keeping the quality of both whole and fresh-cut fruit compared to nonbiodegradable polymers. Alginate is one of the commonly used bio-packaging materials reportedly used to decrease the microbial load and weight loss in carrots [11]. It has increased the shelf life of fresh-cut apples [12] and reduced browning effect during storage of fresh-cut mangoes [13]. During film formation, alginates are sometimes mixed with nanoparticles; one of these studies has exhibited favorable effect of nanoparticles on shiitake mushroom during storage [14].

11.3 Alginates in Food Packaging

Alginate is probably the most widely studied essential biopolymer in recent times and it is the preferred film-forming material also [16]. Alginate is the salt of alginic acid that is nonrepeating copolymers of β-D-mannuronic acid (M) and α-L-guluronic acid (G) linked by 1–4 glycosidic bonds. Figures 11.2 and 11.3 display block structures of M, G and alginates,

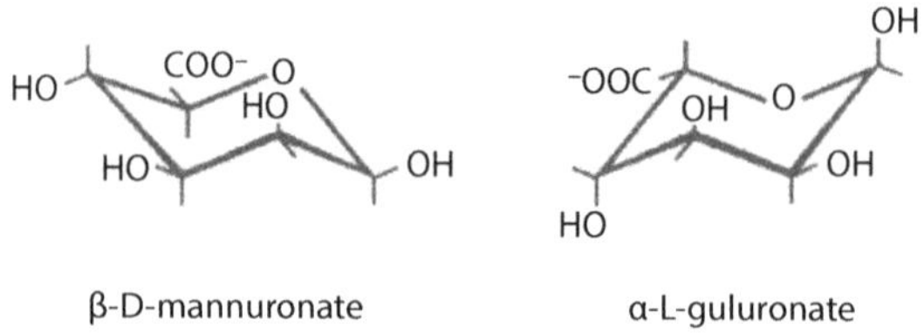

Figure 11.2 Conformations of M- and G-blocks [20].

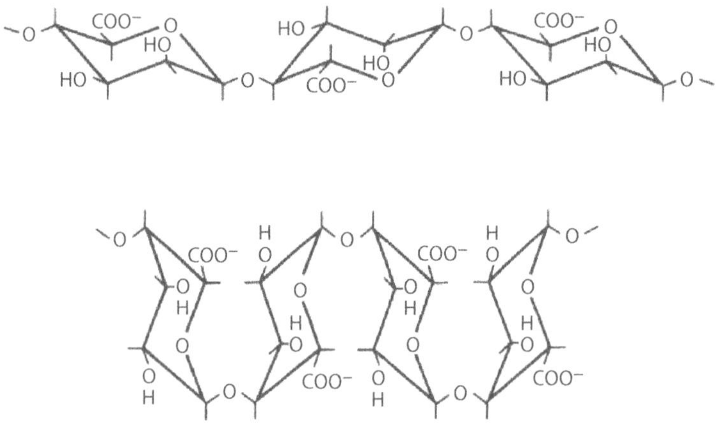

Figure 11.3 Block structures of alginates. Poly-β-D-mannuronate (above) and poly-α-L-guluronate (below) [20].

respectively. Polymer chains of alginates are separated into varied segments of M-blocks, G-blocks, and MG-blocks [17]. The quantity and the prevalence of blocks differ with species of the brown algae, maturity level, origin, and harvesting period of the algae. Alginate is renewable, abundant, biodegradable, biocompatible, and water soluble [18]. Commercially it is available as sodium alginate, commonly used as a thickener, stabilizer, and gelling agent in food like deserts, sauces, stabilizers, and beverages [19]. Table 11.1 exhibits different applications of alginate in food packaging. Apart from food packaging, alginates are also used in drug delivery systems, and encapsulation of alginate has also been done for formulation of herbicides, cells, and microbes [17].

Alginates are generally extracted from brown seaweeds (*Phaeophyceae*) or produced from the extracellular matrix of certain bacteria (*Pseudomonas* and *Azotobacter*) [37]. Jost *et al.*, [17] explained that alginates produced from bacteria contain acetyl group, and those derived from algae have hydroxyl groups. Presence of alginate in seaweeds provides flexibility and mechanical strength to them. Moreover, it acts as a water reservoir,

Table 11.1 Application of alginate in food packaging.

Form of alginate	Additional compounds	Advantage of the process	Reference
Sodium alginate	Montmorillonite/cellulose nanoparticles	Reduction in the water vapor permeability	[21]
	Montmorillonite clay	Reduced water permeability and increased water solubility	[22]
	Polyethyleneimine, biaxially oriented poly(lactic acid)	Increases oxygen barrier properties	[23]
	Natamycin	Increase in soluble matter in water, water vapor permeability and opacity; decrease in tensile strength	[24]
	Silver nanoparticles	Increases shelf life of carrot and pear	[25]
	Partially hydrolyzed sago starch, lemongrass oil, and glycerol	Increased antimicrobial property, flexibility, and mechanical strength	[26]
	$CaCl_2$	Improved mechanical properties, water resistance; decrease in water vapor permeability	[27]
	Lysozyme, nisin, grape fruit seed extract, ethylenediaminetetraacetic acid	Antimicrobial effect against bacterial strains studied	[28]
Calcium alginate	Silver-montmorillonite nanoparticles	Prevents dehydration and microbial spoilage of fresh-cut carrot; increases shelf life	[29]
	Nisin	Suppression of bacterial growth	[30]
	–	Lowered shrinkage loss, drip and degree of off-odor in beef; prolonged muscle color	[31]

(Continued)

Table 11.1 Application of alginate in food packaging. (*Continued*)

Form of alginate	Additional compounds	Advantage of the process	Reference
Food-grade alginate	–	Preserves quality of button mushroom and increases its shelf life	[14]
Alginate cast film	Glycerol and sorbitol	Both plasticizers have positive influence on the mechanical properties	[17]
	Enterocins and high-pressure processing	Delayed growth of *L. monocytogenes* in cooked ham	[32]
	Potassium sorbate	Potential use as systems for release of active substances	[33]
Gelatine/sodium alginate blend	Corn oil and olive oil	Minimizes water loss from sausages; manufactured sausages exhibited high oxygen barrier properties	[34]
Propylene glycol alginate	Soy protein isolate films	Tensile strength increased, percentage elongation at break decreased; decrease in water vapor permeability and water solubility	[35]
Biocomposite film of cellulose and alginate	Unmodified birch pulp, microfibrillated cellulose, nanofibrillated cellulose, nanofibrillated anionic dicarboxylic acid cellulose	Increased mechanical properties of alginate films, excellent grease barrier properties, and reduced WVP	[36]

which prevents seaweeds from dehydration once they get exposed to air [38]. The composition of alginate obtained from different seaweeds varies in accordance with their growth conditions, seasonal changes, and other environmental factors. These variations also can later affect the mechanical properties of alginate [39, 40].

There are several alginate-coated food products (reformed meat, pet food, carb sticks, onion rings and a few others) available in the market and widely distributed [41–43]. Every year, about 30,000 metric tons of sodium alginates is used in food, pharmaceutical, and textile industries as thickening, stabilizing, and jellifying agents [44].

During alginate biofilm formation, calcium ions act as gelling agent and are used to bind alginate chains by means of ionic interaction. Calcium ions bind strongly with guluronate blocks, which leads to interchain organization. This structural alignment is known to be "egg-chain" mechanism [45]. Therefore, alginate films have high water vapor permeability, impermeability to greases and oils, and barrier property to oxygen [9]. During microwave cooking, lack of crispiness in the packed food has been observed; to overcome this problem, active packaging with susceptor (used for converting electromagnetic energy to heat) is developed specifically for microwave heating. Due to its edible property, alginate film with high concentration of salt is used as a susceptor in microwave cooking [46].

11.4 Biosynthesis of Alginate

Alginates are commonly represented as a family of related molecules. The structure of alginates can be explained as unbranched (1-4)-linked polysaccharides consisted of β-D-mannuronate and its α-L-guluronate as its C5-epimer. Though alginate is manufactured mainly from brown seaweeds, some bacteria of the genera *Pseudomonas* and *Azotobacter* can also produce this biopolymer [41, 47–49]. It is known that the polysaccharide arrangements in the structural blocks of the monomer residue in the alginate produced by seaweeds are similar to those synthesized by *Azotobacter vinelandii* [20].

Biosynthesis of alginate from *A. vinelandii* was first studied by Pindar and Bucke [50]. They have explained the biosynthesis procedure in four stages: (i) synthesis of precursor, (ii) transfer and polymerization of cytoplasmic membrane, (iii) transfer and modification of periplasm, and (iv) transport through outer membrane [51]. Mostly, all commercially available alginates are produced from algal source. Later, Szekalska *et al.*, [52] reported on the extraction of alginate from seaweeds, which involves multistage processing such as drying of raw material and treating the raw material with mineral acid, and further purification of the acquired alginic acid to transform it into water-soluble sodium salts. Compared with other polymer composition, these native alginates exhibit poor consistency and purity. Careful purification of alginate is necessary to overcome the protein and immunogenic contaminations, which affect the purity of the same.

Alginates produced by fermentation exhibit effective physicochemical properties in terms of their tensile strength and mechanical properties. Therefore, nowadays, alginates are commercially produced using fermentation. Alginates manufactured from bacteria possess desired material property compared to alginates produced from seaweeds. These material properties of alginates depend upon acetylation, composition of monomer, polymer length, and also type and degree of modifications [53]. Further, bacterial alginates can be produced with "user-defined" material properties by proper understanding of the mechanism of bacterial enzymes and modifications of the same by protein or genetic engineering to "tailor-made" bacterial alginates [54]. For these reasons, though production of bacterial alginate is costly compared with seaweed-derived alginate, the former has more applications than the latter.

Production of alginate by shake-flask method from *A. vinelandii* was optimized by Clementi *et al.*, [44]. Glucose media with varying glucose concentrations, shaking speed, fermentation temperature, and concentrations of acetate and sodium phosphate were studied. After separation from the culture broth, the polymer was characterized by mannuronate fraction at different fermentation times. This similar process is also followed during alginate production from *Macrocystis pyrifera* and *Ascophyllum nodosum*; in the later process, the ratio of guluronic–guluronic blocks is found to be quite small ($GG = 0.037 \pm 0.006$).

Bacteria-derived alginates show well-defined material and chemical properties. Hay *et al.*, [55] reported that alginate produced from *Azotobacter* has higher concentration of G residues, which reflects its rigid nature. This alginate is generally used in the development of desiccation-resistant cysts in pharmaceutical industries. As discussed earlier, engineered bacterial alginate can be produced with desired properties by modification of bacterial enzymes; for example, epimerases from *A. vinelandii* can change the G/M ratio of the alginate [56].

11.5 Application of Alginate in Formation of Biofilm

11.5.1 Preparation of Packaging Films

Wang *et al.*, [57], in their work, explained about commonly followed film-casting procedure also known as the solvent-casting method to prepare packaging films. In this method, they have prepared film using agar/alginate/collagen (A/A/C) where film solution was prepared by dissolving 1 g of each material (agar, alginate, and collagen) in 150 mL of distilled water. In this solution, 0.9 g of glycerol was vigorously mixed for about 25 min at 95°C. The film solution thus prepared was spread onto a glass plate (24 cm × 30 cm) and covered with Teflon layer and dried for 24 h at room temperature. After drying, it was unwrapped from the plate and collected. Prepared film samples were pretreated in a humidity chamber set at 25°C and 50% relative humidity for at least 48 h. Figure 11.4 exhibits the scanning electron micrograph of the alginate cast film.

11.5.2 Role of Alginate in Biofilm Formation

Alginate hydrogels are formed by exterior gelation using calcium ions as cross-linking agents. The cation reacts with guluronate blocks of alginate

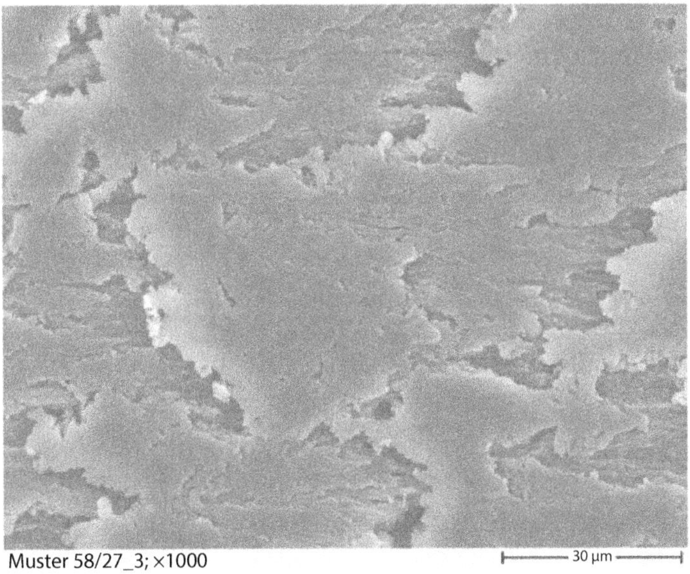

Figure 11.4 Scanning electron micrograph of alginate cast film [17].

chains and binds with them forming the gel network [57]. Alginate, which has high concentration of polyguluronate, produces rigid but fragile gels, whereas those that have higher concentration of polymannuronate or other mixed combination exhibit gels with more elasticity. For efficient gel formation, it is necessary that the blocks of guluronate should be 20 monomer units long; this is due to the prerequisite nature of selective binding of ions [58]. This phenomenon is observed since gelation of alginates depends upon cation binding property; in addition, specific lengths of junction zones are also needed for gelation [20].

Studies have shown that during gel formation, random blocks of alginates help in binding huge amount of water with the help of positive osmotic pressure and limiting elasticity of the gel network [59, 60]. Hence, flexibility of the alginate chain, hydrophilicity, and the number of cross-links (junction zones) inside the gel network all together govern the amount of water trapped inside the gel [33]. A recent study revealed that appropriate binding of calcium with guluronate blocks results in an interchain association known as the egg-box mechanism. In this structure, calcium ions are situated in the cavities formed naturally in between segments of guluronate (Figure 11.5). Therefore, their strength is higher compared to mannuronic acid sequences, though they apparently do not have cross-linking. Considering all these factors, it is expected that alginate with higher content of glucuronic acid shows structural difference compared with mannuronate system. When the content of glucuronate is high, gradually the gel turns out to be extensible, harder, and brittle [45].

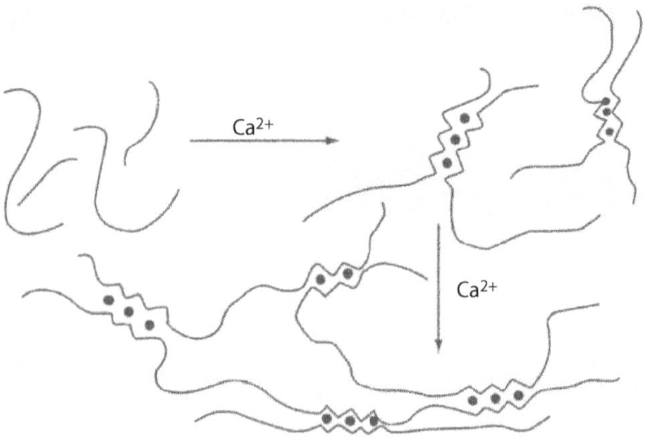

Figure 11.5 Schematic cross-linking of alginate in the presence of calcium counterions (•) complexed with L-guluronic blocks [65].

Viscosity of alginate is the most important factor that governs gel homogeneity. This viscosity can be modified by adjusting the pH of the solution. Studies revealed that at neutral pH, the gel can retain its viscosity properly. Certain properties, for instance, surface tension, viscosity, and density of coating solution, are important to estimate the film thickness [62]. Donnan and Rose [63] conducted physiochemical studies on algal alginates and determined that intrinsic viscosity of the alginates is directly proportional to their molecular size. Properties of starch and alginate together in the film formation were studied by Wu *et al.*, [64]; mixture of these two biopolymers was found to be a favorable combination of biofilm for better mechanical property. The shear-thinning behavior of the alginate film solution depends on the added starch. Therefore, addition of starch in alginate solution is maintained carefully in the industries to obtain better efficiency since increasing the shear rate will reduce the dynamic viscosity [65].

11.6 Packaging Properties of Alginate

Alginates have unique packaging properties such as biodegradability, tensile strength, and water permeability [66]. Characteristics of alginate are defined by their degree of polymerization, composition of the monomer, acetylation, and monomer sequence [55]. Sustainability, packaging function, and performance of the biopolymers depend on their processing techniques and compositional properties [57]. Because of the efficient oxygen barrier properties, alginates can act as a sacrificing agent; they are applied as coating for meat products where they reduce dehydration in fresh meats, and can also protect foods from oxidation [16].

Sirviö *et al.*, [36] reported that the alginate films mixed with cellulose exhibited higher thickness due to the fibrous cellulose structure. This thickness difference was studied by Rhim [27] who reported that the increase in thickness is due to the cross-linkage of alginate with aqueous solution of calcium chloride. Russo *et al.*, [67] have also validated the potential of alginate in the field of biopolymer packaging. In a different study, Yang *et al.*, [68] reported the development of collagen composite biofilm as packaging material with incorporation of sodium alginate. The authors found that addition of sodium alginate in the biofilm improves its tensile strength and elongates the film. Further, sodium alginate in the biofilm with collagen composite results in uniform structural morphology of the film.

11.6.1 Thermostability of Alginate Packaging

Generally during enzyme immobilization, calcium alginate beads are used, which do not harm enzymes, but gradually the immobilized enzymes diffuse out [69]. For bead formation, attachment of the enzyme with alginate before bead preparation is necessary. For example, enzymes like glucoamylase and pullulanase show increased starch hydrolysis along with thermostability of the enzyme during its entrapment in alginate beads [70].

11.6.2 Water Solubility

Water solubility is a widely used parameter for assessment of the resistivity of the film to water, which is considered to be an important characteristic of biopolymers in food packaging [21]. Generally, packaging material that exhibits higher water solubility will have poor resistance to water. Abdollahi *et al.*, [21] reported that water solubility of pure alginate was 99.5%, depicting increased hydrophilic nature of alginate. In order to overcome this, nanoparticles (5%) were incorporated into the film. This resulted in the decrease in water solubility of about 77.49% with the addition of cellulose nanoparticles, and it was further decreased to 61.35% after the addition of MMT.

11.6.3 Water Vapor Permeability

Water vapor permeability (WVP) is another major property for any biobased food packaging. Higher WVP of the films directly affects the food and causes deterioration of the same [21]. Many studies have been reported in this area of alginate coatings exhibiting high oxygen barrier property. The poor moisture barrier of alginate packaging is due to swelling of the same after absorption of moisture vapor from surrounding that increases the water vapor transmission and water uptake [17].

11.6.4 Tensile Strength

Fang *et al.*, [71] studied that alginates possess linear and well-organized chain structure, which helps in the formation of proper connection with calcium ions. This high efficient cross-linking of alginate increases the cohesive force between the chains resulting in high tensile strength of the packaging material. Da Silva *et al.*, [72] have also found that alginates possess high tensile strength compared to other biofilms like pectin. This characteristic behavior of alginate was also studied by Sriamornsak and Kennedy [73].

11.6.5 Oxygen Permeability

Oxygen permeability is another important factor for the shelf life of food products, yet limited reports are available regarding oxygen permeability of biopolymer packaging. Moisture absorption and oxygen permeability are both parallel properties of packaging materials. Moisture absorption depends on the molecular size of the alginate film. Compared to other polymers, alginate packaging being a biopolymer is known to exhibit little elevated permeability to air. Chemical composition and structural differences of the polymers could also be the reason for this elevated permeability. Over a period of time, alginate packaging shows stable oxygen permeability, but the WVP was found to be decreasing [73–75].

11.6.6 Barrier Property

Rojas-Graü *et al.*, [12] reported that alginates exhibit good barrier property toward water molecules and found that 2% of alginate coating applied on fresh-cut "Fuji" apples was successful in avoiding water loss. In a separate study, Montero Calderón [76] observed that no off-flavors were developed in alginate-coated pineapple pieces. This study proves that there is no evidence for deterioration of the product due to retention of water inside the cut pieces. Sirviö *et al.*, [36] studied incorporation of additives having lesser permeability in alginate films. The authors observed an improved barrier property of the films.

For improvement of the barrier property of the packaging material, concentration of alginate coating is needed to be optimized prior to its application into the packaging system [17]. One major concern is to take care of the fragility of the alginate coating. Strength of the alginate in packaging application can be improved by increasing the weight of the molecules or length of the G-blocks. Increasing the M-blocks increases the flexibility and fragility of the films. Solubility of the alginates in acid depends upon its alternating divisions D-mannuronic acid and L-guluronic acid [77, 60]. Application of nano- and microsized cellulose to alginate films also renders strength to the material and thus decreases permeability of water molecule through the film [78].

11.6.7 Antimicrobial Activity

Appendini and Hotchkiss [79] have asserted the importance of maintaining the quality, safety, and freshness of food using various packaging technologies. According to the authors, this includes developing packaging material with antimicrobial property also (Figure 11.6). These technologies improve the shelf life of the product aside from limiting pathogens.

This antimicrobial effect is rendered by inhibiting or retarding the growth of microorganisms present inside the packed food. Microbiological deteriorations of packaged foods are spoilages that occur in the food system when there is more water activity and oxygen consumption or when there is an imbalance in the temperature of the food. Averting oxygen permeability could be a measure to combat this difficulty. Research indicates acceptability in the same, when using an engineered biopolymer with antimicrobial activity as a packaging material. The idea of infusing antimicrobials into the packaging material is to inhibit the growth of microbes on the surface of food. Antimicrobials are out from the packaging material over a particular period; microbial growth kinetics and activity are maintained [79].

In another study, it was revealed that white ginseng when incorporated into alginate films produced antimicrobial effects. Even at times, lactic acid bacteria have even been included in the formation of biofilms to control the growth of food-borne pathogens in ready-to-eat food. Alginate films have been functionalized with the addition of garlic oil, an antibacterial [61]. Biopolymers mixed with ethanol, water, and other solvents also act as antimicrobial elements in packaging [88]. In an interesting find, researchers

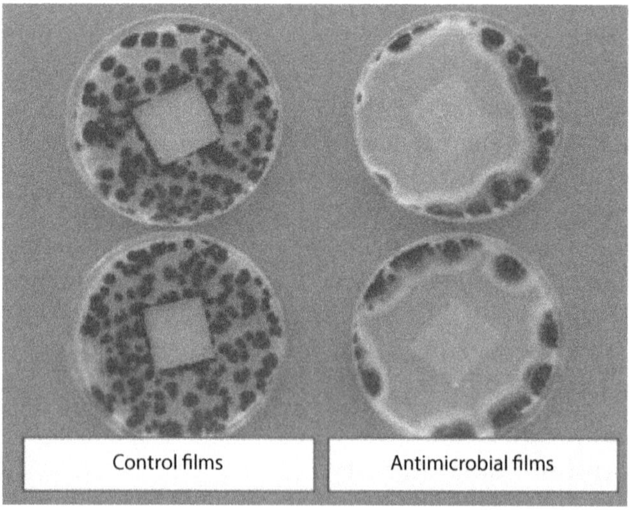

Figure 11.6 Effect of antimicrobial plastic film on *Aspergillus niger*. Agar diffusion method [79].

report on the antibacterial ability of sodium alginate incorporated with nanoparticles on test strains. Also, the film with nanoparticles was found to increase shelf stability of pears and carrots in terms of soluble protein substance and reduction in the weight with storage [25].

Biopolymers employed as antibacterial agents in packaging could be starch or protein based, and have been highlighted in one of the reviews articles [80]. The authors stated utility of benzoic acid, lysozyme, propionic acid, ascorbic acid, lactic acid, and nisin as additives in biofilms for antimicrobial effect [81]. However, to assess the potential of the said material, storage and distribution system must also be accounted for. A recent review on biopolymers put forth that potential application of alginate with antimicrobial agents could be a measure to extend the shelf life of meat and meat products [82]. Effect of antibacterial biopolymer films on shelf life of food against contamination by *Listeria monocytogenes* and *Escherichia coli* arrested growth of these microorganisms [57].

There are some biopolymer packagings that have built-in mechanism of restricting microbial growth, even inside the packaged food; but of course, the added antimicrobial should not contaminate the food product. Unfortunately at times, migration of these elements from the packaging material to food occurs, and therefore appropriate research must be conducted before commercialization [19]. The study of Appendini and Hotchkiss [79] concluded that calcium alginates affected the growth of microbes such as coliform bacterial strain on beef and other natural flora. Such a possibility has been attributed to the presence of calcium chloride in the biofilm encasing the food [83].

Notably, not all biofilms have similar antimicrobial properties, as evidenced by a research highlighting more inhibitory effect of sodium alginate than κ-carrageenan in biofilms. As described, alginate absorbs higher moisture and thus the antimicrobial agents in the films are provided a quicker and easier passage into the food matrix than κ-carrageenan. Such additions have been referred to as "smart packaging" [17]. Incorporation of organic acids in films is reported to protect beef by inhibiting growth of microbes such as *L. monocytogenes*, *Salmonella typhimurium*, and *E. coli* [84]. In fact, certain biofilms have thermoresistant abilities, derived from bacteriocins of *Pediococcus* sp., along with chelating activity particularly against *L. monocytogenes* [85]. Recent studies have demonstrated the efficiency of alginate films in holding bacteriocins [86].

Alginate, polyvinyl alcohol, and zein have been found to be enterocins carrier polymers, thanks to their excellent film-forming property; besides the ability of effective entrapment and delivery of antimicrobials such as lysozyme and nisin in food. Preserving the food item with these combined technologies will result in better antimicrobial effects, which helps in improving the shelf life of the food product [87–89]. In another study, Mecitoglu *et al.*, [90] examined the effect of incorporation of lactoperoxidase (LPS), separated from bovine, into alginate films. The findings revealed immobilization of the enzyme resulting from the treatment, emphasizing the role of antimicrobial components of thiocyanate present in the films. Some authors even showed that antimicrobial films facilitate constant migration of antimicrobial agents from film to food, rendering safety during shipment and storage [91]. For safe production of such biofilms, plasma treatment or electron irradiation has been practiced.

In a similar endeavor, researchers have developed polyelectrolyte structured packaging with antimicrobial properties [19]. Prepared with cationic starch, sodium alginate, and starch, the film promises high antimicrobial potency, as reported. These films showed good thermal stability and increased glass transmission temperature with the increase in sodium alginate content. The experiment proved that polyelectrolyte sodium alginate packaging has good antimicrobial property and can be used as suitable food packaging material.

11.7 Effect of Alginate on the Quality of Food

Alginate employed in the design of food packaging materials has a considerable effect on the food also. According to Cha and Chinnan [10], alginate coatings prepared with glucose oxidase extend the shelf life of raw fish, including the winter flounder fish [92]. Other advantages include delayed purification and consequent prolonged sensory acceptance. In another study, maintenance of color and gloss was observed in alginate-coated cheese [93]. Additionally, alginate packaging reduces the loss of moisture and resists microbial spoilage in food materials, and this packaging could be made as an edible biofilm. Unlike conventional polymers, alginate packaging shows no negative impact on the quality of foods.

There are several other findings as well, reflecting beneficial effects of alginate in food packaging. Recently, Jost *et al.*, [17] described the use of calcium alginate as edible coating for the improvement of quality of pork patties. Alginate can also be used to enhance the color and flavor properties of frozen shrimps [94]. In another study on meat, Williams *et al.*, [31] reported a visible color difference between the alginate-coated meat and uncoated meat, post 144 h from treatment. As described, calcium alginate helped in maintaining the oxymyoglobin for a long time period in meat, which is responsible for meat's red color. Uncoated meat showed no color retention. These authors further stated that the coating decreased the percentage of meat shrinkage, as well as development of off-flavors in meat with storage.

In another work, Bourtoom [95] explains that the functional property of the film can be improved by combining different proteins, lipids, and some synthetic polymers along with the edible film. This synergistic effect of the film gives compatibility and enhances its mechanical properties, which will further improve the shelf life of the food product. Skandamis *et al.*, [96] have verified that infusion of edible oil into meat products extends the shelf life by delaying its lipid peroxidation. Mixing of alginate with essential oils of cinnamon, lemongrass, and palmarosa reportedly rendered extended storability of melons [97]. The benefits of alginate coating, as evidenced, are many. In fact, few alginate compounds are approved to be used as food additives (E 400). Adding plasticizers can improve the flexibility of the biopolymer by decreasing the melting temperature (T_m), melt viscosity, Young's modulus (YM), and glass transition temperature (T_g).

11.8 Interaction between Food and Alginates

The interaction that occurs between the coating and the food matrix is equally important as any other observable effect [97]. Oxidation in the food is one major adverse interaction that happens between alginate packaging and the contained food. Generally, oxidation in the packaged food leads to rancidity of fats therein leading to development of off-flavors. Even residual oxygen can react biochemically with the packed food and cause adverse oxidative effects that further alter the internal temperature of the food product. Sometimes alginate coating reacts with the pigments of the meat product and causes browning and darkening of fresh meats. Under certain circumstances, alginate packaging can affect the respiration of the packed food material also leading to proteolysis and lipolysis, for example, in fresh fruit, vegetables, and meats. Flavor deterioration and discoloration of food due to pigment interaction are some of the most common difficulties in alginate coating [98].

11.9 Environmental Effects on Alginate Packaging

The stability of alginate packaging strongly depends upon the environmental conditions to which the packages are subjected, i.e., pH, temperature, external pressure or humidity, and similar factors that are in its surroundings [99]. When there is an extreme change in the pH of alginate films, there is a reduction in the degree of polymerization.

Thermal stability of the alginate packaging is another important attribute that needs to be investigated while considering it as a packaging material for foods. This can be determined by laboratory techniques. Thermograms of alginate films show that the films will undergo thermal degradation at some point when temperature increases. These films start to lose their native property when temperature crosses 100°C, with major degradation occurring between 100°C and 200°C. During degradation, there is a cross-linking in the polymer linkages [19].

Surface wettability is another major disadvantage regarding biopolymer packaging. Surface wettability of alginate packaging can be determined by measuring the contact angle between the water drop and the surface film. This is to estimate the surface hydrophobicity of the films. The disadvantage from wettability commonly occurs in starch-based films, since they are hydrophilic in nature [19]. Sodium alginate is known to have higher hydrophobicity than starch, and when wet, alginate's surface free energy also increases. As the free energy on the surface increases, the contact angle also increases owing to modified hydrophobic properties [100]. So this explains the concern that increasing alginate content in starch-based films would render concerns about surface wettability of the package. Generally, with increasing alginate content in biofilms, intercellular gel matrix also increases, providing mechanical strength to the packaging. Alginate has been reported to be the material of choice for packing foods that are easily affected by gas penetration [17]. Cross-linking in alginate films is also important since films lacking the same dissolve and disintegrate in water. As a counter measure, water solubility of alginate films can be decreased by cross-linking with multivalent ions [36].

11.10 Market Outlook

Public acceptability to the use of biopolymers in food should also be considered. Educating the masses on beneficial effects and food safety aspects of this coating is the first step. General inclination is towards petroleum-based packaging materials. This is because petroleum-derived products show higher performance than renewable/recyclable-based products. The production cost is also lower in the former case. However, the petro-based packaging unfortunately renders more nuisance than use. Commercializing eco-friendly packaging material that is made up of biodegradable elements has a huge positive impact on countries that depend upon landfill as their major means of waste management. However, considering the lesser environmental impact, development of renewable material-based packages should be encouraged and popularized among people. Over the last few years, there have been many changes and development in the field of food packaging. For affordability, alginate, a commonly expensive biopolymer, could be mixed with other biopolymers; while nanocomposites of biopolymers are under development [100].

Although there are a good number of studies on the development of alginate packaging for foods, there is still a hesitation among the consumer, and awareness in the public is much needed regarding the benefits of biopolymers over plastics. Alginate packaging enhances the utilization of biodegradable package and provides food safety. Moreover, for some food products, alginate packaging helps to retain the preferred quality of the food product.

11.11 Conclusion

Biopolymers are eco-friendly, biodegradable, and less hazardous; no harmful ingredients are added during their preparation, and they safeguard food safety and quality. Numerous developments have been made recently in the fields of biopolymer packaging, manufacturing techniques such as gelation formation, film development, and coating on the food material. Alginate packaging has emerged to be the most promising. The benefits of alginate as packaging material for foods are many, with few challenges compared with petrochemical-based polymers, due to tensility, water permeability, heat, environment abrasion, and a few other external factors. However, research shows the gains outweigh the concerns. Despite the advantages, more studies are needed in biopolymers area for development of packaging material, with incorporation of intelligent packaging systems. Nanotechnology could be employed by applying nanoparticles into the packaging materials, averting microbial spoilage and oxidative deterioration.

References

1. Nisperos-Carriedo, M.O., Edible coatings and films based on polysaccharides, in: *Edible Coatings and Films to Improve Food Quality*, vol. 1, pp. 322–323, Technomic Publ. Co, 1994.
2. Gennadios, A., Weller, C.L., Gooding, C.H., Measurement errors in water vapor permeability of highly permeable, hydrophilic edible films. *J. Food Eng.*, 21, 4, 395–409, 1994.
3. Arvanitoyannis, I.S., *Irradiation of food commodities: Techniques, applications, detection, legislation, safety and consumer opinion*, Academic Press, 2010.
4. Debeaufort, F., Quezada-Gallo, J.A., Voilley, A., Edible films and coatings: Tomorrow's packagings: A review. *Crit. Rev. Food Sci. Nutr.*, 38, 4, 299–313, 1998.
5. Rhim, J.W. and Wang, L.F., Mechanical and water barrier properties of agar/kappa carrageenan/konjac glucomannan ternary blend biohydrogel films. *Carbohydr. Polym.*, 96, 1, 71–81, 2013.
6. Hoffman, A.F., Hydrogels for biomedical applications. *Adv. Drug Deliv. Rev.*, 54, 1, 3–12, 2002.
7. Farris, S., Schaich, K.M., Liu, L., Piergiovanni, L., Yam, K.L., Development of polyion-complex hydrogels as an alternative approach for the production of bio-based polymers for food packaging applications: A review. *Trends Food Sci. Technol.*, 20, 8, 316–332, 2009.
8. De Vlieger, J.J., Green plastics for food packaging, in: *Novel Food Packaging Techniques*, CRC Press, Washington DC, USA, 2003.
9. Xiao, Q., Gu, X., Tan, S., Drying process of sodium alginate films studied by two dimensional correlation ATR-FTIR spectroscopy. *Food Chem.*, 164, 179–184, 2014.
10. Cha, D.S. and Chinnan, M.S., Biopolymer-based antimicrobial packaging: A review. *Crit. Rev. Food Sci. Nutr.*, 44, 4, 223–237, 2004.
11. Amaral, R.D.A., Achaerio, I., Benedetti, B.C., Pujolà, M., The influence of edible coatings, blanching ultrasound treatments on quality attributes shelf-life of vacuum packaged potato strips. *LWT-Food Sci. Technol.*, 85, 449–455, 2017.
12. Rojas-Graü, M.A., Raybaudi-Massilia, R.M., Soliva-Fortuny, R.C., Avena-Bustillos, R.J., McHugh, T.H., Martín-Belloso, O., Apple puree-alginate edible coating as carrier of antimicrobial agents to prolong shelf-life of fresh-cut apples. *Postharvest Biol. Technol.*, 45, 2, 254–264, 2007.
13. Chiumarelli, M., Ferrari, C.C., Sarantópoulos, C.I., Hubinger, M.D., Fresh cut 'Tommy Atkins' mango pre-treated with citric acid coated with cassava (*Manihot esculenta* Crantz) starch or sodium alginate. *Innov. Food Sci. Emerg. Technol.*, 12, 3, 381–387, 2011.
14. Jiang, T., Effect of alginate coating on physicochemical and sensory qualities of button mushrooms (*Agaricus bisporus*) under a high oxygen modified atmosphere. *Postharvest Biol. Technol.*, 76, 91–97, 2013.

15. Johansson, C., Bras, J., Mondragon, I., Nechita, P., Plackett, D., Simon, P., Aucejo, S., Renewable fibers and bio-based materials for packaging applications—A review of recent developments. *BioResour.*, 7, 2, 2506–2552, 2012.
16. Vu, C.H.T. and Won, K., Novel water-resistant UV-activated oxygen indicator for intelligent food packaging. *Food Chem.*, 140, 1–2, 52–56, 2013.
17. Jost, V., Kobsik, K., Schmid, M., Noller, K., Influence of plasticiser on the barrier, mechanical and grease resistance properties of alginate cast films. *Carbohydr. Polym.*, 110, 309–319, 2014.
18. Espitia, P.J.P., Du, W., Avena-Bustillos, R. de J., Soares, N. de F.F., McHugh, T.H., Edible films from pectin: Physical–mechanical and antimicrobial properties—A review. *Food Hydrocoll.*, 35, 287–296, 2014.
19. Şen, F., Uzunsoy, İ., Baştürk, E., Kahraman, M.V., Antimicrobial agent-free hybrid cationic starch/sodium alginate polyelectrolyte films for food packaging materials. *Carbohydr. Polym.*, 170, 264–270, 2017.
20. Gacesa, P. and Russell, N.J., The structure properties of alginate, in: *Pseudomonas Infection Alginates*, pp. 29–49, Springer, Dordrecht, 1990.
21. Abdollahi, M., Alboofetileh, M., Rezaei, M., Behrooz, R., Comparing physico-mechanical thermal properties of alginate nanocomposite films reinforced with organic or inorganic nanofillers. *Food Hydrocoll.*, 32, 2, 416–424, 2013.
22. Alboofetileh, M., Rezaei, M., Hosseini, H., Abdollahi, M., Effect of montmorillonite clay biopolymer concentration on the physical mechanical properties of alginate nanocomposite films. *J. Food Eng.*, 117, 1, 26–33, 2013.
23. Gu, C.H., Wang, J.J., Yu, Y., Sun, H., Shuai, N., Wei, B., Biodegradable multilayer barrier films based on alginate/polyethyleneimine and biaxially oriented poly (lactic acid). *Carbohydr. Polym.*, 92, 2, 1579–1585, 2013.
24. Bierhalz, A.C.K., Da Silva, M.A., Kieckbusch, T.G., Natamycin release from alginate/pectin films for food packaging applications. *J. Food Eng.*, 110, 1, 18–25, 2012.
25. Fayaz, A.M., Balaji, K., Girilal, M., Kalaichelvan, P., Tvenkatesan, R., Mycobased synthesis of silver nanoparticles their incorporation into sodium alginate films for vegetable fruit preservation. *J. Agric. Food Chem.*, 57, 6246–625, 2009.
26. Maizura, M., Fazilah, A., Norziah, M.H., Karim, A.A., Antibacterial activity and mechanical properties of partially hydrolyzed sago starch–alginate edible film containing lemongrass oil. *J. Food Sci.*, 72, 6, 2007.
27. Rhim, J.W., Physical and mechanical properties of water resistant sodium alginate films. *LWT-Food Sci. Technol.*, 37, 3, 323–330, 2004.
28. Cha, D.S., Choi, J.H., Chinnan, M.S., Park, H.J., Antimicrobial films based on Na-alginate κ-carrageenan. *LWT-Food Sci. Technol.*, 35, 8, 715–719, 2002.
29. Costa, C., Conte, A., Buonocore, G.G., Lavorgna, M., Del Nobile, M.A., Calcium-alginate coating loaded with silver-montmorillonite nanoparticles to prolong the shelf-life of fresh-cut carrots. *Food Res. Int.*, 48, 1, 164–169, 2012.

30. Cutter, C.N. and Siragusa, G.R., Reduction of *Brochothrix thermosphacta* on beef surfaces following immobilization of nisin in calcium alginate gels. *Lett. Appl. Microbiol.*, 23, 1, 9–12, 1996.
31. Williams, S.K., Oblinger, J.L., West, R.L., Evaluation of a calcium alginate film for use on beef cuts. *J. Food Sci.*, 43, 2, 292–296, 1978.
32. Marcos, B., Aymerich, T., Monfort, J.M., Garriga, M., Use of antimicrobial biodegradable packaging to control *Listeria monocytogenes* during storage of cooked ham. *Int. J. Food Microbiol.*, 120, 1–2, 152–158, 2007.
33. Zactiti, E.M. and Kieckbusch, T.G., Potassium sorbate permeability in biodegradable alginate films: Effect of the antimicrobial agent concentration and cross-linking degree. *J. Food Eng.*, 77, 3, 462–467, 2006.
34. Liu, L., Kerry, J.F., Kerry, J.P., Application and assessment of extruded edible casings manufactured from pectin and gelatin/sodium alginate blends for use with breakfast pork sausage. *Meat Sci.*, 75, 2, 196–202, 2007.
35. Rhim, J.W., Wu, Y., Weller, C.L., Schnepf, M., Physical characteristics of a composite film of soy protein isolate propylene glycol alginate. *J. Food Sci.*, 64, 1, 149–152, 1999.
36. Sirviö, J.A., Kolehmainen, A., Liimatainen, H., Niinimäki, J., Hormi, O.E., Biocomposite cellulose-alginate films: Promising packaging materials. *Food Chem.*, 151, 343–351, 2014.
37. Stokke, B.T., Draget, K.I., Smidsrod, O., Yuguchi, Y., Urakawa, H., Kajiwara, K., Small-angle X-ray scattering and rheological characterization of alginate gels. 1. Calcium alginate gels. *Macromolecules*, 33, 5, 1853–1863, 2000.
38. Donati, I. and Paoletti, S., Material properties of alginates, in: *Alginates: Biology and Applications*, pp. 1–53, Springer, Berlin Heidelberg, 2009.
39. Andresen, I.-L., Skipnes, O., Smidsrød, O., Østgaard, K., Hemmer, P.C., Some biological functions of matrix components in benthic algae in relation to their chemistry and the composition of seawater. *ACS Symp. Ser.*, 48, 361–381, 1977.
40. Indergaard, M. and Skjåk-Bræk, G., Characteristics of alginate from *Laminaria digitata* cultivated in a high-phosphate environment. *Hydrobiologia*, 151, 1, 541–549, 1987.
41. Cottrell, I. and Kovacs, P., Alginates, in: *Handbook of Water-Soluble Gums and Resins*, p. 143, McGraw-Hill, New York, NY, 1980.
42. Littlecott, G.W., Food gels—The role of alginates. *Food technology in Australia*, 1982.
43. Sime, W.J., Alginates, in: *Food Gels*, P. Harris (Ed.), pp. 53–78, Springer, Dordrecht, 1990.
44. Clementi, F., Moresi, M., Parente, E., Alginate from *Azotobacter vinelandii*, in: *Carbohydrate Biotechnology Protocols*, pp. 23–42, Humana Press, Totowa, New Jersey, USA, 1999.
45. Banks, W. and Greenwood, C.T., *Starch and its components*, Edinburgh University Press, Edinburg, 1975.

46. Albert, A., Salvador, A., Fiszman, S.M.A., Film of alginate plus salt as an edible susceptor in microwaveable food. *Food Hydrocoll.*, 27, 421–426, 2012.
47. Gorin, P.A.J. and Spencer, J.F.T., Exocellular alginic acid from *Azotobacter vinelandii. Can. J. Res.*, 44, 9, 993–998, 1966.
48. Govan, J.R.W., Fyfe, J.A.M., Jarman, T.R., Isolation of alginate-producing mutants of *Pseudomonas fluorescens, Pseudomonas putida* and *Pseudomonas mendocina. Microbiology*, 125, 1, 217–220, 1981.
49. Linker, A. and Jones, R.S., A new polysaccharide resembling alginic acid isolated from pseudomonads. *J. Biol. Chem.*, 241, 16, 3845–3851, 1966.
50. Pindar, D.F. and Bucke, C., The biosynthesis of alginic acid by *Azotobacter vinelandii. Biochem. J.*, 152, 3, 617–622, 1975.
51. Remminghorst, U. and Rehm, B.H.A., *In vitro* alginate polymerization the functional role of Alg8 in alginate production by *Pseudomonas aeruginosa. Appl. Environ. Microbiol.*, 72, 298–305, 2006.
52. Szekalska, M., Puciłowska, A., Szymańska, E., Ciosek, P., Winnicka, K., Alginate: Current use future perspectives in pharmaceutical biomedical applications. *Int. J. Polym. Sci.*, 53, 11, 106, 2016.
53. Moe, S.T., Draget, K.I., Skjäk-Brñk, G., Smidsrùd, O., Alginates, in: *Food Polysaccharides Applications*, A.M. Stephen (Ed.), pp. 245–286, Marcel Dekker, New York, 1995.
54. Salzig, M. and Rehm, B.H.A., Microbial Production of Alginates: Self Assembly and Applications, in: *Microbial Bionanotechnology: Biological Self-Assembly Systems and Biopolymer-Based Nanostructures*, B.H.A. Rehm (Ed.), pp. 125–152, Horizon Scientific Press, Norfolk, United Kingdom, 2006.
55. Hay, I.D., Ur Rehman, Z., Ghafoor, A., Rehm, B.H., Bacterial biosynthesis of alginates. *J. Chem. Technol. Biotechnol.*, 85, 6, 752–759, 2010.
56. Moreno, S., Nájera, R., Guzmán, J., Soberón-Chávez, G., Espín, G., Role of alternative ς factor AlgU in encystment of *Azotobacter vinelandii. J. Bacterial.*, 180, 10, 2766–2769, 1998.
57. Wang, L.F. and Rhim, J.W., Preparation application of agar/alginate/collagen ternary blend functional food packaging films. *Int. J. Biol. Macromol.*, 80, 460–468, 2015.
58. Draget, K.I., Østgaard, K., Smidsrød, O., Homogeneous alginate gels: A technical approach. *Carbohydr. Polym.*, 14, 2, 159–178, 1990.
59. Smidsrød, O., Molecular basis for some physical properties of alginates in the gel state. *Faraday Discuss. Chem. Soc.*, 57, 263–274, 1974.
60. Skjåk-Bræk, G., Grasdalen, H., Smidsrød, O., Inhomogeneous polysaccharide ionic gels. *Carbohydr. Polym.*, 10, 1, 31–54, 1989.
61. Rinaudo, M., Biomaterials based on a natural polysaccharide: Alginate. *J. Biomater. Appl.*, 17, 1, 92–96, 2014.
62. Tavassoli-Kafrani, E., Shekarchizadeh, H., Masoudpour-Behabadi, M., Development of edible films and coatings from alginates and carrageenans. *Carbohydr. Polym.*, 137, 360–374, 2016.

63. Donnan, F.G. and Rose, R.C., Osmotic pressure, molecular weight and viscosity of sodium alginate. *Can. J. Res.*, 28, 3, 105–113, 1950.
64. Wu, Y., Weller, C.L., Hamouz, F., Cuppett, S., Schnepf, M., Moisture loss and lipid oxidation for precooked ground-beef patties packaged in edible starch-alginate-based composite films. *J. Food Sci.*, 66, 3, 486–493, 2001.
65. Barnes, H.A., Hutton, J.F., Walters, K., *An Introduction to Rheology*, vol. 3, Elsevier Science Publishers, Amsterdam, Netherland, 1989.
66. Rehm, B.H.A. and Valla, S., Bacterial alginates: Biosynthesis applications. *Appl. Microbiol. Biotechnol.*, 48, 3, 281–288, 1997.
67. Russo, R., Malinconico, M., Santagata, G., Effect of cross-linking with calcium ions on the physical properties of alginate films. *Biomacromolecules*, 8, 10, 3193–3197, 2007.
68. Yang, H., Guo, X., Chen, X., Shu, Z., Preparation and characterization of collagen food packaging film. *J. Chem. Pharm. Res.*, 6, 6, 740–745, 2014.
69. Smidsrod, O. and Draget, K.I., Alginate gelation technologies. Food colloids: Proteins, lipids and polysaccharides. E. Dickinson, and B. Bergenstahl. *RSC Adv*, 1, 279–293, 1997.
70. Roy, I. and Gupta, M.N., Hydrolysis of starch by a mixture of glucoamylase and pullulanase entrapped individually in calcium alginate beads. *Enzyme Microb. Technol.*, 34, 1, 26–32, 2004.
71. Fang, Y., Al-Assaf, S., Phillips, G.O., Nishinari, K., Funami, T., Williams, P.A., Binding behavior of calcium to polyuronates: Comparison of pectin with alginate. *Carbohydr. Polym.*, 72, 2, 334–341, 2008.
72. Da Silva, M.A., Bierhalz, A.C.K., Kieckbusch, T.G., Alginate pectin composite films cross-linked with Ca^{2+} ions: Effect of the plasticizer concentration. *Carbohydr. Polym.*, 77, 4, 736–742, 2009.
73. Sriamornsak, P. and Kennedy, R.A., Swelling and diffusion studies of calcium polysaccharide gels intended for film coating. *Int. J. Pharm.*, 358, 1–2, 205–213, 2008.
74. Aider, M., Chitosan application for active bio-based films production potential in the food industry: Review. *LWT-Food Sci. Technol.*, 43, 6, 837–842, 2010.
75. Rojas-Graü, M.A., Raybaudi-Massilia, R.M., Soliva-Fortuny, R.C., Avena-Bustillos, R.J., McHugh, T.H., Martín-Belloso, O., Apple puree-alginate edible coating as carrier of antimicrobial agents to prolong shelf-life of fresh-cut apples. *Postharvest Biol. Technol.*, 45, 2, 254–264, 2007.
76. Montero Calderón, M., *Internal quality profile and influence of packaging conditions on fresh-cut pineapple*, Universitat de Lleida, Escola Tècnica Superior d'Enginyeria Agrària, Lleida, Spain, 2010.
77. McHugh, D.J.A., *Guide to the seaweed industry*, FAO Fisheries the Technical Paper 441, Rome, Italy, 2003.
78. Huq, T., Salmieri, S., Khan, A., Khan, R.A., Le Tien, C., Riedl, B., Lacroix, M., Nanocrystalline cellulose (NCC) reinforced alginate based biodegradable nanocomposite film. *Carbohydr. Polym.*, 90, 4, 1757–1763, 2012.

79. Appendini, P. and Hotchkiss, J.H., Review of antimicrobial food packaging. *Innov. Food Sci. Emerg. Technol.*, 3, 2, 113–126, 2002.
80. Valencia-Chamorro, S.A., Palou, L., del Río, M.A., Pérez-Gago, M.B., Antimicrobial edible films and coatings for fresh and minimally processed fruits and vegetables: A review. *Crit. Rev. Food Sci. Nutr.*, 51, 9, 872–900, 2011.
81. Varela, P. and Fiszman, S.M., Hydrocolloids in fried foods. A review. *Food Hydrocoll.*, 25, 8, 1801–1812, 2011.
82. Wanstdts, K.G., Seideman, S.C., Donnely, L.S., Quenzer, N.M., Sensory attributes of precooked, calcium alginate-coated pork patties. *J. Food Sci.*, 44, 732–735, 1981.
83. Cuq, B., Gontard, N., Guilbert, S., Edible films and coatings as active layers, in: *Active Food Packaging*, pp. 111–142, Springer, Boston, MA, 1995.
84. Kerry, J.P., O'grady, M.N., Hogan, S.A., Past, current potential utilisation of active intelligent packaging systems for meat muscle-based products: A review. *Meat Sci.*, 74, 1, 113–130, 2006.
85. Katz, F., Smart packaging adds a dimension to safety. *Food Technol.*, 53, 106, 1999.
86. Millette, M.C.L.T., Le Tien, C., Smoragiewicz, W., Lacroix, M., Inhibition of *Staphylococcus aureus* on beef by nisin-containing modified alginate films beads. *Food Control*, 18, 7, 878–884, 2007.
87. Natrajan, N. and Sheldon, B.W., Efficacy of nisin-coated polymer films to inactivate *Salmonella typhimurium* on fresh broiler skin. *J. Food Prot.*, 63, 9, 1189–1196, 2000.
88. Leistner, L., Basic aspects of food preservation by hurdle technology. *Int. J. Food Microbiol.*, 55, 181–186, 2000.
89. Buonocore, G.G., Conte, A., Corbo, M.R., Sinigaglia, M., Del Nobile, M.A., Mono-multilayer active films containing lysozyme as antimicrobial agent. *Innov. Food Sci. Emerg. Technol.*, 6, 4, 459–464, 2005.
90. Mecitoglu, C., Yemencioglu, A., Arslanoglu, A., Elmac, Z.S., Koreal, F., Cetin, A.E., Incorporation of partially purified hen egg white lysozyme into zein films for antimicrobial food packaging. *Food Res. Int.*, 39, 12–21, 2006.
91. Quintavalla, S. and Vicini, L., Antimicrobial food packaging in meat industry. *Meat Sci.*, 62, 3, 373–380, 2002.
92. Field, C.E., Pivarnick, L.F., Barnett, S.M., R, A., Utilization of glucose oxidase for extending the shelf-life of fish. *J. Food Sci.*, 51, 66–70, 1986.
93. Kampf, N. and Nussinovitch, A., Hydrocolloid coating of cheeses. *Food Hydrocoll.*, 14, 6, 531–537, 2000.
94. Earle, R.D. and Snyder, C.E., Method of preparing frozen seafood. U.S. Patent 3,255,021, issued June 7, 1966.
95. Bourtoom, T., Edible films and coatings: Characteristics and properties. *IFRJ*, 15, 3, 237–248, 2008.
96. Skandamis, P.N. and Nychas, G.-J.E., Preservation of fresh meat with active and modified atmosphere packaging conditions. *Int. J. Food Microbiol.*, 79, 1–2, 35–45, 2002.

97. Rojas-Graü, M., Oms-Oliu, G., Soliva-Fortuny, R., Martín-Belloso, O., The use of packaging techniques to maintain freshness in fresh-cut fruits vegetables: A review. *Int. J. Food Sci. Technol.*, 44, 5, 875–889, 2009.
98. Skurtys, O., Acevedo, C., Pedreschi, F., Enronoe, J., Osorio, F., Aguiler, J.M., *Food hydrocolloid edible films and coatings*, Nova Science Publishers, Inc., US, 2010.
99. Brownlee, I.A., Seal, C.J., Wilcox, M., Dettmar, P.W., Pearson, J.P., Applications of alginates in food, in: *Alginates: Biology Applications*, pp. 211–228, Springer, Berlin Heidelberg, 2009.
100. Lange, J. and Wyser, Y., Recent innovations in barrier technologies for plastic packaging—A review. *Packag. Technol. Sci.*, 16, 4, 149–158, 2003.

12

Potential Application of Alginates in the Beverage Industry

S. Vijayalakshmi, S.K. Sivakamasundari, J.A. Moses and C. Anandharamakrishnan*

Computational Modeling and Nanoscale Processing Unit, Indian Institute of Food Processing Technology, Thanjavur, India

Abstract

Alginates are polysaccharides derived from brown seaweeds, namely *Laminaria* sp., *Macrocystis* sp., and *Ascophyllum* sp. They are used as stabilizers, thickeners, and emulsifiers in the beverage industry. Alginate-fortified beverages and fermented drinks have also been reported to have effects on gastric emptying and glycemic modulation. They can also be used to enhance bioavailability of beneficial compounds by acting as a potential carrier. Alginates can also be used in alcoholic beverages to induce and stabilize frothing. Calcium alginate microencapsulated yeast has been widely employed in wine industries to improve flavor and color, reduce browning, and also improve overall quality. Recently, alginates have been used to microencapsulate probiotic bacteria and have found applications in the production of fermented dairy products like yogurt with better viability during storage and under gastrointestinal conditions. This chapter summarizes the potential applications of alginates in the beverage industry focusing on recent advancements in research and commercialization.

Keywords: Alginates, fortified beverage, fermented drink, microencapsulated yeast, probiotic

12.1 Introduction

Hydrocolloids and gums are used in the food system to improve the texture, enhance stability, or improve its nutritional quality. In general,

**Corresponding author*: anandharamakrishnan@iicpt.edu.in

Shakeel Ahmed (ed.) Alginates, (233–262) © 2019 Scrivener Publishing LLC

hydrocolloids are used to form gel, and also as a thickening and stabilizing agent, film-forming agent, and foam stabilizer, to control syneresis, to stabilize emulsion or suspension, to prevent crystal growth, encapsulation of cells, and bioactive compounds, to improve bake and freeze–thaw stability, and to retain water [1, 2]. The main reason behind the wide application of hydrocolloids is their ability to modify the rheology of food systems. The addition of hydrocolloids in food systems influences its flow behavior (viscosity) and textural property [1]. The commonly used hydrocolloids are alginates, carrageenan, xanthan gum, gum arabic, carboxy methyl cellulose, hydroxypropyl methyl cellulose, pectin, gellan gum, and guar gum. Alginate is one of the prevalently used hydrocolloids in the food and beverage industry.

Alginate is basically extracted from the cell wall of brown algae. The cell wall of brown algae comprises of laminaran, sulfated hexouronoxylofucans, cellulose, fucodian and alginate. Alginate is the major composition of brown algae that contributes to about 45% of the dry weight of the seaweeds. Alginates are commercially available in acid and salt forms. The acidic form of alginates is known as alginic acid, and it is a linear polyuronic acid. The alginate salts are the major cell wall components of the brown algae that constitute up to 40–47% of its dry weight [3, 4]. The alginates are widely used in food industry for their ability to cross-link with ions (Ca^+), gel formation, water retention, ability to modify the viscosity, and also the ability to stabilize. The other advantage of using alginates is their ability to form heat-stable gels even at room temperature. Sources of alginates, their extraction, and their various applications in beverage industry are discussed in detail in this chapter.

12.2 Alginate Source

Alginates can be derived from the cell walls of brown algae/seaweed and from few soil bacteria. The alginate occurs in the form of capsular polysaccharide in *Pseudomonas aeruginosa* [5, 6]. Though alginates are also produced by bacterial species, the commercial production of alginates is from brown seaweeds. The commercial extraction of alginates is from brown seaweed species such as *Laminaria hyperborea, Laminaria digitata, Durvillea antarctica, Macrocystis pyriera, Ascophyllum nodosum, Lessonia nigrescens, Ecklonia maxima*, and *Sargassum* spp. [6–9]. The high quality of alginates can be obtained from cultivated brown seeds. Table 12.1 represents the various grades of alginates extracted from different brown seaweed sources.

Table 12.1 Grades of alginates extracted from various seaweeds in their country of harvest (modified from [13]).

Species	Grade	Prominent harvest location
Laminaria hyperborea	High	France, Ireland, UK, Norway
Lessonia negrescens	Medium-high	Chile, Peru
Laminaria japonica	Medium	China, Japan
Macrocystis	Low	USA, Mexico, Chile
Durvillaea antarctica	Low	Australia
Flavicans	High	Chile, Peru
Ecklonia maxima	Medium	South Africa
Ascophyllum nodosum	Low	France, Ireland, Iceland, Norway, UK

The chemical structure and the corresponding functional property of the alginates derived from brown algae vary with the species used for extraction. Extraction of alginates from *Macrocystis* leads to medium viscosity or highly viscous alginates based on the extraction procedure employed, whereas the alginates extracted from *Sargassum* yield low viscous alginates. Alginates with the ability to form strong gels can be derived from *L. hyperborea* and *Durvillaea*, and those that form gels of soft to medium strength are extracted from *L. digitata*. Hence, the producers of alginates use a mixture of seaweeds to produce alginates of desired good quality. The quality of alginates and their composition and structural sequence are based on the taxonomy, species, age and type of seaweed, season and environmental conditions [10–12].

12.3 Extraction of Alginates

The alginates are commercially available as alginic acid, salts of alginates (sodium, calcium, potassium, and ammonium), and propylene glycol alginate. The chemistry behind extraction process of alginates from cell wall is the conversion of insoluble alginate salts in seaweeds to soluble form of acid and salts in water and alcohol accordingly.

The extraction of seaweed starts with the shredding and milling of seaweeds with 0.1–0.2 M mineral acid to extract the salts of alginates present in the cell wall. The acidification of milled seaweeds also helps in removal of contaminant glycans like fucan and laminarian [9, 14, 15]. The alginates extracted will be in the insoluble alginic acid form. This alginic acid gets converted to sodium alginate (soluble form of alginate) by treating the extract with alcohol, followed by sodium carbonate or sodium hydroxide. The sodium alginate solution is further separated from the algal residues by employing extensive separation procedures such as sifting, floatation, centrifugation, and filtration. The extracted sodium alginate is recovered by three methods: directly precipitated using alcohol, by alginic acid, and calcium alginate method.

In the alginic acid method, the dilute sodium alginate solution is further treated with mineral acid to form alginic acid, which can be easily separated from water. Then the collected alginic acid can be further treated with alcohol and sodium carbonate to form sodium alginates. In the second process of recovery using calcium salt, the sodium alginate solution is treated with calcium carbonate to form insoluble calcium carbonate fibers. The suspended calcium alginate fibers are then treated with acid to form alginic acid. The formed alginic acid as described in the alginic acid recovery process is further converted to sodium form of alginate by treating with alcohol and sodium carbonate. The sodium alginate recovered from the alcohol is extruded into pellets, dried, and milled to powder form [9, 16].

The soluble salts of alginate, i.e., potassium and ammonium salts, can be produced in a similar manner as that of sodium salt extraction using potassium carbonate and ammonium hydroxide, respectively, during the alkali treatment step. The calcium alginate and alginic acid can be extracted from the appropriate stages of sodium alginate recovery process, and after which the insoluble calcium fibers and alginic acid are thoroughly washed, dried, and milled. Figure 12.1 represents the commercial extraction procedure of various forms of alginates.

Propylene glycol alginate is an ester of alginic acid and has different uses and properties form the sodium alginate. The production of propylene glycol alginate involves reaction of alginic acid (that is partially treated with sodium carbonate) with propylene oxide (Figure 12.2) in a pressurized vessel at 80°C for 2 hours [17].

12.4 Physical, Chemical and Functional Properties of Alginate

The physical and chemical properties of the extracted alginates depend on their chemical composition and structure. Alginates are basically unbranched binary copolymers comprising of (1-4)-linked β-D-mannuronic acid (M) and α-L-glucuronic acid (G) residues, wherein the composition and sequence vary with the source from which they are extracted [9]. The chemical and physical properties of alginates are dependent on the distribution

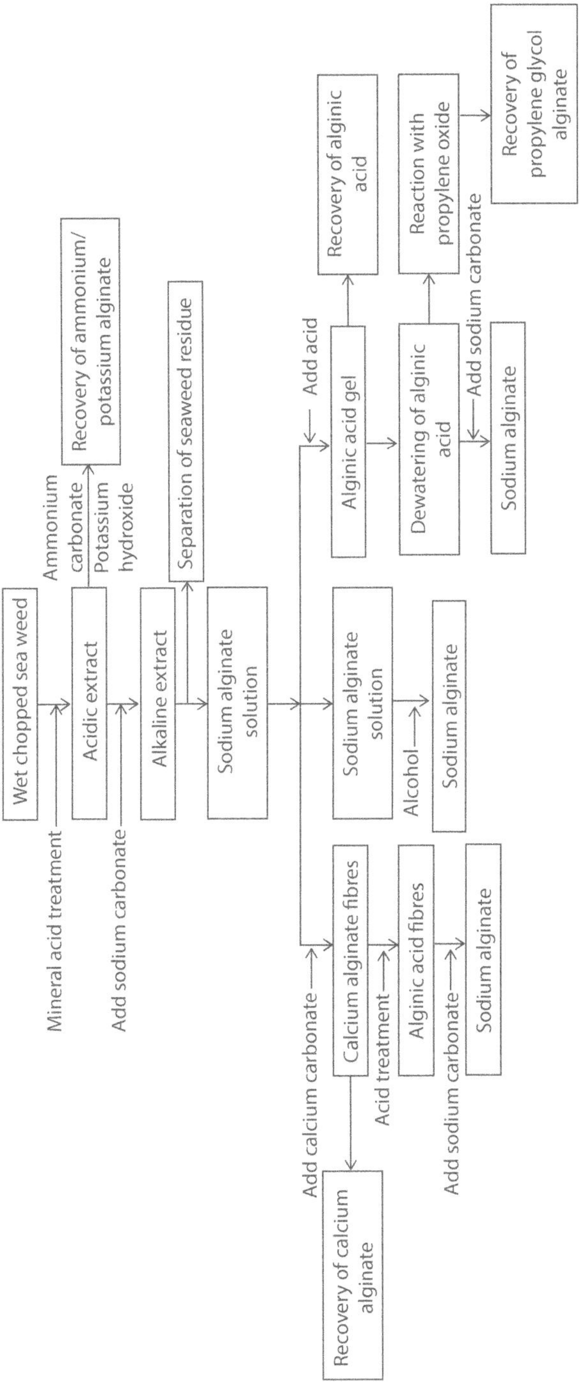

Figure 12.1 Schematic representation of commercial extraction of alginates.

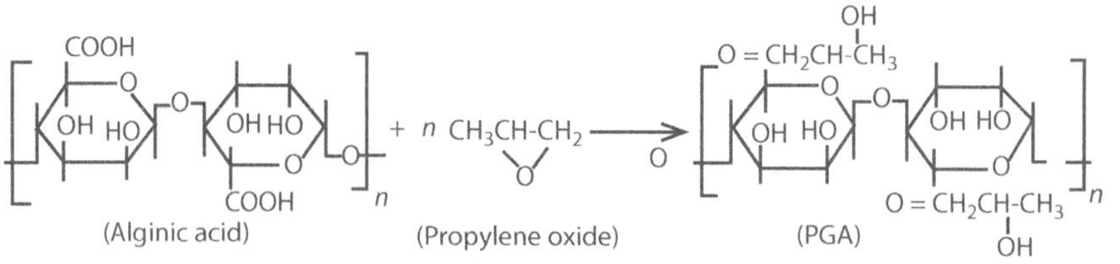

Figure 12.2 Production of propylene glycol alginate: Reaction between alginic acid and propylene oxide [18].

and proportion of β-D-mannuronic and α-L-guluronic acid present in them [19, 20]. Figure 12.3 indicates the structural configuration of alginates based on M and G. The average molecular weight of the alginate is in the range of 32 to 400 kg/mol irrespective of the G/M composition and its sequence in the alginates derived from various sources [5].

The solubility of alginate depends on the pH of the solvent (Table 12.2). The solubility also depends on the type of alginate used. For example, calcium and magnesium alginates are insoluble in water, whereas sodium, potassium, and ammonium alginates are soluble in water but insoluble in alcohol. This is because of the effect of electrostatic charges of the uronic acid residues and also the ionic strength of the solute. The hardness of water also influences the solubility of alginates.

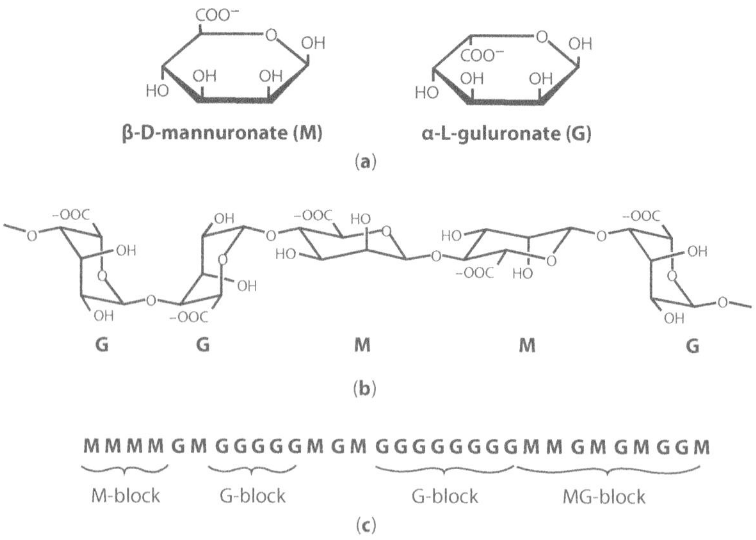

Figure 12.3 Structural characteristics of alginates: (a) monomers, (b) chain conformation, (c) block [18].

Table 12.2 Solubility of alginates in various pH solutions [18].

	Acidic conditions	Alkaline conditions	In solution with divalent cations
Type of alginate	***Fruit juice, liquor, salad dressing, etc.***	***Kansui, etc.***	***Hard water, milk, etc.***
Alginic acid	Insoluble	Soluble	Insoluble
Sodium alginate	Insoluble	Soluble	Insoluble
Potassium alginate	Insoluble	Soluble	Insoluble
Calcium alginate	Insoluble	Insoluble	Insoluble
Ammonium alginate	Insoluble	Soluble	Insoluble
PGA	Soluble	Soluble*	Soluble

*In neutral and alkaline conditions, PGA decomposes and forms an alginate salt.

The rheological property of alginates is based on its G/M ratio, molecular weight, concentration, and pH of the medium. The major rheological properties influenced by addition of alginates are viscosity and textural property. The thickening and gelling property of alginates influences the rheology of beverages. The gelling property of alginates is due to its affinity toward ions [21]. The affinity for multivalent cations also depends on the basic composition of alginates. The selectivity and affinity of alginates toward similar ions are determined by nonspecific electrostatic interaction and chelation of its G-block structure. The characteristic feature of formation of egg-box structure (Figure 12.4) with Ca^{2+} ion is based on the formation of linkages between ions and glucuronate residues [22, 23]. The swelling and complex formation of alginates in presence of suitable ions and solvents help in the impregnation of sensitive compounds for further application in food and beverage industry [24].

The uses of alginates in various food and beverage products depend on the characteristic nature of the alginates used, i.e., salts of alginates. Sodium alginate forms a viscous colloidal solution in water and is insoluble in alcohol, ether, and chloroform. This property of sodium alginates makes it suitable for application as a thickener, stabilizer, and emulsifier in various food products. Calcium alginates are insoluble in water and acid, but are soluble in alkaline solutions. Hence, calcium alginates can be used as a thickening and stabilizing agent in ice cream, cheese products, canned fruits, and sausage casings. Potassium alginate is widely used as a thickener and stabilizer in dairy products, canned fruits, and sausage casing, and it is also used as an emulsifier due to its viscous nature when dissolved in water. Potassium alginate is also insoluble in alcohol like sodium alginate.

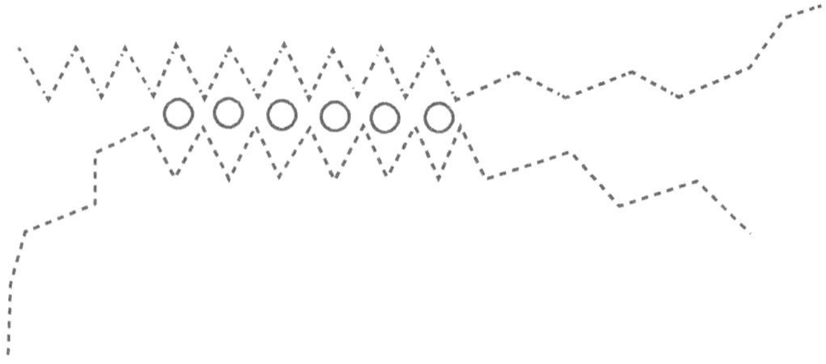

Figure 12.4 Egg box complex of Ca^{2+} ion and alginate (modified from [25]).

12.5 Uses as a Food Additive/Ingredient

In general, hydrocolloids are added to beverages to ensure suspension of particles, increase viscosity, improve fruit pulp stabilization, prevent oil separation, induce satiating mouth feel, and enable fiber addition. It is also used in fruit-fermented dairy products to control its viscosity, ensure excellent suspension of fruits, maintain fruit structure, increase gel network formation, and also stabilize the final product [2]. Alginates are widely used hydrocolloids derived from brown seaweed [5]. Various applications of alginates in food industry are as an emulsifier, stabilizer, thickener, flavor enhancer, firming agent, formulation and processing aid, and texture modifier [26]. The various food applications of alginates are achieved by addition at a level of 0.5–1.5% [27].

The application of alginates is mainly based on their three characteristic properties: to modify the viscosity (dissolve in water and thicken the solution), ability to form gels, and to form films (sodium or calcium alginate) and fibers (calcium alginate). The gelation and physical properties of gel are affected by factors like viscosity, molecular weight, structure (M/G ratio), sequence of M-G blocks, and concentration of alginate [28]. The ability of sodium, potassium, and ammonium salts of alginic acid to dissolve in cold water makes its application in food industry vital, as a thickener, gelling agent, stabilizer, film formation, and controlled release of bioactive compounds or food ingredients [8]. The various applications of alginates in food industries are represented in Table 12.3 and its permissible limit in various foods are listed in Table 12.4.

Alginates find major application in beverage industry as a thickener and stabilizer. In fruit and vegetable juices, alginates are used to prevent layering and sedimentation on processing. In products like Leben, they are used to prevent milk protein agglutination and sedimentation. Alginates are also used to prevent oil separation and also to avoid protein sedimentation. Similarly in vegetable protein beverages, alginates are added to prevent coagulation of milk protein and other solid particles and also to avoid flotation of fat. The edible coating by sodium alginate of fruit and vegetables helps in ensuring excellent suspension of particles in suspended beverages [29]. The sodium alginate acts as an effective clarifying in wine and also helps in the removal of unwanted color.

The use of alginates such as propylene glycol in yogurt improves its texture, body, and appearance as it leads to stabilization of milk proteins under characteristic acid condition of yogurt. Propylene glycol alginate and sodium alginate also find application in the suspension of fruit pulps

Table 12.3 Various applications of alginates and its purpose of use (Source: modified from [18]).

Application of alginate	Purpose
Premium beer foam stabilizer	PGA usage allows better head retention and protects against foam-negative contaminants.
Restructured foods	• Use in reformation of food materials (e.g., onion rings, pimento pieces in olives). • Endows food product with thermostability and desired consistency.
Further uses of PGA (propylene glycol alginates)	PGA is acid-stable and resists loss of viscosity. Has unique suspension and foaming properties. Wide range of applications includes • Soft drinks • Dressings/condiments • Milk drinks • Sorbet • Ice cream • Noodles/pasta
Bakery products	• Proves bakery cream with freeze/thaw stability and reduced synergies. • Improves shelf life and moisture retention in bread and cake mixes. • Allows cold solubility in instant flan preparations.
Fruit preserves	• Commonly used as gelling, thickening, and stabilizing agents in jams, marmalades, and fruit sauces. • Alginate-pectin gels are heat reversible and give higher gel strength than either individual component.
Ice cream	• Allows correct viscosity of ice creams, while avoiding crystallization and shrinkage, also secures heat shock resistance and allows homogeneous melting without whey separation. • Used in combination with other stabilizers for further effects (e.g., increased thickening and slow melting with guar/locust bean gum)
Other	• Desserts (e.g., mousses, instant puddings, ripple syrups) • Emulsions and sauces (e.g., low fat mayonnaise, tomato ketchup, salad dressings, low fat spreads), extruded foods (e.g., noodles and pasta)

Table 12.4 Permissible limit for alginates in different food products (Source: modified from [18]).

Food use category	Ammonium alginate	Calcium alginate	Potassium alginate	Sodium alginate	Propylene glycol alginate
Alcoholic beverages		0.4%			
Baked goods		0.002%			0.5%
Cheese					0.9%
Condiments and relishes				1.0%	0.6%
Confections and frostings	0.4%	0.4%	0.1%	0.3%	0.5%
Egg products		0.6%			
Fats and oils	0.5%	0.5%			1.1%
Frozen dairy desserts					0.5%
Gelatins and puddings		0.25%	0.7%	4.0%	0.6%
Gravies and sauces		0.4%	0.4%		0.5%
Hard candy				10.0%	
Jams and jellies	0.4%	0.5%			0.4%
Pimento ribbon for stuffed olives				6.0%	
Processed fruits and fruit juices			0.25%	2.0%	
Seasonings and flavors					1.7%
Sweet sauces	0.5%	0.5%			
All other food categories	0.1%	0.3%	0.01%	1.0%	0.3%

in fruit drinks, where they prevent the sedimentation of pulp. Alginates are also used in the production of chocolate milk to keep the cocoa suspended. They are also used to thicken and stabilize whipped cream.

Recently, the application of propylene glycol alginate in beverage industries is on the rise due to its readily water-soluble nature. Propylene glycol alginate is highly soluble in water and forms viscous liquid solutions in both hot and cold water, thereby making its application suitable as a thickener, stabilizer, and suspending agent. But it lacks the ability to form gels, insoluble fibers, and films. Propylene glycol alginate is also seen to be more stable in acidic solutions and calcium-rich foods due to the esterification of its reactive carboxylic acid groups [8]. Hence, it is more prevalently used at lower concentrations used in acidic beverage and food applications [5, 27]. Propylene glycol alginate cannot dissolve in milk at temperatures below boiling point. Based on the degree of esterification and stability at low pH, the wide applications of propylene glycol alginate include as a stabilizer to stabilize beer foam, meringues, noodles, oil–vinegar emulsion in salad dressings, fruit and vegetable drinks, flavor oils, and fermented milk drinks [8]. The use of propylene glycol alginate in beer and soft drink production has been essential for its unique suspension and foaming applications [27]. Beer is, in general, preferred to be consumed with thick stable foam and the ones with poor foam are considered to be of low quality. The addition of propylene glycol at low concentrations was sufficient to provide stable longer-lasting foam.

Other major food applications of alginates include the following: edible coating, encapsulating agent, and in the formation of restructured food. Alginate coating can be used to extend shelf life and reduce pathogenic surface microbes [30]. The alginate edible coating or film acts as a protective layer to food or active ingredient by retarding food spoilage and also by extending the effectiveness of active ingredient [31]. Alginates are used as carriers of bioactive or functional food ingredients such as antibrowning agents, flavors, colors, spices, nutrients, and antimicrobial agents [30]. The gelling properties of alginates are used to produce fruit substitutes with the characteristic flavor encapsulated in calcium alginate complex. This application of alginate meets a great demand in bakery industries. Edible coatings made up of alginates also help in the prevention of phase separation, loss of water, and syneresis [8]. Alginates are also used in preparation of restructured foods due to their ability to form instant gels, fibers, and films (skin) [8]. The thermostability of alginates also enables the microwaving of alginate-based products.

12.6 Alginate as Stabilizer

Sodium alginates used in various beverages as a stabilizer are reported to enhance their functional property. Pelkman *et al.*, [32] reported that a novel beverage made up of calcium gelled alginate pectin reduced the energy and food intake by stimulating endogenous satiety signaling in overweight and obese women. Sodium alginate fibers are reported to act as a dietary fiber that influences gastric emptying and nutrient absorption. Thereby, it is reported to influence the satiety in men and women subjects [32–35]. Paxman *et al.*, [36] substantiated the health benefits of addition of alginates. The authors reported that sodium alginate-based drink reduces the uptake of cholesterol and glucose in overweight males. El Khoury *et al.*, [37] reported that sodium alginate added to chocolate milk reduced appetite and glucose release, and it is also reported to influence the food intake and insulin production in healthy men. Dietary alginate consumed is also reported to beneficially influence the growth of *Bifidobacterium*, a probiotic microbe present in the intestine. Alginate is also reported to promote the production of organic acid and inhibit the harmful putrefactive activity of intestinal bacteria [38].

The apparent viscosity and sensory properties of beverages such as tomato juice and coffee are reported to be affected by the type of hydrocolloids used. Sodium alginate of varying concentrations is also used to study its influence on beverages. The increase in the concentration affected both the apparent viscosity and sensory properties of tomato juice and coffee. Hence, the addition of sodium alginate concentration influences the product quality [39].

The highly viscous and shear thinning or pseudoplastic flow property of the alginate solution makes it application in liquid food systems more feasible. The presence of calcium ion in beverages such as milk makes the application of sodium and potassium salts of alginates less preferable. In order to avoid sodium and potassium alginate interaction with calcium ions, sequestrants are used [40]. The increase in the concentration of alginates is also said to alter the viscosity of milk-based beverages. Sodium alginate is prevalently used in chocolate milk, eggnog, and drinkable and fruit-flavored yogurts as a clarifying agent and stabilizer, to produce a smoother and stable product. In wines, sodium alginate is used to clarify wine and to reduce tannins, clouding substances and nitrogenous compounds. A combination of alginate-phosphate is used in chocolate-milk drinks as a stabilizer.

The esterified nature of propylene glycol alginate makes it a good stabilizing agent and viscosity modifier in fruit beverages. The ester groups of propylene glycol alginate make the reactive sites unavailable for association with calcium ions, unlike that of the calcium ion sensitive sodium and potassium alginates. The stability of propylene glycol alginate in acidic solutions makes its application suitable as an emulsifier, stabilizer, and suspending agent in various beverages such as fermented dairy/milk, fruit and vegetable juices, and fruit–milk beverages. Propylene glycol alginate in combination with other hydrocolloids such as xanthan gum, carrageenan, and pectin shows a potential application in fruit juice beverages wherein particle suspension, fortification of calcium, and viscosity control are necessary.

The major problem faced by the fruit juice industry is the separation of fruit pulp into a clear upper layer and a thick pulpy bottom layer. The application of propylene glycol alginate is proved to be boon to juice and soft drink industries, wherein the fruit pulp used in the formulation is stabilized. A concentration of 0.1% of propylene glycol alginate is proved to act as an effective stabilizer without affecting its taste and texture. It is also reported that the addition of propylene glycol alginate improves the taste and texture of fruit juice as expected by the consumer [41].

The formulated acid milk beverage formulations involve milk (raw or powdered) or soybean milk, acid such as lactic or malic acid, sweeteners, stabilizers, and flavor and coloring agents. The protein content, in general, is greater than 1% in acid milk beverage. Hence, the problem of precipitation and layering of protein occurs during production and storage of formulated acid milk beverage. This mandates the need for appropriate stabilizers like propylene glycol alginates. The ability of propylene glycol to form stable solutions in acidic conditions makes its application in the formulation more preferable and feasible. A combination of stabilizers along with the propylene glycol alginate is also used by formulated acid milk beverage manufacturers, where 60–70% of the stabilizer used comprises of propylene glycol alginate. The use of stabilizers helps in preventing the layering and sedimentation of proteins through the period of shelf life [42].

Cheong *et al.*, [43] studied the effect of modified starch, propylene glycol alginate, sucrose laurate, and sucrose stearate on creaming index, droplet size, cloudiness, and conductivity of soursop beverage emulsion. The propylene glycol alginate used individually or in combination with modified starch influences the physical stability of soursop beverage emulsion.

Propylene glycol alginate possesses a greater foam stabilizing property due to the electrostatic interaction between the carboxyl group of alginate and protein in the bubble wall. This interaction is said to improve the foam stability in beverages such as beer [44]. Sahai and Sher [45], in their invention, report that addition of propylene glycol alginate (ester alginate) improved the foaming capacity and stability of foaming fruit juices and is also reported to influence the viscosity, clarity, and color of the juice. Ester alginate, when added at a level of about 0.05% to 1% in foaming juice compositions, is reported to achieve stable foams and improved product quality. The foam stabilizer available in the market mainly comprises of propylene glycol alginate and its combination with modified/resistant starch. The combination of propylene glycol alginate with modified starch or protein finds it application as a stabilizer in low malt/protein beverages.

12.7 As Encapsulating Wall Material

Alginate is generally preferred as an encapsulating wall material due to its water absorption ability [46]. The gelling capacity of alginates is based on their ability to form complexes with multivalent cations. Propylene glycol alginate lacks the gelling property due to the esterification of carboxyl groups of alginate. Sodium and potassium salts of alginic acid are, in general, said to form gels in the presence of calcium in food system. The gelling property of alginate is beneficially used to preserve bioactive compounds by encapsulating them. The desired compound or substances such as flavor, color, antimicrobial compounds, enzymes, and yeast cells are encapsulated with calcium alginate.

The cross-linking or gel formation of alginate is carried out by two methods: external ionic gelation through extrusion (Figure 12.5) and internal ionic gelation (Figure 12.6) [47]. In the diffusion setting method, the desired compound or ingredient in the alginate solution is dropped into the calcium chloride solution, wherein in immediate contact with calcium ions, the cross-linking of alginates starts and encapsulation of ingredients or compounds takes place [48, 49]. The gelled particles or capsules formed are 500 μm to 3 mm in diameter, and the size of calcium capsules obtained is based on the diameter of the needle used for extrusion, concentration, and viscosity of alginate solution and the distance between the syringe and calcium chloride solution [50]. The use of pressurized extrusion system and nebulizers can help in reducing the particle size to less than 300 μm and less than 1 μm, respectively [51]. The application of calcium capsules in food and beverage industries is highly desirable with a size of approximately 100 μm [52]. Hence, the microencapsulated compounds can be used to improve the characteristic quality and property of food and beverages. They can also be used to retain the flavor and color of beverages,

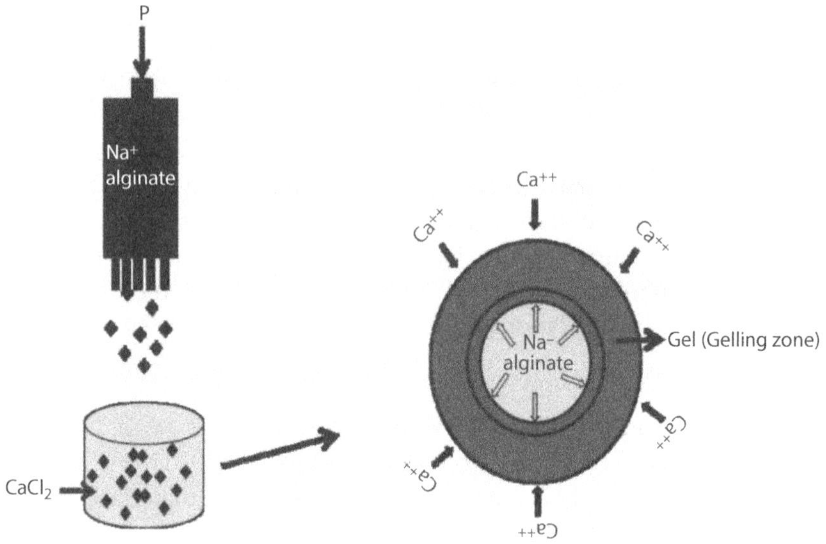

Figure 12.5 External ionic gelation through extrusion (Source: modified from [9]).

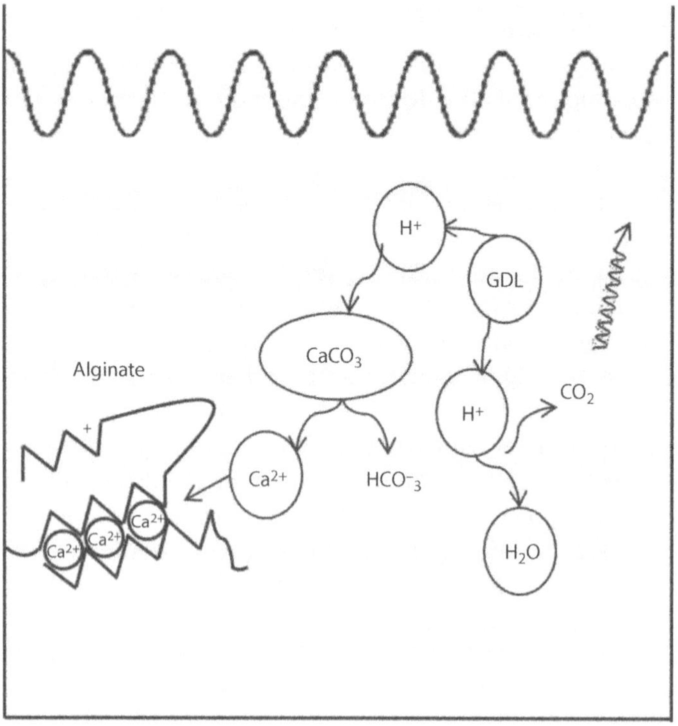

Figure 12.6 Internal setting method (Source: modified from [9]).

in controlled release of enzymes, and in retention of yeast cell activity in alcoholic beverages.

In the internal ionic gelation method, the calcium ion is released in a controlled fashion from an inert calcium source within the alginate solution–vegetable oil emulsion [52]. The release of calcium ions can be attained by changing the pH, using a limitedly soluble calcium salt source, or using chelating agents [50]. This technique is prevalently used in microencapsulation of probiotics [53, 54]. The microencapsulated probiotics can be incorporated in beverages such as milk-based beverages and yogurts. The size of the microcapsules produced by this method is in the range of 20–25 μm to 2 mm, and it depends on the homogeneity and agitation speed [53].

The production of microcapsules for immobilization of living cells and for inclusion of desired food compounds/ingredients without the use of any organic solvent is preferred and is of great interest and need [55, 56].

12.7.1 Immobilization of Biocatalysts

Immobilization of live cells or enzymes helps in the retention of their viability and metabolic activity. In general, sodium, calcium, and barium salts of alginic acid are used for cell entrapment and immobilization of enzymes for their application in beverage industries especially wine and beer production [57]. Fruit pieces and skins are also used as an immobilization medium for cells in wine making to improve its flavor and quality [58–61]. The yeast cells immobilized in alginate beads are used in alcoholic drinks for ethanol production and for second fermentation of champagne. According to Hill [62], the calcium alginate immobilized yeast is suitable for bottle fermentation of sparkling wines. Immobilization of yeast cells for use in wine production enhances the taste, aroma, and overall quality of produced wine [63].

The inventors Medina *et al.*, [64] developed a novel application of immobilized yeast cells to control the browning of drinks and to also prevent any changes caused by the color change. The yeast cells were immobilized using natural alginate or carrageenan. This invention was mainly focused on the application of immobilized yeast cells in food and beverage industries to control the browning reaction.

Immobilization of enzymes using alginate also finds a major application in food and beverage industry. The immobilized enzymes are used for the conversion of glucose to fructose, production of amino acids for its application in foods, for continuous production of yogurt, and for conversion of starch to ethanol in brewing of beer. The desired reaction such as sugar to ethanol or starch ethanol can be easily controlled by removal of the immobilized enzymes or cells as soon as the desired alcohol production is achieved. The use of immobilized cells is also said to ease the process of yeast cell separation [65].

12.7.2 Probiotics

Probiotics are beneficial live microorganisms that are consumed to improve the immunity and overall health of humans [66]. The commonly known probiotic foods such as fermented milks and yogurt, cheese, ice cream, and dairy desserts face major problem in the retention of viability of microbes. The factors that influence the viability of probiotic bacteria include titratable acidity, pH, oxygen concentration, storage temperature, lactic and acetic acid concentrations, and interaction with other microbes in the product [67].

Microencapsulation technique favors the viability of probiotic cultures by protecting them from environmental conditions and also helping in their controlled release. The encapsulation of probiotics in alginates is preferred due to the ease of production or handling, nontoxic nature, biocompatibility, increased viability, and low cost of production [68]. The microencapsulation with sodium alginate is also proved to form a versatile gel matrix, a biocompatible and nontoxic protection layer for probiotic microbes from pH, heat, oxygen, and others factors when the food is exposed to different processing and storage conditions [46]. Among all the microencapsulation techniques such as spray drying, spray congealing, and spray freezing, the ionic gelation method is preferred due to its simple and inexpensive procedure [69]. As discussed earlier, the microencapsulation of probiotics using alginate in the desired size for food application can be achieved by both external ionic and internal ionic gelation methods (Figure 12.7). But time, cost, and complexity involved in the internal ionic gelation method make it less preferable than the external ionic method.

The use of microencapsulated probiotic bacterial culture protects the microbes from detrimental conditions in fermented milk products [53, 70]. The process of microencapsulation of probiotics with alginates helps to improve its viability in fermented milks [71–73] and in gastrointestinal tract [74–78]. Alginate is also preferred as an encapsulating agent because of its biocompatibility, simplicity, availability, cheapness, and good intestinal digestibility [53, 70].

Chandramouli *et al.*, [79] reported that the calcium alginate encapsulated *Lactobacillus acidophilus* was viable in the acidic condition (pH 2.0). It is also reported that a maximum viability of 106 CFU/mL was achieved when the culture was encapsulated with 2% calcium alginate. In another study by Kim *et al.*, [80], *L. acidophilus* ATCC 43121 encapsulated with calcium

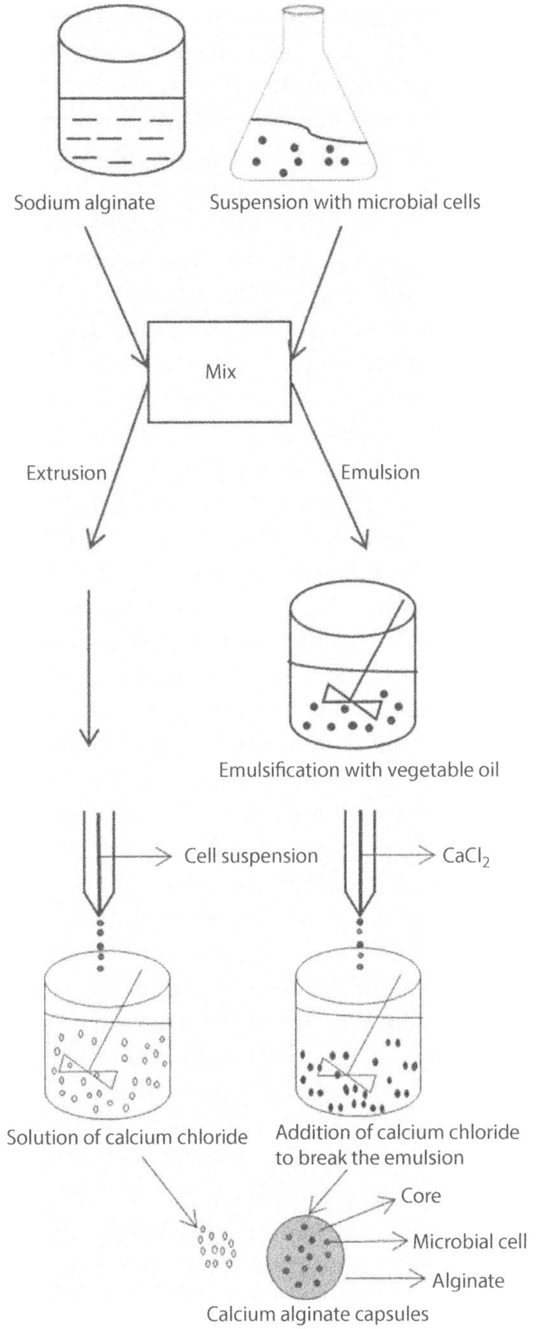

Figure 12.7 Microencapsulation of probiotic culture by external and internal ionic gelation technique (Source: modified from [25]).

alginate by extrusion method showed that the cells were viable when exposed to gastrointestinal tract condition and various thermal treatments.

Multilayer microencapsulation of probiotics is a recent advancement that is used to increase the viability of microbes. Electrostatic interactions between polymers are used in the preparation of multilayer encapsulations of polyelectrolytes [81, 82]. Chitosan, polyamino acid, and whey protein form a strong complex with alginates that are stable in the presence of chelating agents/sequestrants [47]. The association between materials like chitosan, polyamino acid, and whey protein with calcium alginate leads to a formation of stable double-walled microcapsules [83].

Lee *et al.*, [84] developed multilayered microparticles of *Lactobacillus bulgaricus* KFRI 673 using alginate and chitosan and studied their survival in gastric and intestinal juices and their stability under various storage conditions. The multilayer encapsulated culture was seen to have achieved maximum viability and stability in gastrointestinal juices and storage conditions. In a similar study, Gbassi *et al.*, [85] encapsulated *Lactobacillus plantarum* with whey protein and alginate and reported that maximum probiotic viability was achieved under stimulated gastric system test analysis. Similarly, Sathyabama and Vijayabharathi [86] studied the stability and viability of multilayer (chicórea and beet sugar + alginate) encapsulated *Enterococcus fecium* under stimulated gastrointestinal system and proved that the multilayer encapsulation was effective in retaining the vitality of the probiotic microbe. Martin *et al.*, [54] examined the stability and viability of single (sodium alginate) and multilayer (sodium alginate + starch) encapsulated *Lactobacillus fermentum* CECT5716 and reported that the log reduction was lower in multilayered capsules.

Krasaekoopt and Watcharapoka [56] encapsulated *L. acidophilus* and *Lactobacillus casei* with calcium alginate and chitosan, which are then incorporated in yogurt and fruit juices to study their stability and viability. It was reported that the encapsulated cells were viable and stable for a storage period of 4 weeks in both yogurt and fruit juices under refrigerated condition.

The incorporation of calcium alginate encapsulated probiotic cultures *L. acidophilus* LA-5 and *Bifidobacterium lactis* Bb-12 in Iranian yogurt drink (Doogh) is examined for their stability on storage (for 42 days at 4°C) under stimulated gastrointestinal conditions. The viability of encapsulated *L. acidophilus* and *B. lactis* was 4.0–5.5 log cycles higher than that in the control drink with unencapsulated microbes by the end of 42-day storage period. The survival rate of encapsulated *L. acidophilus* and *B. lactis* was 26.3% and 34.0% higher under the gastrointestinal conditions, respectively [87]. Table 12.5 represents various studies performed with multilayer encapsulation of probiotics.

Table 12.5 Multilayer encapsulation of various probiotic microbes with multilayer wall materials.

Microorganisms	**Special treatment**	**Reference**
Lactobacillus spp.	Alginate + chitosan and corn starch	[88]
Lactobacillus gasseri and *Bifidobacterium bifidum*	Alginate + chitosan	[83]
	Alginate + resistant starch	[89]
Lactobacillus casei and *Bifidobacterium lactis*	Alginate + resistant starch	[90]
BAL	Alginate + starch	[91]
Bifidobacterium animalis subsp. *lactis Bb12*	Alginate + chitosan-acryl-eze	[92]
Bifidobacterium	Alginate + poly-L-lysine	[93]
Lactobacillus casei	Alginate + starch	[94]
Lactobacillus acidophilus and *Lactobacillus casei*	Alginate + resistant starch	[95]

12.7.3 Improvement of the Alginate Encapsulation: Prebiotics Addition

The alginates used in beverages are established to act as dietary fibers that help in the growth of beneficial microorganisms in the intestine. This nature of alginate makes it an effective prebiotic substance. The use of alginates as a stabilizer, thickener, emulsifier, and coating material in beverages is also reported to contribute to many health benefits like regulating the glucose and insulin release, reducing food intake, and also enhancing the growth of intestinal beneficial microbes.

The combination of alginate with prebiotic compounds such as resistant starch, inulin, and fructo-oligosaccharides enhances the protection of probiotic microbes in the food systems due to its synergistic and symbiotic relationship [96, 97]. The three-dimensional microcrystal structure of prebiotics along with the complex alginate structure contributes to an effective protective layer for bacterial cells [98].

Homayouni *et al.*, [90] showed that alginate and starch combination improved the stability of probiotic microbial cultures due to the formation of prebiotic starch–alginate complex and effective microencapsulation. The addition of prebiotics ensures protection of the probiotic microbes and also aids in their growth in the stimulated digestive system.

In another study, Nazzaro *et al.*, [99] developed a functional fermented carrot juice with probiotic microbes (*L. rhamnosus* DSM20711 and *L. bulgaricus*

ATCC 11842) and prebiotic components (inulin and fructo-oligosaccharides). The cultures were reported to grow and survive in the carrot juice at acidic pH condition during a storage period of 4 weeks at 4°C. In another study by Nazzaro *et al.*, [100], carrot juice was incorporated with alginate-inulin-xanthan prebiotic gum encapsulated *L. acidophilus* DSM 20079, and its growth and stability during storage and under gastrointestinal conditions were examined. Encapsulation enhanced the cell viability and stability of *L. acidophilus* during storage and under gastrointestinal conditions.

The use of prebiotic alginate encapsulated probiotic microbe in dairy-based beverages not only protects and enhances the viability of probiotic microbes but also improves the overall functional property of the product.

12.8 Conclusion

Alginates are hydrocolloids derived from certain bacterial species and brown seaweeds/algae. Commercially available alginic acid and its salts are derived from brown seaweed sources such as *Laminaria*. Alginates, due to their physical and chemical properties, find various applications in food and beverage industries. Alginates are prevalently used as a stabilizer, thickener, suspending agent, emulsifier, encapsulating wall material, and a prebiotic source in various beverage products. Alginates find their application in beverage industries as an additive and ingredient. Salts of alginates such as sodium and potassium alginates find their application as a thickener and stabilizer in many non-acidic and calcium-free beverages. Calcium alginate is, in general, preferred as an encapsulating agent for protecting live microbes, flavor, color, and other bioactive compounds to enhance its viability and stability in beverage products, to which it is added, during processing and storage. Propylene glycol alginate, an esterified form of alginic acid, is known for its water solubility and viscosity-modifying property. Propylene glycol alginate is prevalently used in beverages as a stabilizer, thickener, emulsifier, and suspension agent, whereas it cannot be used as a gelling agent due to the esterification of carboxyl groups. Thus, application of alginates is established to improve the rheological and textural properties and also the functional property of beverages.

References

1. Saha, D. and Bhattacharya, S., Hydrocolloids as thickening and gelling agents in food: A critical review. *J. Food Sci. Technol.*, 47, 6, 587–597, 2010.

2. e-PG Pathashala. Role played by hydrocolloids in food industry. http://epgp.inflibnet.ac.in/epgpdata/uploads/epgp_content/food_technology/food_additives/10.role_played_by__hydrocolloids_in_food__industry/et/2685_et_m10.pdf. 2018.
3. Arasaki, S. and Arasaki, T., *Low calorie, high nutrition: Vegetables from the sea; To help you look and feel better*, p. 60, Japan Publications, Tokyo, 1983.
4. Rasmussen, R.S. and Morrissey, M.T., Marine biotechnology for production of food ingredients. *Adv. Food Nutr. Res.*, 52, 237–292, 2007.
5. Draget, K.I., Smidsrød, O., Skjak-Bræk, G., Alginates from algae. *Biopolym Online*, 2005.
6. Pawar, S.N. and Edgar, K.J., Alginate derivatization: A review of chemistry, properties and applications. *Biomaterials*, 33, 11, 3279–3305, 2012.
7. Hay, I.D., Rehman, Z.U., Moradali, M.F., Wang, Y., Rehm, B.H., Microbial alginate production, modification and its applications. *Microb. Biotechnol.*, 6, 6, 637–650, 2013.
8. Saltmarsh, M. and Saltmarsh, M. (Eds.), *Essential guide to food additives*, Royal Society of Chemistry, 2013.
9. Nalamothu, N., Potluri, A., Muppalla, M.B., Review on marine alginates and its applications. *J. Pharm. Res.*, 4, 10, 2014.
10. Kelly, B.J. and Brown, M.T., Variations in the alginate content and composition of *Durvillaea antarctica* and *D. willana* from southern New Zealand. *J. Appl. Phycol.*, 12, 3–5, 317–324, 2000.
11. Jothisaraswathi, S., Babu, B., Rengasamy, R., Seasonal studies on alginate and its composition II: *Turbinaria conoides* (J. Ag.) *Kütz. (Fucales, Phaeophyceae). J. Appl. Phycol.*, 18, 2, 161, 2006.
12. Rioux, L.E., Turgeon, S.L., Beaulieu, M., Rheological characterisation of polysaccharides extracted from brown seaweeds. *J. Sci. Food Agric.*, 87, 9, 1630–1638, 2007.
13. Market and Market analysis. Alginates & derivatives market—Global trends & forecast to 2019, 2015.
14. Fleury, N. and Lahaye, M., Studies on by-products from the industrial extraction of alginate. *J. Appl. Phycol.*, 5, 6, 605–614, 1993.
15. Hernández-Carmona, G., McHugh, D.J., Arvizu-Higuera, D.L., Rodríguez-Montesinos, Y.E., Pilot plant scale extraction of alginate from *Macrocystis pyrifera*. 1. Effect of pre-extraction treatments on yield and quality of alginate. *J. Appl. Phycol.*, 10, 6, 507–513, 1998.
16. Vauchel, P., Leroux, K., Kaas, R., Arhaliass, A., Baron, R., Legrand, J., Kinetics modeling of alginate alkaline extraction from *Laminaria digitata*. *Bioresour. Technol.*, 100, 3, 1291–1296, 2009.
17. McHugh, D.J., Production, properties and uses of alginates. pp. 58–115. 1987.
18. Technical Evaluation Report. Alginates—Handling/Processing. Compiled by OMRI for the USDA National Organic Program. 1–23. https://www.ams.usda.gov/sites/default/files/media/Alginates%20TR%202015.pdf, 2015.

19. Nyvall, P., Corre, E., Boisset, C., Barbeyron, T., Rousvoal, S., Scornet, D., Kloareg, B., Boyen, C., Characterization of mannuronan C-5-epimerase genes from the brown alga *Laminaria digitata. Plant Physiol.*, 133, 2, 726–735, 2003.
20. Rinaudo, M., Main properties and current applications of some polysaccharides as biomaterials. *Polym. Int.*, 57, 3, 397–430, 2008.
21. Drury, J.L., Dennis, R.G., Mooney, D.J., The tensile properties of alginate hydrogels. *Biomaterials*, 25, 16, 3187–3199, 2004.
22. Siew, C.K., Williams, P.A., Young, N.W., New insights into the mechanism of gelation of alginate and pectin: Charge annihilation and reversal mechanism. *Biomacromolecules*, 6, 2, 963–969, 2005.
23. Gomez, C.G., Lambrecht, M.V.P., Lozano, J.E., Rinaudo, M., Villar, M.A., Influence of the extraction-purification conditions on final properties of alginates obtained from brown algae (*Macrocystis Pyrifera*). *Int. J. Biol. Macromol.*, 44, 4, 365–371, 2009.
24. Borgogna, M., Skjåk-Bræk, G., Paoletti, S., Donati, I., On the initial binding of alginate by calcium ions. The tilted egg-box hypothesis. *J. Phys. Chem. B*, 117, 24, 7277–7282, 2013.
25. Etchepare, M.de-A., Barin, J.S., Cichoski, A.J., Jacob-Lopes, E., Wagner, R., Fries, L.L.M., Menezes, C.R., Microencapsulation of probiotics using sodium alginate. *Ciênc Rural, Santa Maria*, 45, 7, 1319–1326, 2015.
26. Truong, V.D., Walter, W.M., Giesbrecht, F.G., Texturization of sweet potato puree with alginate: Effects of tetrasodium pyrophosphate and calcium sulfate. *J. Food Sci.*, 1054–1059, 1995.
27. Brownlee, I.A., Allen, A., Pearson, J.P., Dettmar, P.W., Havler, M.E., Atherton, M.R., Onsøyen, E., Alginate as a source of dietary fiber. *Crit. Rev. Food Sci. Nutr.*, 45, 6, 497–510, 2005.
28. d'Ayala, G.G., Malinconico, M., Laurienzo, P., Marine derived polysaccharides for biomedical applications: Chemical modification approaches. *Molecules*, 13, 9, 2069–2106, 2008.
29. Sidley Chemicals. Application of sodium alginate in food. http://www.visitchem.com/application-of-sodium-alginate-in-food/, 2018.
30. Rößle, C., Brunton, N., Gormley, R.T., Wouters, R., Butler, F., Alginate coating as carrier of oligofructose and inulin and to maintain the quality of fresh-cut apples. *J. Food Sci.*, 76, 1, 2011.
31. Datta, S., Janes, M.E., Xue, Q.G., Losso, J., La Peyre, J.F., Control of *Listeria monocytogenes* and *Salmonella anatum* on the surface of smoked salmon coated with calcium alginate coating containing oyster lysozyme and nisin. *J. Food Sci.*, 73, 2, 2008.
32. Pelkman, C.L., Navia, J.L., Miller, A.E., Pohle, R.J., Novel calcium-gelled, alginate-pectin beverage reduced energy intake in nondieting overweight and obese women: Interactions with dietary restraint status. *Am. J. Clin. Nutr.*, 86, 6, 1595–1602, 2007.

33. Solah, V.A., Kerr, D.A., Adikara, C.D., Meng, X., Binns, C.W., Zhu, K., Devine, A., Prince, R.L., Differences in satiety effects of alginate-and whey protein-based foods. *Appetite*, 54, 3, 485–491, 2010.
34. Kristensen, M. and Jensen, M.G., Dietary fibres in the regulation of appetite and food intake. Importance of viscosity. *Appetite*, 56, 1, 65–70, 2011.
35. Peters, H.P., Koppert, R.J., Boers, H.M., Ström, A., Melnikov, S.M., Haddeman, E., Schuring, E.A., Mela, D.J., Wiseman, S.A., Dose-dependent suppression of hunger by a specific alginate in a low-viscosity drink formulation. *Obesity*, 19, 6, 1171–1176, 2011.
36. Paxman, J.R., Richardson, J.C., Dettmar, P.W., Corfe, B.M., Daily ingestion of alginate reduces energy intake in free-living subjects. *Appetite*, 51, 3, 713–719, 2008.
37. El Khoury, D., Goff, H.D., Berengut, S., Kubant, R., Anderson, G.H., Effect of sodium alginate addition to chocolate milk on glycemia, insulin, appetite and food intake in healthy adult men. *Eur. J. Clin. Nutr.*, 68, 5, 613, 2014.
38. Terada, A., Hara, H., Mitsuoka, T., Effect of dietary alginate on the faecal microbiota and faecal metabolic activity in humans. *Microb. Ecol. Health Dis.*, 8, 6, 259–266, 1995.
39. Pangborn, R., Gibbs, Z.M., Tassan, C., Effect of hydrocolloids on apparent viscosity and sensory properties of selected beverages. *J. Texture Stud.*, 9, 4, 415–436, 1978.
40. Yanes, M., Durán, L., Costell, E., Effect of hydrocolloid type and concentration on flow behavior and sensory properties of milk beverages model systems. *Food Hydrocol.*, 16, 6, 605–611, 2002.
41. Sidley Chemicals. Application of PGA in fruit juice. http://www.visitchem.com/application-propylene-glycol-alginate-fruit-juice/, 2018.
42. Sidley Chemicals. Application of propylene glycol alginate (PGA) in the formulated acid milk beverage. http://www.visitchem.com/propylene-glycol-alginate-pga-food/, 2018.
43. Cheong, K.W., Mirhosseini, H., Hamid, N.S.A., Osman, A., Basri, M., Tan, C.P., Effects of propylene glycol alginate and sucrose esters on the physicochemical properties of modified starch-stabilized beverage emulsions. *Molecules*, 19, 8691–8706, 2014.
44. Jackson, G., Roberts, R.T., Wainwright, T., Mechanism of beer foam stabilization by propylene glycol alginate. *J. Inst. Brew.*, 86, 1, 34–37, 1980.
45. Sahai, D. and Sher, A.A., Foaming juice compositions. US Patent: WO2012/076578, Assigned to Nestec S.A., 2012.
46. Goh, C.H., Heng, P.W.S., Chan, L.W., Alginates as a useful natural polymer for microencapsulation and therapeutic applications. *Carbohydr. Polym.*, 88, 1, 1–12, 2012.
47. Gombotz, W.R. and Wee, S., Protein release from alginate matrices. *Adv. Drug Deliv. Rev.*, 31, 3, 267–285, 1998.

48. Yeo, Y., Baek, N., Park, K., Microencapsulation methods for delivery of protein drugs. *Biotechnol. Bioprocess Eng.*, 6, 4, 213–230, 2001.
49. Smrdel, P., Bogataj, M., Zega, A., Planinsek, O., Mrhar, A., Shape optimization and characterization of polysaccharide beads prepared by inotropic gelation. *J. Microencapsul.*, 25, 2, 90–105, 2008.
50. Burey, P., Bhandari, B.R., Howes, T., Gidley, M.J., Hydrocolloid gel particles: Formation, characterization, and application. *Crit. Rev. Food Sci. Nutr.*, 48, 5, 361–377, 2008.
51. Burgess, D.J. and Hickey, A.J., Microsphere technology and applications. *Encyclopedia of Pharmaceutical Technology*, 10, 1–29, 1994.
52. Heidebach, T., Först, P., Kulozik, U., Microencapsulation of probiotic cells by means of rennet-gelation of milk proteins. *Food Hydrocol.*, 23, 7, 1670–1677, 2009.
53. Mortazavian, A., Razavi, S.H., Ehsani, M.R., Sohrabvandi, S., Principles and methods of microencapsulation of probiotic microorganisms. *Iran J. Biotechnol.*, 5, 1, 1–18, 2007.
54. Martin, M.J., Lara-Villoslada, F., Ruiz, M.A., Morales, M.E., Effect of unmodified starch on viability of alginate-encapsulated *Lactobacillus fermentum* CECT5716. *LWT-Food Sci. Technol.*, 53, 2, 480–486, 2013.
55. Patil, J.S., Kamalapur, M.V., Marapur, S.C., Kadam, D.V., Ionotropic gelation and polyelectrolyte complexation: The novel techniques to design hydrogel particulate sustained, modulated drug delivery system: A review. *Dig. J. Nanomater. Biostruct.*, 5, 1, 241–248, 2010.
56. Krasaekoopt, W. and Watcharapoka, S., Effect of addition of inulin and galactooligosaccharide on the survival of microencapsulated probiotics in alginate beads coated with chitosan in simulated digestive system, yogurt and fruit juice. *LWT-Food Sci. Technol.*, 57, 2, 761–766, 2014.
57. Colagrande, O., Silva, A., Fumi, M.D., Recent applications of biotechnology in wine production. *Biotechnol. Prog.*, 10, 1, 2–18, 1994.
58. Kourkoutas, Y., Komaitis, M., Koutinas, A.A., Kanellaki, M., Wine production using yeast immobilized on apple pieces at low and room temperatures. *J. Agric. Food Chem.*, 49, 3, 1417–1425, 2001.
59. Kourkoutas, Y., Douma, M., Koutinas, A.A., Kanellaki, M., Banat, I.M., Marchant, R., Room and low temperature continuous wine making using yeast immobilized on quince pieces. *Process Biochem.*, 39, 2, 143–148, 2003.
60. Kourkoutas, Y., Komaitis, M., Koutinas, A.A., Kaliafas, A., Kanellaki, M., Marchant, R., Banat, I.M., Wine production using yeast immobilized on quince biocatalyst at temperatures between 30 and 0°C. *Food Chem.*, 82, 3, 353–360, 2003.
61. Tsakiris, A., Bekatorou, A., Psarianos, C., Koutinas, A.A., Marchant, R., Banat, I.M., Immobilization of yeast on dried raisin berries for use in dry white wine-making. *Food Chem.*, 87, 1, 11–15, 2004.
62. Hill, F., Immobilization of yeast in alginate beads for production of alcoholic beverages. US Patent 5,070,019. Assigned to Huels AG, 1991.

63. Reddy, L.V.A. and Reddy, O.V.S., Production and characterization of wine from mango fruit (*Mangifera indica* L). *World J. Microbiol. Biotechnol.*, 21, 8-9, 1345–1350, 2005.
64. Medina, C.M., Garcia, M.J.C., Mayén, R.M., Mérida, G.J., Millán, P.M.C., Moreno, V.J., Moyano, C.M.L., Ortega, R.J.M., Zea, C.L., Berlanga, F.T., López, T.A., Maestre, D.R., Muñoz, R.D., Peinado, A.R., Novel applications of gels that contain immobilised yeasts. Patent: WO2003070930A1. Assigned to Universidad de Corodoba, 2003.
65. Jain, T. Ethanolic fermentation with immobilized yeast. Patent WO2014159159A1. Assigned to The Board of Regents of the University of Idaho, 2013.
66. Sanders, M.E., Probiotics: Considerations for human health. *Nutr. Rev.*, 61, 3, 91–99, 2003.
67. De Castro-Cislaghi, F.P., Carina Dos Reis, E.S., Fritzen-Freire, C.B., Lorenz, J.G., Sant'Anna, E.S., *Bifidobacterium* Bb-12 microencapsulated by spray drying with whey: Survival under simulated gastrointestinal conditions, tolerance to NaCl, and viability during storage. *J. Food Eng.*, 113, 2, 186–193, 2012.
68. Burgain, J., Gaiani, C., Linder, M., Scher, J., Encapsulation of probiotic living cells: From laboratory scale to industrial applications. *J. Food Eng.*, 104, 4, 467–483, 2011.
69. de Vos, P., Bučko, M., Gemeiner, P., Navrátil, M., Švitel, J., Faas, M., Strand, B.L., Skjak-Braek, G., Morch, Y.A., Vikartovská, A., Lacík, I., Multiscale requirements for bioencapsulation in medicine and biotechnology. *Biomaterials*, 30, 13, 2559–2570, 2009.
70. Krasaekoopt, W., Bhandari, B., Deeth, H., Evaluation of encapsulation techniques of probiotics for yoghurt. *Int. Dairy J.*, 13, 1, 3–13, 2003.
71. Adhikari, K., Mustapha, A., Grün, I.U., Fernando, L., Viability of microencapsulated bifidobacteria in set yogurt during refrigerated storage. *J. Dairy Sci.*, 83, 9, 1946–1951, 2000.
72. Krasaekoopt, W., Bhandari, B., Deeth, H.C., Survival of probiotics encapsulated in chitosan-coated alginate beads in yoghurt from UHT-and conventionally treated milk during storage. *LWT-Food Sci. Technol.*, 39, 2, 177–183, 2006.
73. Sultana, K., Godward, G., Reynolds, N., Arumugaswamy, R., Peiris, P., Kailasapathy, K., Encapsulation of probiotic bacteria with alginate–starch and evaluation of survival in simulated gastrointestinal conditions and in yoghurt. *Int. J. Food Microbiol.*, 62, 1–2, 47–55, 2000.
74. Rao, A.V., Shiwnarain, N., Maharaj, I., Survival of microencapsulated *Bifidobacterium pseudolongum* in simulated gastric and intestinal juices. *Can. Inst. Food Sci. Technol. J.*, 22, 4, 345–349, 1989.
75. Wenrong, S. and Griffiths, M.W., Survival of bifidobacteria in yogurt and simulated gastric juice following immobilization in gellan-xanthan beads. *Int. J. Food Microbiol.*, 61, 17–25, 2000.

76. Lee, K.Y. and Heo, T.R., Survival of *Bifidobacterium longum* immobilized in calcium alginate beads in simulated gastric juices and bile salts solution. *Appl. Environ. Microbiol.*, 66, 869–873, 2003.
77. Hansen, L.T., Allan-Wojtas, P.M., Jin, Y.L., Paulson, A.T., Survival of Ca-alginate microencapsulated *Bifidobacterium* spp. in milk and simulated gastrointestinal conditions. *Food Microbiol.*, 19, 1, 35–45, 2002.
78. Krasaekoopt, W., Bhandari, B., Deeth, H., The influence of coating materials on some properties of alginate beads and survivability of microencapsulated probiotic bacteria. *Int. Dairy J.*, 14, 8, 737–743, 2004.
79. Chandramouli, V., Kailasapathy, K., Peiris, P., Jones, M., An improved method of microencapsulation and its evaluation to protect *Lactobacillus* spp. in simulated gastric conditions. *J. Microbiol. Methods*, 56, 1, 27–35, 2004.
80. Kim, S.J., Cho, S.Y., Kim, S.H., Song, O.J., Shin, I.S., Cha, D.S., Park, H.J., Effect of microencapsulation on viability and other characteristics in *Lactobacillus acidophilus* ATCC 43121. *LWT-Food Sci. Technol.*, 41, 3, 493–500, 2008.
81. Kreft, O., Prevot, M., Möhwald, H., Sukhorukov, G.B., Shell-in-shell microcapsules: A novel tool for integrated, spatially confined enzymatic reactions. *Angew. Chem. Int. Ed.*, 46, 29, 5605–5608, 2007.
82. Jun-Xia, X., Hai-Yan, Y., Jian, Y., Microencapsulation of sweet orange oil by complex coacervation with soybean protein isolate/gum Arabic. *Food Chem.*, 125, 4, 1267–1272, 2011.
83. Chávarri, M., Marañón, I., Ares, R., Ibáñez, F.C., Marzo, F., del Carmen Villarán, M., Microencapsulation of a probiotic and prebiotic in alginate-chitosan capsules improves survival in simulated gastro-intestinal conditions. *Int. J. Food Microbiol.*, 142, 1–2, 185–189, 2010.
84. Lee, J.S., Cha, D.S., Park, H.J., Survival of freeze-dried *Lactobacillus bulgaricus* KFRI 3673 in chitosan coated calcium alginate microparticles. *J. Agric. Food Chem.*, 52, 7300–7305, 2004.
85. Gbassi, G.K., Vandamme, T., Ennahar, S., Marchioni, E., Microencapsulation of *Lactobacillus plantarum* spp in an alginate matrix coated with whey proteins. *Int. J. Food Microbiol.*, 129, 1, 103–105, 2009.
86. Sathyabama, S. and Vijayabharathi, R., Co-encapsulation of probiotics with prebiotics on alginate matrix and its effect on viability in simulated gastric environment. *LWT-Food Sci. Technol.*, 57, 1, 419–425, 2014.
87. Mortazavian, A.M., Ehsani, M.R., Azizi, A., Razavi, S.H., Mousavi, S.M., Sohrabvandi, S., Reinheimer, J.A., Viability of calcium-alginate-microencapsulated probiotic bacteria in Iranian yogurt drink (Doogh) during refrigerated storage and under simulated gastrointestinal conditions. *Aust. J. Dairy Technol.*, 63, 1, 24–29, 2015.
88. Iyer, C., Phillips, M., Kailasapathy, K., Release studies of *Lactobacillus casei* strain Shirota from chitosan-coated alginate-starch microcapsules in *ex vivo* porcine gastrointestinal contents. *Lett. Appl. Microbiol.*, 41, 6, 493–497, 2005.
89. Fahimdanesh, M., Mohammadi, N., Ahari, H., Zanjani, M.A., Hargalani, F.Z., Behrouznasab, K., Effect of microencapsulation plus resistant starch on

survival of *Lactobacillus casei* and *Bifidobacterium bifidum* in mayonnaise sauce. *Afr. J. Microbiol. Res.*, 6, 40, 6853–6858, 2013.
90. Homayouni, A., Azizi, A., Ehsani, M.R., Yarmand, M.S., Razavi, S.H., Effect of microencapsulation and resistant starch on the probiotic survival and sensory properties of synbiotic ice cream. *Food Chem.*, 111, 1, 50–55, 2008.
91. Jankowski, T., Zielinska, M., Wysakowska, A., Encapsulation of lactic acid bacteria with alginate/starch capsules. *Biotechnol. Tech.*, 11, 1, 31–34, 1997.
92. Liserre, A.M., Re, M.I., Franco, B.D., Microencapsulation of *Bifidobacterium animalis* subsp. lactis in modified alginate-chitosan beads and evaluation of survival in simulated gastrointestinal conditions. *Food Biotechnol.*, 21, 1, 1–16, 2007.
93. Cui, J.H., Goh, J.S., Kim, P.H., Choi, S.H., Lee, B.J., Survival and stability of bifidobacteria loaded in alginate poly-l-lysine microparticles. *Int. J. Pharm.*, 210, 1–2, 51–59, 2000.
94. Chan, E.S., Preparation of Ca-alginate beads containing high oil content: Influence of process variables on encapsulation efficiency and bead properties. *Carbohydr. Polym.*, 84, 4, 1267–1275, 2011.
95. Bigdelian, E. and Razavi, S.H., Evaluation of survival rate and physicochemical properties of encapsulated bacteria in alginate and resistant starch in mayonnaise sauce. *J. Bioproces Biotechniq*, 4, 5, 166, 2014.
96. Chen, K.N., Chen, M.J., Liu, J.R., Lin, C.W., Chiu, H.Y., Optimization of incorporated prebiotics as coating materials for probiotic microencapsulation. *J. Food Sci.*, 70, 5, 2005.
97. Siró, I., Kapolna, E., Kapolna, B., Lugasi, A., Functional food. Product development, marketing and consumer acceptance—A review. *Appetite*, 51, 3, 456–467, 2008.
98. Tárrega, A., Rocafull, A., Costell, E., Effects of blends of short and long-chain inulin on the rheological and sensory properties of prebiotic low fat custards. *LWT-Food Sci. Technol.*, 43, 3, 556–562, 2010.
99. Nazzaro, F., Fratianni, F., Sada, A., Orlando, P., Synbiotic potential of carrot juice supplemented with *Lactobacillus* spp. and inulin or fructooligosaccharides. *J. Sci. Food Agric.*, 88, 13, 2271–2276, 2008.
100. Nazzaro, F., Fratianni, F., Coppola, R., Sada, A., Orlando, P., Fermentative ability of alginate-prebiotic encapsulated *Lactobacillus acidophilus* and survival under simulated gastrointestinal conditions. *J. Funct. Foods*, 1, 3, 319–323, 2009.

13

Alginates in Comestibles

Ashwini Ravi[1], S. Vijayanand[1], Velu Rajeshkannan[2], S. Aisverya[3], K. Sangeetha[3], P.N. Sudha[3] and J. Hemapriya[4*]

[1]Bioresource Technology Lab, Department of Biotechnology, Thiruvalluvar University, Sekkadu, Vellore, Tamilnadu, India
[2]Department of Microbiology, Bharathidasan University, Tiruchirapalli, Tamil Nadu, India
[3]PG & Research, Department of Chemistry, DKM College for Women (Autonomous), Vellore, Tamilnadu, India
[4]PG & Research, Department of Microbiology, DKM College for Women (Autonomous), Vellore, Tamilnadu, India

Abstract

Alginates are polysaccharide compounds derived from the cell walls of brown algae of Class Phaeophyceae and some earth bacterial species. The alginates derived from these organisms are used in food industries as food additives to improve, modify, and stabilize the texture of certain food products. Alginates are classified as synthetic, nonagricultural substances that can be used as an ingredient in processed food products, and they have been listed at 7CFR section 205.605(b) for their use in food products. They have been approved and demonstrated safe for use throughout the European Union (EU), and they have been commonly consumed in western countries. The alginates have been designated E numbers as per EU. The number E400 was designated for alginic acid, and the numbers E401–E404 were designated for salts of alginic acid. Esters of alginic acid with Propylene Glycol Alginates (PGA) has been designated E405. Though alginates have no nutritional value, their properties such as viscosity enhancement, gel-forming ability, and stabilization of aqueous mixtures, dispersions, and emulsions make their effective use in food industries. They were also found to increase the shelf life of products and protect them from pathogens. They have been used as anti-browning agents, colorants, flavors, nutrients, spices, and antimicrobial compounds in food industry. They act as a protective barrier to moisture, oxygen, lipid oxidation, and loss of volatile aromas and flavors in food products. They also

**Corresponding author*: micropriya80@gmail.com

Shakeel Ahmed (ed.) Alginates, (263–280) © 2019 Scrivener Publishing LLC

prevent water loss, syneresis, and phase separation, and increase the functionality of fresh-cut and prepackaged fruits in other food products. The present chapter deals with the application of alginates in food industry.

Keywords: Alginates, food additives, emulsions, flavors, antimicrobial agents

13.1 Introduction

Hydrocolloids are macromolecular hydrophilic substances that will either dissolve in water forming colloidal solution or swell in water forming gels. They are highly used in the food industry as emulsifiers, thickeners, and stabilizers. Hydrocolloids are divided into four types based on the source of extraction: hydrocolloids obtained from plants, fermentation, chemically modified sources, and animals. Alginates are one of the hydrocolloids used widely in food industry that have been isolated from marine algae. These polysaccharides are either purified as a sole component or modified from the original alginate as propylene glycol alginate [1].

Alginates are polyuron polysaccharides found in the cell wall and intercellular spaces of brown algae of order Phaeophyceae. They have been identified in the year 1881 and found to be applied in various industries such as textile industry, paper coating, pharmaceuticals, and food industry [2]. Alginates are composed of mannuronate units called M blocks and guluronate units called G blocks, which during extraction and purification can be modified into alginates containing MM blocks, GG blocks, or MG blocks. This modification of alginates is very essential as it determines the physical properties of the compound [3]. Alginates have been used in food industry as additives, emulsifiers, thickeners, gelling agents, stabilizers, food coating, etc. They have been found to have no health concerns and have no acceptable daily intake limits, but PGA was found to have acceptable daily intake (ADI) of 25 mg/kg body weight per day [1]. Alginates are used in various food products such as diet products, baked goods, frozen foods, mayonnaise, salad dressings, ice creams, foams, processed cheese, meat, canned vegetables, soups, noodles, pasta, etc. In Asian countries, alginates have not been used as additive but as food [4].

Apart from their use as a single compound, they have also been used in combinations such as sodium alginates, calcium alginates, starch–alginate (SA) combinations, alginate–stearic acid (SAS) combinations, tocopherol-coated SA and tocopherol-coated SAS, etc. [5]. The present chapter deals with the application of alginates in food industries.

13.2 Alginates in Agricultural Marketing

Alginates have been approved by the FDA in the group of compounds that are used for oral applications. Alginates are nontoxic and biodegradable when applied orally, but when taken intravenously, they were found to cause several immunogenic ailments [6]. However, they are approved by several food committees for consumption. Table 13.1 demonstrates the approval of alginates by various food committees around the world.

Table 13.1 clearly depicts that all forms of alginate have been approved in several nations of the world. As mentioned earlier, they have also been used in food as thickeners, stabilizers, emulsifiers, gelling agents, etc.

Table 13.1 Approval of alginates by various food committees.

Name of food committee	Code	Type of alginate	Reference
European Economic Community	400	Alginic acid	[7]
	401	Sodium alginate	
	402	Potassium alginate	
	403	Ammonium alginate	
	404	Calcium alginate	
	405	PGA	
Code of Federal Regulations	9005-38-3	Sodium alginate	[8]
	9005-30-0	Calcium alginate	
Japanese Agricultural Standard	INS 401	Sodium alginate	[9]
	INS 400	Alginic acid	
FAO/WHO–Codex Alimentarius–International Food Standards	INS 400	Alginic acid	[10]
	INS 401	Sodium alginate	
	INS 402	Potassium alginate	
	INS 403	Ammonium alginate	
	INS 404	Calcium alginate	

(Continued)

Table 13.1 Approval of alginates by various food committees. (*Continued*)

Name of food committee	Code	Type of alginate	Reference
Government of Canada	A5	Algin	[11]
	A6	Alginic acid	
	A7	Ammonium alginate	
	C1	Calcium alginate	
	P6	Potassium alginate	
	P12	PGA	
	S2	Sodium alginate	
USDA	GRAS 184.1011	Alginic acid	[12]
	GRAS 184.1724	Sodium alginate	
	GRAS 184.1610	Potassium alginate	
	REG 173.310	Ammonium alginate	
	GRAS 184.1187	Calcium alginate	
	REG 172.858	PGA	
IFOAM	INS 400	Alginic acid	[13]
	INS 401	Sodium alginate	
	INS 402	Potassium alginate	

13.3 Use of Alginates in Food Industry

Food substances are those compounds that are used to enhance the properties of food by enhancing flavor and color, preserving them, or changing their consistency. Several hydrocolloids, especially marine polysaccharides such as agar, xanthan, carrageenans, alginates, etc., have been used for ages as additives in food industry. The present topic describes the use of alginates in food industries for various purposes.

13.3.1 Thickeners and Gelling Agents

The thickening ability of any hydrocolloid is due to the nonspecific entanglement of confirmationally disordered polymer chains. Thickening was found to occur usually at overlap concentrations of the colloid with solvent [14, 15]. In dilute concentrations, colloids can move freely and do not cause thickening effect, whereas in concentrated solutions, the contact between colloids is increased, thereby their interaction and also their thickening behavior. In case of no polymer interaction, the thickening ability of colloid is affected by its molecular weight. Therefore, it can be clearly said that the thickening of any hydrocolloid depends on intrinsic velocity, the molecular weight of the colloid, and its concentration [16]. Similarly, the gelling ability of any compound is due to the formation of junction zones due to the three-dimensional interlinking of polymer with solvent interstices [17]. The gelation is formed by either of three mechanisms called as ionotropic gelation, cold set gelation, and heat set gelation [18]. Ionotropic gelation is caused by the presence of ions in which the phenomenal gelation is occurring in alginates, whereas the other two methods are followed by other colloids such as agar, starch, etc. [19–21].

The use of alginates in foods is long-standing in history and they are mainly used as thickeners, gelling agents, stabilizers, etc. [22]. The use of alginates is due to their physical properties such as viscosity enhancement, gel-forming ability, stabilization of mixtures, and interactions with food components such as protein, fat, or fiber. Apart from these, the flexibility of controlling alginate production makes them an efficient candidate [23, 24]. The most commonly used alginate in food is sodium alginate apart from other alginates, and alginates of different viscosity have been prepared from variable degree of polymerization such as 80% to >750%.

As a gelling agent, alginates have several advantages in food industry. The gelling property of alginate is affected by adding calcium ions. Pure alginic acid swells in cold water 200–300 times rather than dissolving in them, and it immediately dissolves in hot water, but by acidification, in the presence of calcium ions, it forms a gel in hot water [25]. This intrinsic property of alginates in the presence of calcium ions enables them to be produced as homogenous thermostable gels under controlled conditions. Gels of alginate are formed by egg-box mode where cooperative ion binding occurs with G blocks of adjacent alginate molecules. Apart from these, G-type alginates react highly with calcium ions than the M-type alginates [26–28].

Several studies on the thickening and gelling ability of alginates have also been studied. Radiation of alginates as powder with different doses showed that at decreasing temperature, the viscosity of alginates increases [22]. A similar study on irradiating sodium alginate powder by gamma radiation shows that with irradiation dose of 0.5% to 1.0% and more than 30%, the viscosity and gel strength of alginate decreased, respectively. However, gels were formed at all doses of irradiation [29]. Gelling properties of sodium alginate studied with various concentrations of calcium showed that the highest rupture force was obtained at concentration of 60 g/kg sodium alginate with 100 g/kg calcium chloride. Similarly, the highest deformation in rupture point was found at 60 g/kg sodium alginate with 10 g/kg calcium chloride. In contrast, the smallest deformation of rupture was observed at 30 g/kg sodium alginate with 70 g/kg $C_6H_{10}O_6$ Ca and 650 g/kg sucrose. It has also been found from the study that the increase in calcium chloride concentration at particular alginate concentration and gelation time leads to rupture formation [30]. The effect of collagen fiber and gelatin on gelling properties of alginates showed that there was lower shrinkage level on alginates and mechanical properties were similar to those of pure alginates [31]. The effect of sodium alginate on drying and rehydration of konjac noodles showed that sodium alginate of 1% to 1.5% showed better springiness; however, the cohesiveness varied between 0.66 and 0.71, whereas sodium alginate of 0.25% and 0.5% showed no effect on springiness and cohesiveness on konjac noodles [32]. Similarly, covalent and noncovalent interactions of konjac glucomannan, xanthum gum, and alginate were studied. It has been found from the study that primary covalent structures were not affected, whereas the noncovalent structures of ternary mixtures caused sedimentation in the presence of sodium alginate. However, on heating the solution and at concentrations greater than 8% sodium alginate, the ternary mixtures had a shift to higher sedimentation coefficients [33]. As thickeners and gelling agents, alginates have been used in ice creams, soups, sauces, dressings, ketchup, mayonnaise, margarine, milkshakes, fruit juices, liquors, desserts, jams, puddings, whipped cream, pie fillings, mashed potatoes, and restricted foods [3].

13.3.2 Stabilizers and Emulsifiers

Stabilizer is a single component or mixture of components offering long-term stability to the food by involving adsorptive mechanism or any other covalent interactions, whereas emulsifier is a single chemical or mixture of components that gives temporary stabilization and also aids in promoting emulsification [34]. Stabilizers protect food substance by minimizing water migration, prevent oil separation, provide suspension, increase viscosity, and prevent ice crystal formation, whereas emulsifiers reduce surface tensions and interfacial tensions and stabilize oil in water emulsions through the steric and mechanical stabilization mechanisms [35, 36].

There have been several studies on using alginate as an emulsifier and stabilizer. Beneficiary lipids that are highly unstable are converted to nanoemulsions for better stability and release. In order to obtain such nanoemulsions, they have to be made stable. Alginates have been used as a stabilizer for such nanoemulsions and have been found that stable emulsions were formed at alkaline pH, temperature 50 °C at an electrolyte concentration of 100 mM [37]. Alginates were also found to enhance the stability of dietary protein emulsions [38]. Further study on alginates extracted from six types of alginofit called *S. crassifolium*, *S. polycystum*, *Padina*, *S. echinocarpum*, *S. duplicatum*, and *S. binderi* in cake making showed that good quality of cake has been obtained from 0.75% alginate from *S. duplicatum*. They were also found to have fat and protein content of 18.98% and 63.11%, respectively [39]. Sodium alginate was found to increase stabilization of ice cream than other stabilizers and also found to have better whipping of the mix when combining sodium alginate with calcium caseinate. The increased whipping was found due to the formation of sodium caseinate and calcium alginate [40]. In soursop beverage emulsions, PGA, either as sole compound or as mixtures at $p < 0.005$, was found to increase higher cloudiness, larger droplet size, and also low creaming stability [41]. In addition to this, alginates were also found to stabilize the wheat dough during proofing and also retard crumb firming [42].

13.3.3 Texturizers

Texturizers are compounds that protect the food component by increasing its texture. Alginates exclusively as texturizers have been extensively studied, and they have been used as a texturizer in food industry [43]. In a study by Nicomrat *et al.*, [44], alginate has been used as a texturizer in moo yor batter. Sodium alginate was found to affect the rheological properties; decrease cooking loss and moisture content; increase hardness, chewiness, springiness, gumminess, and cohesiveness; and improve product texture. They were found to be more suitable as texturizers in moo yor batter than other agents.

13.3.4 Encapsulation

Encapsulation is the process to entrap one substance to another and the substance that is entrapped can be of solid, liquid, or gas. Encapsulation produces particles with a diameter in the range of millimeters to nanometers. The substance that is entrapped is called the core material, active phase, or pay load phase, whereas the substance that is used for encapsulation is called the shell, carrier material, or matrix. The encapsulates are categorized into two types, viz., the reservoir type, which has a shell over the active agent, and the matrix type, which has active compound dispersed over the carrier material in relatively small droplets. It may also be found as homogenous deposition over encapsulate. Encapsulation in food industry helps to improve flavors in food, protects sensitive and expensive nutrients, prevents loss of food quality by external factors and also masks the undesirable taste of some nutrients that affects the free nutrients [5–49]. Several hydrocolloids have been used in encapsulation, but alginate hydrogels are the most exploited [50, 51].

Alginates have been extensively used in encapsulation of probiotics. Microencapsulation of probiotics by sodium alginate is done by two methods, viz., extrusion method in which sodium alginate solution is dripped into solution of calcium salt causing ionic gelation, and emulsification method in which gelation of alginate occurs in water–oil emulsion [52]. Several probiotic organisms such as *Lactobacillus casei* [53], *Bifidobacterium bifidum* [54], *Lactobacillus parcasei* [55], *Lactobacillus acidophilus* [56, 57], *Lactobacillus bulgaricus* [58], *Lactobacillus plantarum* [59], *Enterococcus fecium* [60], and *Lactobacillus fermentum* [61] have been encapsulated by various alginate sources. Apart from that, they have also been encapsulated with several other compounds such as starch, chitosan, carrageenan, gellan gum, locust bean gum, etc. [62]. Apart from these, several other compounds such as pitaya fruit juice [63], human insulin [64], nisin [65], ascorbic acid, [66], etc. have also been encapsulated by alginates.

13.3.5 Food Coating

Edible coatings are used for extending the life of fresh cut fruits and vegetables. These coatings also make them safe to be eaten along with the food compounds and do not cause any harmful effects to the food stuff [66, 67]. Several compounds such as hydrocolloids, lipids, and composites are used as food coating material. Alginate is one among the polysaccharide-based hydrocolloids that is used in food coating [68]. Alginates have been used in coating several fruits and vegetables such as fresh cut apples [69, 70], William pear [71], pumpkin [72], fresh cut nectarine [73], etc. Alginates have also been used in coating fresh and frozen meats such as beef [74], chicken pieces [75], pork [76], beef steak [77], bream fish [78], kilka fish [79], and lamb meat [80]. They are also used in extending the life span of Coalho cheese [81].

13.4 Use of Alginates for Pets

Apart from their variable application in human consumables, alginates have also been used as animal feed [82] and as a binder in feed for fishes [83]. Sodium alginate has no maximum level of usage, whereas potassium alginate has a usage level of 40,000 mg/kg feed on dry matter [84]. Alginates have been microencapsulated by animal feed supplements for controlled release in gastrointestinal tract [85]. It has also been found that sodium alginate oligosaccharides from brown algae inhibit colonization of *Salmonella enteritidis* in broiler chicken [86].

13.5 Effect of Dietary Alginates

In vitro treatment of dietary alginates was found to inhibit the digestive enzymes and was found to be used in obesity treatment. Adding alginates to bread vehicle was found to inhibit lipases rather than cooking and digestion and was found to be used as an effective treatment against obesity [87]. In another study by Jensen *et al.*, [88], alginate supplementation was found to improve weight loss in obese subjects who completed a 12-week dietary intervention. It has been found that alginate does not affect the hematological indices, plasma biochemistry parameters, urinalysis parameters, blood glucose, plasma insulin concentrations, and breath hydrogen concentrations. They were also found not to cause any allergic response in the tested individuals [89]. In addition to these, alginate increases the gut microbe Bifidobacteria and decreases the levels of harmful microflora Enterobacteriaceae and lecithinase-negative clostridia. There was no detectable change in other microorganisms during the period of alginate intake [90].

Besides their beneficial effects in humans, they were also found to increase the immune response in shrimp and increase the survival rate of fishes *Epinephelus bruneus* and *Epinephelus coioides* infected with *Streptococcus iniae* and *Streptococcus* sp. and iridovirus, respectively [91–93]. In addition to these, the alginates are found to increase the survival rate of Nile tilapia against *Streptococcus agalactiae* infection [94]. Apart from these, they were found to increase the gut flora in Indian shrimp *Fenneropenaeus indicus* [95]. Alginates were found to increase the meat quality of pig longissimus muscle during ageing [96] and increase the body weight of gilthead sea bream [97], common pheasant chicks [98], ad libitum-fed growing pigs [99], and rats [100].

13.6 Alginate Safety

Gastroesophageal reflux symptoms such as heartburn and regurgitation were found to occur in patients with nonerosive reflux disease (NERD). Treating those patients with oral supplementation of sodium alginate was found to indicate relief of gastroesophageal reflux symptoms. They were found safe than the chemical compound omeprazole in relieving heartburn or regurgitation symptoms of NERD patients [101]. Similarly, they were found to be effective in the treatment of gastroesophageal reflux disease in preterm infants [102]. In another study by European food safety authority (EFSA), the novel food alginate–konjac–xanthan polysaccharide (PGX) was found safe for consumption at a rate of 1.8 g/kg body weight per day [103]. Apart from its safe use for human consumption, alginates have also been found to be safe for pets and non-food-producing animals and fish [104].

13.7 Conclusion

The physical properties of alginates, which are noninteractivity, viscosity, and formation of gelling agents, make them useful in various food-related applications such as thickening agents, stabilizers, emulsifiers, additives, encapsulation, etc. They were also found to not cause any harm to human consumption when taken in considerable amount and also found to be useful for animals. Also, another advantage is their use in medical applications. These advantages of alginates make them a highly efficient candidate for application in several fields including food industry.

References

1. Scherz, H., Hydrocolloids: Stabilizers, Thickening and Gelling Agents in Food Products, Food Chemistry/Food Quality, in: *Food Chemical Society GDCh*, vol. 2, Behr's Verlag GmbH, Hamburg, 1996.
2. Onsoyen, E., Commercial application of alginates. *Carbohydr. Eur.*, 46, 26–31, 1996.
3. Draget, K.I., Smidsrød, O., Skjåk-Bræk, G., Alginates from algae, in: *Polysaccharides and Polyamides in the Food Industry. Properties, Production, and Patents*, A. Steinbüchel and S.K. Rhee (Eds.), pp. 1–30, Verlag, Wiley, 2005.
4. Brownlee, I.A., Allen, A., Pearson, J.P., Dettmar, P.W., Havler, M.E., Atherton, M.R., Onsoyen, E., Alginate as a source of dietary fiber. *Crit. Rev. Food Sci. Nutr.*, 45, 6, 497–510, 2005.

5. Wu, Y., Weller, C.L., Hamouz, F., Cuppett, S., Schnepf, M., Moisture Loss and Lipid Oxidation for precooked ground-beef patties packaged in edible starch-alginate-based composite films. *J. Food Sci.*, 66, 3, 486–493, 2001.
6. Soon-Shiong, P. and Heintz, R.E., Insulin independence in a type 1 diabetic patient after encapsulated islet transplantation. *Lancet*, 16, 343(8903), 950–1.
7. Commission Regulation (EU) No 257/2010 of 25 March 2010 setting up a programme for the re-evaluation of approved food additives in accordance with Regulation (EC) No 1333/2008 of the European Parliament and of the Council on food additives. OJ L 80, 26.03.2010, p. 19.
8. https://www.accessdata.fda.gov/scripts/cdrh/cfdocs/cfCFR/CFRSearch.cfm?fr=184.1724, https://www.accessdata.fda.gov/scripts/cdrh/cfdocs/cfCFR/CFRSearch.cfm?fr=184.1187
9. Japanese Agricultural Standard for Organic Processed Foods, www.maff.go.jp/e/jas/specific/pdf/834_2012-3.pdf
10. http://www.fao.org/gsfaonline/docs/CXS_192e.pdf, Codex Alimentarius.
11. Government of 2006 - 016 Canada, https://www.canada.ca/en/health-canada/services/food-nutrition/food-safety/food-additives/lists-permitted/4-emulsifying-gelling-stabilizing-thickening-agents.html
12. USDA, https://www.fda.gov/food/ingredientspackaginglabeling/foodadditivesingredients/ucm091048.htm
13. The IFOAM norms for organic production and processing, Version 2005, 2009.
14. Philips, G.O., Wedlock, D.J., Williams, P.A., Molecular origin of hydrocolloid functionality, in: *Gums and Stabilizers for the Food Industry*, vol. 3, G.O. Philips, P.A. Williams, D.J. Wedlock (Eds.), pp. 3–5, IRL Press, Oxford, 1986.
15. Philips, G.O. and Williams, P.A., Introduction to food hydrocolloids, in: *Handbook of Hydrocolloids*, pp. 1–19, Woodhead Publ Ltd, New York, 2000.
16. Sworn, G., Hydrocolloid thickeners and their applications, in: *Gums and Stabilizers for the Food Industry*, vol. 12, G.O. Philips and P.A. Williams (Eds.), pp. 13–22, RSC Publishing, Oxford, 2004.
17. Oakenfull, D., Gelling agents. *Crit. Rev. Food Sci. Nutr.*, 26, 1, 1–31, 1987.
18. Burey, P., Bhandari, B.R., Howes, T., Gidley, M., Hydrocolloid gel particles: Formation, characterization and application. *Crit. Rev. Food. Sci. Nutr.*, 48, 5, 361–377, 2008.
19. Draget, K.I., Alginates, in: *Handbook of Hydrocolloids*, G.O. Philips and P.A. Williams (Eds.), pp. 379–393, Woodhead Publ Ltd, New York, 2000.
20. Glicksman, M., *Food Hydrocolloids*, vol. 1, pp. 4–21, CRC Publ, Boca Raton, Florida, 1982.
21. Nishinari, K. and Zhang, H., Recent advances in the understanding of heat set gelling polysaccharides. *Trends Food Sci. Technol.*, 15, 305–312, 2004.
22. Aliste, A.J., Vieira, F.F., Del Mastro, N.L., Radiation effects on agar, alginates and carrageenan to be used as food additives. *Radiat. Phys. Chem.*, 57, 305–308, 2000.

23. Neiser, S., Draget, K.I., Smidsrød, O., Gel formation in heat treated bovine serum albumin—Sodium alginate systems. *Food Hydrocoll.*, 12, 2, 127–132, 1997.
24. Neiser, S., Draget, K.I., Smidsrød, O., Interactions in bovine serum albumin-calcium alginate gel systems. *Food Hydrocoll.*, 13, 445–458, 1998.
25. Draget, K.I., Moe, S.T., Skjak-Brek, G., Smidsrod, O., Alginate, in: *Food Polysaccharides and Their Application*, A.M. Stephen, G.O. Phillips, P.A. Williams (Eds.), Taylor & Francis Group, LLC, Boca Raton, 2006.
26. Onsoyen, E., Alginates, in: *Thickening and Gelling Agents for Food*, A. and Irneson (Eds.), pp. 1–24, Blackie, London, 1992.
27. Imerson, A., Applications of alginates, in: *Gums and Stabilisers for the Food Industry*, vol. 5, G.O. Phillips, D. Wedlock, P.A. Williams (Eds.), pp. 553–562, Pergamon Press, Oxford, 1990.
28. Simc, W.J., Alginates, in: *Food Gels*, P. Harris (Ed.), pp. 53–78, Elsevier Applied Science, London, 1990.
29. King, K., Changcs in thc functional propcrtics and molccular wcight of sodium alginate following γ irradiation. *Food Hydrocoll.*, 8, 2, 83–96, 1994.
30. Manev, Z., Denev, P., Zsivanovits, G., Ludneva, D., Structural mechanical and gelling properties of alginate beads. *Bulg. J. Agri. Sci.*, 19, 4, 770–774, 2013.
31. Sato, A.C.K. and Cunha, R.L., *Effect of collagen fiber and gelatin on gelling properties of alginate gels*, Department of Food Engineering, Faculty of Food Engineering, University of Campinas (UNICAMP), Campinas, Brazil.
32. Impaprasert, R., Piyarat, S., Sophontanakij, N., Sakulnate, N., Paengkanya, S., Borompichaichartkul, C., Srzednicki, G., Rehydration and textural properties of dried konjac noodles: Effect of alkaline and some gelling agents. *Horticulturae*, 3, 1, 20, 1–10, 2017.
33. Harding, S.E., Smith, I.H., Lawson, C.J., Gahler, R.J., Wood, S., Studies on macromolecular interactions in ternary mixtures of konjac glucomannan, xanthan gum and sodium alginate. *Carbohydr. Polym.*, 83, 329–338, 2011.
34. Dickinson, E. and Stainsby, G., Emulsion stability, in: *Advances in Food Emulsions and Foams*, E. Dickinson and G. Stainsby (Eds.), pp. 1–44, Elsevier, 1988.
35. Wang, G., Stabilization from natural, in: *Gums/Starches*, D. Berry (Ed.). *Food Product Design*, 20, 3, 1–4, 2010.
36. Garti, N. and Reicmann, D., Hydrocolloids as food emulsifiers and stabilizers. *Food Str.*, 12, 411–426, 1993.
37. Gupta, S.S. and Ghosh, M., Formulation development and process parameter optimization of lipid nanoemulsions using an alginate-protein stabilizer. *J. Food Sci. Technol.*, 52, 5, 2544–2557, 2015.
38. Daniells, S., Alginates to boost stability of dairy—Protein emulsions, food-navigator.com. 1–2.
39. Mushollaeni, W. and Endang Rusdiana, S., Optimizing the use of alginate from *Sargassum* and *Padina* as natural emulsifier and stabilizer in cake. *J. Agric. Food. Tech.*, 2, 7, 108–112, 2012.

40. Goodman, C., Technical control of ice cream with sodium alginate. *Ice Cream Rev.*, 18, 42–48, 1935.
41. Cheong, K.W., Mirhosseini, H., Hamid, N.S.A., Osman, A., Basri, M., Tan, C.P., Effects of propylene glycol alginate and sucrose esters on the physicochemical properties of modified starch-stabilized beverage emulsions. *Molecules*, 19, 6, 8691-8706, 2014.
42. Kohajdová, Z., Karovičová, J., Schmidt, S., Significance of emulsifiers and hydrocolloids in bakery industry. *Acta Chim. Slovac.*, 2, 1, 46–61, 2009.
43. Vilgis, T.A., Hydrocolloids between soft matter and taste: Culinary polymer physics. *Int. J. Gastronomy Food Sci.*, 1, 1, 46–53, 2012.
44. Nicomrat, K., Chanthachum, S., Adulyatham, P., Effect of texturizing agents on quality of Moo yor in a model system. *Int. Food Res. J.*, 23, 2, 675–681, 2016.
45. Augustin, M.A., Sanguansri, L., Margetts, C., Young, B., Microencapsulation of food ingredients. *Food Austr.*, 53, 6, 220–223, 2001.
46. Zuidam, N.J. and Nedović, V.A., *Encapsulation Technologies for Active Food Ingredients and Food Processing*, pp. 3–29, Springer Science+Business Media, LLC, 2010.
47. Mothes, H., Encapsulated flavours for sophisticated products. *Eur. Food Drink Rev.*, Autumn, 20–21, 1997.
48. LaBell, F., Energy drinks with electrifying flavour. *Prep. Foods*, 168, 3, 77, 1999.
49. Meyers, M.A., New encapsulated ingredients for fortified foods, Food Ingredients Europe, Conference proceedings. 118–123, 1998.
50. Tønnesen, H.H. and Karlsen, J., Alginate in drug delivery systems. *Drug Develop. Ind. Pharm.*, 28, 6, 621–630, 2002.
51. Zimmermann, H., Shirley, S.G., Zimmermann, U., Alginate-based encapsulation of cells: Past, present, and future. *Curr. Diabetes Rep.*, 7, 4, 314–320, 2007.
52. Gombotz, W.R. and Wee, S., Protein release from alginate matrices. *Adv. Drug Deliv. Rev.*, 31, 3, 267–285, 1998.
53. Krasaekoopt, W. and Watcharapoka, S., Effect of addition of inulin and galactooligosaccharide on the survival of microencapsulated probiotics in alginate beads coated with chitosan in simulated digestive system, yogurt and fruit juice. *LWT-Food Sci. Technol.*, 57, 761–766, 2014.
54. Zanjani, M.A.K., Tarzi, B.G., Sharifan, A., and Mohammadi, N., Microencapsulation of probiotics by calcium alginate–gelatinized starch with chitosan coating and evaluation of survival in simulated human gastro-intestinal condition. *Iran J. Pharm. Res.*, 13, 3, 843–852, 2014.
55. Spigno, G., Garrido, G.D., Guidesi, E., Elli, M., Spray-drying encapsulation of probiotics for ice-cream application. *Chem. Eng. Trans.*, 43, 49–54, 2015.
56. Shah, N.P. and Ravula, R.R., Microencapsulation of probiotic bacteria and their survival in frozen fermented dairy desserts. *Aust. J. Dairy Technol.*, 55, 139–44, 2000.

57. Chandramouli, V., Kailasapathy, K., Peiris, P., Jones, M., An improved method of microencapsulation and its evaluation to protect *Lactobacillus* spp. in simulated gastric conditions. *J. Microbiol. Methods*, 56, 1, 27–35, 2004.
58. Lee, K.Y. and Heo, T.R., Survival of *Bifidobacterium longum* in calcium alginate beads in simulated gastric juices and bile salt solution. *Appl. Environ. Microbiol.*, 66, 869–873, 2004.
59. Gbassi, G.K., Vandamme, T., Ennahar, S., Marchioni, E., Microencapsulation of *Lactobacillus plantarum* spp in an alginate matrix coated with whey proteins. *Int. J. Food Microbiol.*, 129, 1, 103–105, 2009.
60. Sathyabama, S., Ranjithkumar, M., Bruntha devi, P., Vijayabharathi, R., Brindha Priyadarshini, V., Co-encapsulation of probiotics with prebiotics on alginate matrix and its effect on viability in simulated gastric environment. *Food Sci. Technol.*, 57, 419–425, 2014.
61. Martin, M.J., Effect of unmodified starch on viability of alginate encapsulated *Lactobacillus fermentum* CECT5716. *LWT–Food Sci. Technol.*, 53, 480–486, 2013.
62. Totosaus, A., Ariza-Ortega, T.J., Pérez-Chabela, M.L., Lactic acid bacteria microencapsulation in sodium alginate and other gelling hydrocolloids mixtures. *J. Food Nutr. Res.*, 52, 2, 107–120, 2013.
63. Rodríguez-Sánchez, J.A., Cuatzo-Lozano, M.I., Pérez-Loredo, M.G., Abarca-Sarro, D.I., Navarro, Y.G., Alginate encapsulation as a preservation method of pitaya fruit juice (*Stenocereus* spp.). *J. Food Sci. Eng.*, 7, 127–134, 2017.
64. Silva, C.M., Ribeiro, A.J., Figueiredo, I.V., Goncalves, A.R., Veiga, F., Alginate microspheres prepared by internal gelation: Development and effect on insulin stability. *Int. J. Pharm.*, 311, 1–10, 2006.
65. Wan, J., Gordon, G.B., Muirhead, K., Hickey, M.W., Coventry, M.J., Incorporation of nisin in microparticles of calcium alginate. *Lett. Appl. Microbiol.*, 24, 153–158, 1997.
66. Marcela, F., Lucía, C., Esther, F., Elena, M., Microencapsulation of L-ascorbic acid by spray drying using sodium alginate as wall material. *JEAS*, 6, 1–8, 2016.
67. Baldwin, E.A., Edible coating for fresh fruits and vegetables: Past, present and future, in: *Edible Coating and Films to Improve Food Quality*, J.M. Krochta, E.A. Baldwin, M.O. Nisperos-Carreido (Eds.), pp. 25–64, Techonomic publishing company, Inc., Lancaster, Pennsylvania, USA, 1994.
68. Raghav, P.K., Agarwal, N., Saini, M., Edible coating of fruits and vegetables: A review. *IJSRME*, 1, 1, 188–204, 2016.
69. Ghavidel, R.A., Davoodi, M.G., AdibAs, A.F., Tanoori, T., Sheykholeslami, Z., Effect of selected edible coatings to extend shelf-life of fresh-cut apples. *Int. J. Agri. Crop Sci.*, 6, 16, 1171–1178, 2013.
70. Moreira, M.R., Cassani, L., Martín-Belloso, O., Soliva-Fortuny, R., Effects of polysaccharide-based edible coatings enriched with dietary fiber on quality attributes of fresh-cut apples. *J. Food Sci. Technol.*, 52, 12, 7795–7805, 2015.

71. Moraes, K.S., Fagundes, C., Melo, M.C., Andreani, P., Monteiro, A.R., Conservation of *Williams* pear using edible coating with alginate and carrageenan. *Ciênc. Tecnol. Aliment. Campinas*, 32, 4, 679–684, 2012.
72. Jansrimanee, S. and Lertworasirikul, S., Effect of sodium alginate coating on osmotic dehydration of pumpkin. *Int. Food Res. J.*, 24, 5, 1903–1909, 2017.
73. Chiabrando, V. and Giacalone, G., Effect of chitosan and sodium alginate edible coatings on the postharvest quality of fresh-cut nectarines during storage. *Fruits*, 71, 2, 79–85, 2016.
74. Berlin, A., Calcium alginate films and their application for meats used for freezing. *Mayasnaya Ind. S.S.S.R.*, 28, 44, 1957.
75. Mountney, G.J. and Winterm, A.R., The use of calcium alginate for coating cut up poultry. *Poultry Sci.*, 40, 1, 28–34, 1961.
76. Yu, X.L., Li, X.B., Xu, X.L., Zhou, G.H., Coating with sodium alginate and its effects on the functional properties and structure of frozen pork. *J. Muscle Foods*, 19, 4, 333–351, 2008.
77. Allen, L., Nelson, A.I., Steinberg, M.P., McGill, J.N., Edible corn-carbohydrate food coatings. *Food Technol.*, 17, 104, 1963.
78. Song, Y., Liu, L., Shen, H., You, J., Luo, Y., Effect of sodium alginate-based edible coating containing different anti-oxidants on quality and shelf life of refrigerated bream (*Megalobrama amblycephala*). *Food Control*, 22, 608–615, 2011.
79. Khanedan, N., Motalebi, A.A., Khanipour, A.A., Koochekian sabour, A., Seifzadeh, M., Hasanzati rostami, A., Effects of different concentrations of Sodium alginate as an edible film on chemical changes of dressed Kilka during frozen storage. *Iran J. Fish Sci.*, 10, 4, 654–662, 2011.
80. Koushki, M.R., Azizi, M.H., Koohy-Kamaly, P., Azizkhani, M., Effect of calcium alginate edible coatings on microbial and chemical properties of lamb meat during refrigerated storage. *JFQHC*, 2, 1, 6–10, 2015.
81. Medeiros, B.G.S., Souza, M.P., Pinheiro, A.C., Bourbon, A.I., Cerqueira, M.A., Vicente, A.A., Carneiro-da-Cunha, M.G., Physical characterisation of an alginate/lysozyme nano-laminate coating and its evaluation on 'Coalho' cheese shelf life. *Food Bioprocess Technol.*, 7, 1088–1098, 2014.
82. Edwards, M., Hanniffy, D., Heesch, S., Hernández-Kantún, J., Moniz, M., Queguineur, B., Ratcliff, J., Soler-Vila, A., Wan, A., Soler-Vila, A., Moniz, M., *Macroalgae Fact-sheets*, p. 40, Irish Seaweed Research Group, Ryan Institute, NUI Galway, 2012.
83. Storebakken, T., Binders in fish feeds: I. effect of alginate and guar gum on growth, digestability, feed intake and passage through the gastrointestinal tract of rainbow trout. *Aquaculture*, 47, 1, 11–26, 1985.
84. Rychen, G., Aquilina, G., Azimonti, G., Bampidis, V., Bastos, M.L., Bories, G., Chesson, A., Cocconcelli, P.S., Flachowsky, G., Kolar, B., Kouba, M., Lopez-Alonso, M., Lopez Puente, S., Mantovani, A., Mayo, B., Ramos, F., Saarela, M., Villa, R.E., Wallace, R.J., Wester, P., Lundebye, A.K., Nebbia, C., Renshaw, D., Innocenti, M.L., Gropp, J., Scientific Opinion on the safety and efficacy of sodium and potassium alginate for pets, other non food-producing

animals and fish. *EFSA J.*, 15, 7, 4945, 1–13, 2017, https://doi.org/10.2903/j.efsa.2017.4945.

85. Kumar, B.N.P., Reddy, K.K., Mallikarjuna, P.V.R., Characterisation of blended sodium alginate microcapsules for controlled release of animal feed supplements in the git. *JFAV*, 2, 3, 161–166, 2012.
86. Yan, G.L., Guo, Y.M., Yuan, J.M., Liu, D., Zhang, B.K., Sodium alginate oligosaccharides from brown algae inhibit *Salmonella enteritidis* colonization in broiler chickens. *Immunol. Health Dis.*, 90, 7, 1441–1448, 2011.
87. Houghton, D., Wilcox, M.D., Chater, P.I., Brownlee, I.A., Seal, C.J., Pearson, J.P., Biological activity of alginate and its effect on pancreatic lipase inhibition as a potential treatment for obesity. *Food Hydrocoll.*, 49, 18–24, 2015.
88. Jensen, M.G., Kristensen, M., Astrup, A., Effect of alginate supplementation on weight loss in obese subjects completing a 12-wk energy-restricted diet: a randomized controlled trial. *Am. J. Clin. Nutr.*, 96, 1, 5–13, 2012.
89. Anderson, D.M.W., Brydon, W.G., Eastwood, M.A., and Sedgwick, D.M., Dietary effects of sodium alginate in humans. *Food Addit. Contam.*, 8, 3, 237–248, 1991.
90. Terada, A., Hara, H., Mitsuoka, T., Effect of dietary alginate on the faecal microbiota and faecal metabolic activity in humans. *Microb. Ecol. Health Dis.*, 8, 6, 259–266, 1995.
91. Cheng, W. and Yu, J.S., Effects of the dietary administration of sodium alginate on the immune responses and disease resistance of Taiwan abalone, *Haliotis diversicolor supertexta*. *Fish Shellfish Immunol.*, 34, 3, 902–908, 2013.
92. Harikrishnan, R., Kim, M.C., Kim, J.S., Han, Y.J., Jang, I.S., Balasundaram, C., Heo, M.S., Immunomodulatory effect of sodium alginate enriched diet in kelp grouper *Epinephelus bruneus* against *Streptococcus iniae*. *Fish Shellfish Immunol.*, 30, 2, 543–549, 2011.
93. Yeh, S.P., Chang, C.A., Chang, C.Y., Liu, C.H., Cheng, W., Dietary sodium alginate administration affects fingerling growth and resistance to *Streptococcus* sp. and iridovirus, and juvenile non-specific immune responses of the orange-spotted grouper, *Epinephelus coioides*. *Fish Shellfish Immunol.*, 25, 1-2, 19–27, 2008.
94. Van Doan, H., Hoseinifar, S.H., Tapingkae, W., Khamtavee, P., The effects of dietary kefir and low molecular weight sodium alginate on serum immune parameters, resistance against *Streptococcus agalactiae* and growth performance in Nile tilapia (*Oreochromis niloticus*). *Fish Shellfish Immunol.*, 62, 139–146, 2017.
95. Kokilam, G., Vasuki, S., Babitha, D., Influence of dietary supplementation of sodium alginate on gut flora and biochemical composition in *Fenneropenaeus indicus*(Indian major shrimp). *Adv. Appl. Sci. Res.*, 7, 2, 167–173, 2016.
96. Park, B.Y., Kim, J.H., Cho, S.H., Hwang, I.H., Jung, O.S., Kim, Y.K., Lee, J.M., Yun, S.G., Effects of a dietary chitosan-alginate-Fe(II) complex on meat quality of pig longissimus muscle during ageing. *Asian-Aust. J. Anim. Sci.*, 18, 3, 414–419, 2005.

97. Ayala, M.D., García-Alcázar, A., Arizcun, M., Abdel, I., López-Albors, O., Effect of sodium alginate dietary in body parameters and muscle growth of gilthead sea bream, *Sparus aurata* L. *Turk. J. Fish Aquatic Sci.*, 12, 627–633, 2012.
98. Hanzal, V., Divisova, M., Murawska, D., Janiszewski,P., The effect of dietary bio-alginate supplementation of the growth rate and body weights of common pheasant (*Phasianus colchicus*) chicks. *Pol. J. Natur. Sci.*, 31, 3, 363–371, 2016.
99. Silva, C.S., Bosch, G., Bolhuis, J.E., Stappers, L.J.N., Van Hees, H.M.J., Gerrits, W.J.J., Kemp, B., Effects of alginate and resistant starch on feeding patterns, behaviour and performance in ad libitum-fed growing pigs. *Animal*, 8, 12, 1917–1927, 2014.
100. Yoshie, Y., Suzuki, T., Shirai, T., Hirano, T., Effect of sodium alginate on fat contents and digestive organs of rats fed with fat-free diet. *Fish Sci.*, 61, 4, 668–671, 1995.
101. Chiu, C.T., Chang Gung Memorial Hospital, Efficacy and safety of sodium alginate oral suspension to treat non erosive gastro-esophageal reflux disease, Clinialtrial.gov., 2018.
102. Corvaglia, L., Aceti, A., Faldella, G., Efficacy and safety of sodium alginate for GERD in preterm infants. *Aliment Pharmacol. Ther.*, 33, 979–985, 2011.
103. Turck, D., Bresson, J.L., Burlingame, B., Dean, T., Fairweather-Tait, S., Heinonen, M., Hirsch-Ernst, K.I., Mangelsdorf, I., McArdle, H.J., Naska, A., Neuh€auser-Berthold, M., Nowicka, G., Pentieva, K., Sanz, Y., Siani, A., Sjodin, A., Stern, M., Tome, D., Vinceti, M., Willatts, P., Engel, K.H., Marchelli, R., Peoting, A., Poulsen, M., Schlatter, J.R., Turla, E., Van Loveren, H., Scientific Opinion on the safety of alginate-konjac-xanthan polysaccharide complex (PGX) as a novel food pursuant to Regulation (EC) No 258/97. *EFSA J.*, 15, 5, 4776, 1–24, 2017.
104. http://www.icrt-idtf.com/en/index.php?act=show&id=44, International database Transport (for) Feed.

Part 4

ALGINATES FUTURE PROSPECTS

14

Alginates: Current Uses and Future Perspective

Ashwini Ravi[1], S. Vijayanand[1], G. Ramya[1], A. Shyamala[1], Velu Rajeshkannan[2], S. Aisverya[3], P.N. Sudha[3] and J. Hemapriya[4*]

[1]*Bioresource Technology Lab, Department of Biotechnology, Thiruvalluvar University, Sekkadu, Vellore, Tamilnadu, India*
[2]*Department of Microbiology, Bharathidasan University, Tiruchirapalli, Tamil Nadu, India*
[3]*Biomaterial Research Lab, Department of Chemistry, DKM College for Women, (Autonomous), Vellore, Tamilnadu, India*
[4]*PG & Research, Department of Microbiology, DKM College for Women (Autonomous), Vellore, Tamilnadu, India*

Abstract

Exopolysaccharides of marine origin have been found to be used extensively in various fields, and their demand is continuously increasing due to their cheap extraction process and several advantageous characteristics. The production of these biopolymers was accounted for only 3–5% during 1980 to 2000. But the emerging markets in several developing and developed countries as well created the need of exploring and using more marine polymers. Alginate is an extensively used marine biopolymer that is derived from the cell wall of brown algae and some bacterial species. Alginates have been found to be exploited in the field of pharmaceutics and food industries due to their flexible physical, chemical, and physiological properties. They have been used in drugs, protein delivery, tissue regeneration, wound dressing, and wound healing in pharmaceutical industry. In food industry, they have been used as emulsifiers, stabilizers, texturizers, etc. and in paper industry to improve crumpling and resistance of paper. Also, they find their application in textile, cosmetics, and welding industries as thickeners and as a binder of flux, respectively. These advantageous characteristics and the wide use of alginate make them a promising candidate for future applications. In this chapter, the future perspectives of alginate have been discussed in detail.

Keywords: Alginates, drug delivery, emulsifiers, thickeners, cell culture systems

**Corresponding author*: micropriya80@gmail.com

Shakeel Ahmed (ed.) Alginates, (283–312) © 2019 Scrivener Publishing LLC

14.1 Introduction

Marine organisms have been considered as a valuable source for diverse range of materials that are gaining importance in several fields of development, especially food and medicine. These explorations and use of various marine-originated compounds lead to the development of a new field of biotechnology called blue biotechnology [1]. Among various compounds extracted or obtained from marine organisms, polysaccharides possess a great stand in them [2]. Polysaccharides are a natural, nontoxic, biocompatible substance formed by several polymeric carbohydrate molecules linked together by glycosidic linkages [3, 4]. Though polysaccharides are available in plenty of organisms, those from marine origin have been concentrated by scientists and industrialists since they are the primary constituents of marine organisms, which make them to be obtained easily, and they are available in abundant quantity, which aids in increased production [5]. The main contributor of polysaccharides from marine sources are the sea weeds. During the past few years, polysaccharides from sea weeds are gaining importance due to their diverse structural availability and various biological properties [6]. The polysaccharides obtained from sea weeds include chitosan, fucans, ulvans, carrageenans, laminarins, alginates, agar, hyaluronic acid, xanthan, xyloglucan, pullulan, dextran, etc. [7].

Among various polysaccharides, alginates are the second most common exopolysaccharides produced by bacteria and are abundantly produced in brown sea weeds of Class Phaeophyceae [8, 9]. Like other polysaccharides, alginates form the cell walls and intracellular matrix of brown seaweeds, thereby providing them with mechanical strength and flexibility [10]. Alginates were found to be present in both acid and salt forms and are found to be formed by unbranched exopolysaccharide units of mannuronic acid and guluronic acid [11]. It is the salt form of alginate that is found in the cell wall of brown seaweed and it constitutes 40–45% of dry matter, whereas the acid form of alginate is a linear polyuronic acid called alginic acid [12, 13]. The exploration of alginates has found their application in food industry as gelling agents, viscosity agents, etc. and also in pharmaceutical industry as wound dressings, tissue engineering, etc. [14, 15]. In the present chapter, the biogenesis of alginates along with its applications and future perspectives has been discussed in detail.

14.2 Sources of Alginate Synthesis

The polysaccharide alginate has been majorly extracted from two main sources, viz., brown seaweed and bacteria. The brown sea weeds of class Phaeophycaea and two bacterial species *Azotobacter* and *Pseudomonas* were found to be the major producers.

14.2.1 Brown Seaweeds

Seaweeds are macro algae that are found in various habitats from fresh waters to sea zones, and like land plants, they are capable of producing their own food. The sea weeds have been classified into three phylums, viz., Ochrophyta, Chlorphyta, and Rhodophyta. Of these phyla, brown alga belongs to the category Ochrophyta. The classification of brown algae is given below [16]:

Kingdom	:	Chromista
Phylum	:	Ochrophyta
Class	:	Phaeophyceae
Order	:	Dictyotales Fucales Laminariales Desmarestiales Scytothamnales
Family	:	Dictyotaceae Durvillaeacea Hormosiraceae Sargassaceae Seirococcaceae Xiphophoraceae Alariaceae Laminariaceae Lessoniaceae Desmarestiaceae Splachnidiaceae Scytothamnaceae

The brown algae were found to represent 2000 species in 256 genera and are found in fresh water, ocean, and also in brackish waters. The characteristic color of the brown sea weeds is due to its pigment carotenoid fucoxanthin and in some cases, phaeophycean tannins. The storage reserve includes laminarins and the cell wall is made up of cellulose, usually containing the mucopolysaccharide alginic acid. Therefore, brown algae are considered the major reserve for alginic acid production. The various species from which alginate has been extracted is given in Table 14.1.

Table 14.1 Algae used for the extraction of alginates.

S. No.	Species	Reference
1.	*Laminaria digitata*	[17–19]
2.	*Sargassum* sp.	[20, 21]
3.	*Sargassum filipendula*	[22, 23]
4.	*Sargassum duplicatum*	[24]
5.	*Sargassum crassifolium*	[24, 25]
6.	*Sargassum muticum*	[26]
7.	*Sargassum vulgare*	[27]
8.	*Sargassum augustifolium*	[28]
9.	*Undaria pinnatifida*	[29]
10.	*Sargassum baccularia*	[30]
11.	*Sargassum binderi*	[30]
12.	*Sargassum siliquosum*	[30]
13.	*Turbinaria conoides*	[30]
14.	*Colpomenia sinuosa*	[31]
15.	*Lobophora variegata*	[31]
16.	*Chnoospora implexa*	[31]
17.	*Padina gymnospora*	[31]

(*Continued*)

Table 14.1 Algae used for the extraction of alginates. (*Continued*)

S. No.	Species	Reference
18.	*Sargassum tenerrium*	[31]
19.	*Dictyota dichotoma*	[31]
20.	*Sargassum turbinarioides*	[32]
21.	*Sargassum polycystum*	[25]
22.	*Sargassum echinocarpum*	[25]
23.	*Padina* sp.	[25]
24.	*Sargassum wightii*	[33, 34]
25.	*Macrocystis pyrifera*	[19]
26.	*Durvillaea anthartica*	[35]
27.	*Lessonia flavicans*	[35]
28.	*Ecklonia maxima*	[35]
29.	*Ascophyllum nodosum*	[35]
30.	*Laminaria pallida*	[36]
31.	*Laminaria hyperborea*	[37]
32.	*Laminaria saccharina*	[38]
33.	*Lessonia trabeculata*	[39]
34.	*Lessonia nigrescens*	[40]

Alginates of commercial purposes are usually obtained from *Laminaria* sp., *Lessonia* sp., *Ecklonia maxima, Durvillaea antarctica,* and *Ascophyllum nodosum.* In later years, *Sargassum* sp. has been explored for the production of alginates. It is clear from the table that *Sargassum* sp. has been concentrated more for the production of alginates, whereas the least producers are considered to be *Undaria pinnatifida.*

14.2.2 Bacteria

The production of alginic acid by bacteria made a breakthrough in production of alginic acid by brown sea weeds [41]. Though brown algae are abundantly present in the environment, it has several disadvantages such as high production costs, high price, seasonal changes, and other environmental impacts associated with utilization of these brown sea weeds. These reasons have made researchers to look out for different sources for the production of alginic acid. In later years, it was Linker and Jones [42] and Gorin and Spencer [43] who found that bacteria *Pseudomonas aeruginosa* and *Azotobacter vinelandii* respectively produces polyuronic compounds that resemble alginic acid. After this identification, improvisation in production of alginic acid by these bacterial species has been explored. It has also been found that several other bacterial species, such as *Pseudomonas mendonica, Pseudomonas putida, Pseudomonas phaseolicola, Pseudomonas savastanoi, Pseudomonas glycinae*, and *Pseudomonas fluorescens* produce alginic acid [44–47].

14.3 Synthesis of Alginate

As described earlier, alginate is a linear exopolysaccharide comprising D-mannuronic acid and L-guluronic acid linked by β (1→4) and α (1→4) glycosidic bond. These mannuronate and guluronate units of alginate are called M and G blocks, respectively [10] (Figure 14.1).

Alginate biosynthesis has been divided into four stages, viz., synthesis of precursor substrate, polymerization and cytoplasmic membrane transfer, periplasmic modification, and export through the outer membrane. The biosynthesis of alginates initiates with mannose in sea weeds, which is then converted to mannuronate and then alginate. Whereas in bacteria, the precursor of alginate biosynthesis is fructose-6-phosphate, which is converted to Guanosine diphosphate (GDP) mannuronic acid. It is then

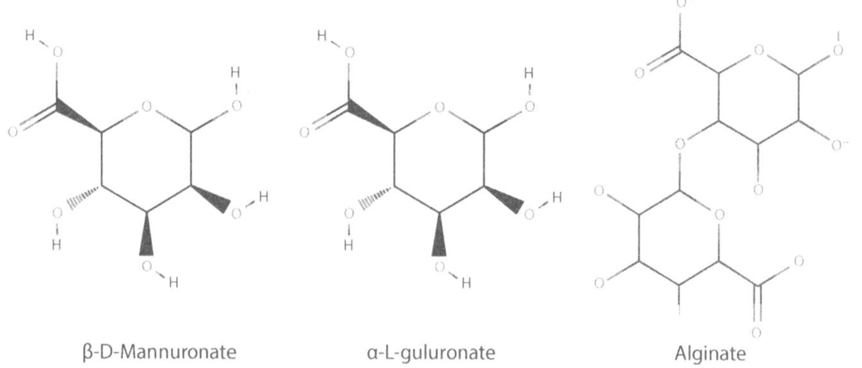

Figure 14.1 Structure of mannuronate, guluronate, and β (1→4) mannuronic acid and α (1→4) guluronic acid.

polymerized to polymannuronic acid. During this, epimerization of mannuronic acid units with guluronic acid units occurs. The variation in epimerization, i.e., the concentration of M and G blocks, alters the structure of alginates. After epimerization, the alginate is transported to the outer membrane. During this process, several phosphate molecules are involved, which is maintained by the alginate synthesis genes.

14.3.1 Alginate Biosynthesis Gene

The production of alginate in bacteria has been studied extensively, and the genes involved in its synthesis have been mapped by various techniques. These genetic studies revealed that alginate biosynthesis in bacteria has been controlled by alg genes. The alg genes have been found to be predominantly present in members of *Pseudomonas*, *Azotobacter*, and also azomonas [48]. To date, 25 alg genes have been identified to be involved in the production of alginates [49, 50].

The genes involved in production of alginates have been categorized into three types, namely, structural or biosynthetic genes, regulatory genes, and genome switching genes. The structural or biosynthetic category includes 13 genes, which are named as algD, algA, alg8, alg44, algK, algE, algG, algX, algI, algJ, algL, algF, and algC. The genes algE and algX were formerly called as alg76 and alg60, respectively. Genetic mapping studies showed that the genes algA through algF were found in a row of 34 minutes whereas the gene algC alone was found to be present 10 minutes far from these genes (Figure 14.2a) [51, 52].

The regulatory genes include algZ, algR, algQ, algB, and algP of which algZ to algP is present continuously for 9 minutes. algP is interrupted from

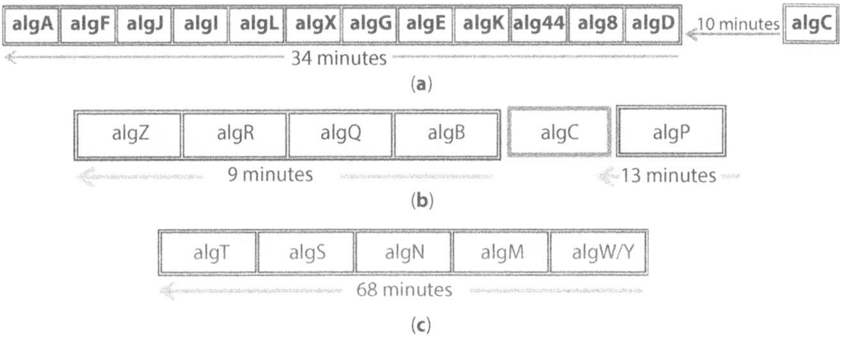

Figure 14.2 Gene order of alginate synthesis genes—(a) biosynthesis genes, (b) regulatory genes, and (c) genotypic switching genes.

the rest of the genes by algC, which is the biosynthetic gene. The algP was found to be present in 13 minutes (Figure 14.2b). The final sets of genes are the genotypic switching genes that include algT and muc genes. The muc genes are categorized into mucA, mucB, mucC, and mucD, which are also called as algS, algN, algM, and algW/Y, respectively. These genes are present in a row for 68 minutes (Figure 14.2c).

14.4 Properties of Alginates

The activity of alginate is due to its physical, chemical, and biological properties, which are discussed below.

14.4.1 Molecular Weight

Alginates have been used extensively in food industries as well as pharmaceuticals mainly because of its gelling property. The gelling property of alginate is determined by its viscosity, which again is determined by its molecular weight [53, 54]. Like other polysaccharides, alginates are also a polydisperse molecule present in various molecular weights [10]. With the wide variety of extraction sources, there are more than 200 types of alginates that are commercially available in the market. It is therefore very much essential to know about their molecular weight before using it for any application [55].

The molecular weight of alginates is affected by the G or M blocks present in them. It has been found that low molecular weight alginates have lower G blocks, and therefore they cannot be used as an effective gelling agent [56]. As of now, molecular weights ranging from 48,000 to 2,700,000 g/mol have been found from a variety of species such as *Laminaria digitata, L. hyperborean*, etc. [57]. Several techniques in identifying the molecular weight of alginates have been used. Majority of the works concentrate on sedimentation diffusing viscosity, sedimentation intrinsic viscosity determination, and light scattering [53, 58]; a work by Sperger *et al.*, [59] showed that they can be determined by solid-state nuclear magnetic resonance (NMR) even in the presence of water molecules. Similarly, several methods were there to find the intrinsic viscosity of alginates; Huggin's method was found to be the best method followed by the Martin, Kraemer, Schulz–Blashke, and Baker–Philippoff method [57]. It has been found from another study that low molecular weight alginate can be used in poly(methylene-co-guanidine) capsules [60]. Similarly, another study by Mancini *et al.*, [61] showed that the amount of alginate, irrespective of its molecular weight, is enough for the structure retention of mayonnaise. In this way, molecular weight affects the function of alginate.

14.4.2 Solubility

The solubility of alginate is essential for its application in various industries. Alginates are soluble in water and insoluble in organic solvents, whereas alginic acid is insoluble in water and soluble in organic solvents. The solubility of any polysaccharide is affected by external factors such as pH, ionic strength, high temperatures, and reducing agents. For alginates, the solubility is affected by pH, ionic strength, and salts; the pH of the solution affects the alginate by altering its uronic acid units. It has been found that the decreased solubility of alginate during the commercial process of extraction is due to the removal of a compound called ascophyllan. This compound binds with the alginate and makes it more soluble even in low pH [62]. As mentioned earlier, alginic acid is insoluble in water. Therefore, the composition of alginic acid was studied by Haug *et al.*, [63] by dissolving it in oxalic acid. It has been found that even at a low pH of 2.85, oxalic acid can dissolve alginic acid to 80–90%.

As discussed earlier, the solubility of alginate is affected by ionic strengths, which cause salting out of alginates and inorganic salts causing precipitation. In case of mixed solution, the solubility is affected by dielectric constant. Since alginates are readily soluble in water, to make use of them in various applications, several hydrophobic molecules are added to their backbone. Such hydrophobic alginates are produced by oxidation of hydroxyl groups or by reacting them with tetrabutylammonium carboxylates [64, 65].

14.4.3 Stability

Alginates can remain stable when not exposed to sunlight without any considerable changes in their molecular weight, and also, they function for several months in a dry, cool place. But alginic acid cannot be kept stable over a period of time [66]. There are several factors that affect the stability of alginates such as acidic hydrolysis, alkaline oxidation, and bacterial degradation. These factors affect the stability of alginate by degrading its molecular structure, thereby reducing its molecular weight. When the molecular weight is affected, the function of alginate is also affected. Therefore, in order to maintain its molecular weight and stability, it has to be protected.

Several studies demonstrated the stability of alginates in their various forms such as impressions, hydrogels, etc. since they are used in dental casings and wound dressings. The stability of alginate has been checked in alginate impressions by neodymium ions concentration, which showed that they can be used as an effective material for dental casing [67]. Similarly, in another study, extended pouring of gypsum on alginate impression showed that it performs as better casting without any dimensional changes [68]. When alginate was tested for its stability with commonly used disinfectants used in dental surgery, glutaraldehyde and hypochlorite solutions, they were found to remain unaffected for 60 minutes and other disinfectant sprays were found to cause no harm [69, 70]. The conformation changes of alginate impression with stainless steel were found to be not affected for 30 minutes, and it remains unchanged even at 100 hours of time, making them a perfect candidate for its use in dental casings [71].

14.4.4 Ionic Binding Property

The ion binding ability of alginate determines the gel formation. The alginates are known to bind well with earth metals such as Ca^{2+}, Mg^{2+}, Sr^{2+}, and Ba^{2+}. This binding ability of alginate with the divalent or sometimes the monovalent earth metals is because of its configuration in the G blocks. It has been found in previous studies that G blocks have the ability to bind with metals, and therefore, the polymer with more GG blocks rather than the M blocks or MG blocks tends to have less binding and therefore less gelling ability [72–74]. This phenomenon of binding metals with G blocks in gel formation has been explained in an "egg box model." This model explains the phenomenon of binding based on the ligands present in G blocks and also the steric interferences of them with G blocks. However, this model is based completely on the intuitive understanding; it is still used as the base model for ion binding properties of alginate [56, 75, 76].

This egg box model has been studied to better understand the interaction of earth metals with alginates and has been found that interaction of Ca^{2+}-induced alginates was unable to produce chains but is able to produce sufficient short bondings with the oligoguluronate, which resulted in less gelling properties with respect to its interaction with non-Ca^{2+}-limited regimes. This study made clear that the oligoguluronate oligomers either sequester calcium by binding with the oligoguluronate sequence or between the free oligoguluronates [77]. Later, in the year 2013, Borgogna *et al.*, [78] proposed a tilted egg box model, which says that addition of calcium even in smaller amounts to alginates causes multicomplex binding modality. This was in contrast with the monocomplex model proposed by Fang *et al.*, [79]. The alginate has been bound to Cu^{2+} to study its gelling abilities and has been proved that they can function better than calcium for studying its ion binding properties [80]. Ca-alginate has been used in several applications such as removal of heavy metal from water bodies and various medical applications [81–84]. Apart from the ultimate binder, calcium, other metals such as Fe have been used for binding with alginates and have been used in human dermal fibroblasts [85].

14.4.5 Gel Formation Ability

Alginates are best known for their property as a gelling agent. They are obtained from nature as Ca, K, Na, and Mg salts. Of the other forms, Na-alginate is used as the best gelling agent. Since Na- alginates are soluble in both hot and cold waters, they are used as thickeners, emulsifiers, and gel-forming agents in food industry [86]. Addition of calcium to alginic acid causes them to be converted into gels. But this is crucial since it results in instantaneous precipitation rather than forming gels. Therefore, it is very much essential to be done in a controlled environment where the pH is lowered gradually below the pKa of guluronic acid and addition of slowly hydrolyzing lactones is entertained [86].

As discussed earlier, the gelling of alginate is instantaneous once the addition of calcium is started. It has been found that calcium lactate was found to have the shortest gelling time of 3 minutes, whereas calcium carbonate was found to have the longest gelling time of more than 3 days at 20°C and 60°C, respectively [87]. It has been denoted before that alginates with G blocks have better gelling ability when binding with ions, but as far as its gelling ability for absorption or immobilization is concerned, M block was found to have better gel swelling ability than the G blocks, and also the gel swelling ability of G blocks was found to be increased by addition of sodium ions into the fiber [88]. In the field of medicine, Ca-treated alginate gels were found to be used in bone-like apatite formations [89, 90].

14.4.6 Biological Properties

Alginates do not have any nutritional values, but their physical properties of gel formation and reacting with polymers of food components such as proteins under favorable conditions make them to be useful in maintaining the structure of certain processed foods and pet foods. Therefore, they have been used as thickeners, softeners, and gelling agents in food industry [86]. As far as its medicinal application is concerned, the use of alginate in the field of medicine first initiated with transplantation of encapsulated pancreatic islets of Langerhans. But this was found to cause several immunogenic ailments [91]. Since then, several studies report that alginates with M blocks were found to be immunogenic, whereas other reports show no such activity. But still, several works on the use of alginate in medicinal field are still being explored [92]. Houghton *et al.*, [93] showed that alginates can be used as a weight loss agent since they give zero calories.

14.5 Application of Alginates

Alginates have found their application in several industries, which have been discussed in brief in the table below:

Properties	Applications	Reference
In food industry		
Thickeners and gelling agents	Ice creams, soups, sauces, dressings, ketchup, mayonnaise, margarine, milkshakes, fruit juices, liquors, desserts, jams, puddings, whipped cream, pie fillings, mashed potatoes, and restricted foods	[10]
Stabilizers	Ice cream, wheat dough	[94, 95]
Emulsifiers	Beverages	[96]
Texturizers	Desserts	[97]
Encapsulates	Probiotic	[98–101]
Food coatings	Fruits, vegetables, and meat	[102–104]
Binders	In pet foods	[105, 106]
In pharmaceutics		
Delivery	Drugs and proteins	[107–109]
Cell culture		[110–112]
Tissue regeneration		[113]

Wound healing		[114]
Wound dressing		[115]
Other applications		
Thickeners	Printing paints	[116]
Bioremediation	Waste water treatment	[117–119]
Bioremediation	Tannery effluent treatment	[120, 121]
Fertilizers		[122, 123]

14.6 Future Perspectives of Alginates

The property of alginates in forming gels by binding with various ions and its sensitivity to various environmental factors such as temperature and pH make it useful in both food and medicinal industries. Though they have been exploited in both these industries, their applications by varying their physical forms are still explored.

14.6.1 3D-Based Cell Culture Systems

A wide range of cell culture studies such as tissue engineering, drug delivery, and cell culture analysis are carried out in 3D cell cultures. Alginates are used as a cell scaffold in the development of 3D cultures of neural cells such as astroglioma cells, astrocytes, microglia and neurons [124], and human osteoblast cells [125]. Cancer stem cell (CSC) model has become one of most eminent models for assumption of tumorigenesis. A large number of limitations were found in the ancient methods where 2D cell culture model was employed. 3D cell culture plays an efficient role in CSC research and promotes discovery of various anticancer drugs. A CSC-related gene was found to exhibit higher expression level in 3D cell culture model, which has been enriched with alginates [126]. Alginates are found to be a promising candidate in the field of drug delivery. But they also pave the way for the development of devices used for drug delivery and monitoring the effect of chemicals on 3D cells. Drops of sodium alginate immersed in calcium chloride are used in designing 3D environment in electrowetting on dielectric digital microfluids [127].

A work done by Chen *et al.*, [166] clearly describes about the encapsulation of mouse-induced pluripotent stem cells using alginate as core with the thin shell of Poly(L-lysine)-graft-Poly(ethylene glycol) PLL-g-PEG. This microsphere encapsulation methods enable us to analyze and quantify the cell growth effectively and feasibly using high throughput analysis. Development of 3D alginate tumor spheroid culture system for murine osteosarcoma (OS) allows the researchers in investigating the pathobiology of OS. The alginate bead transplantation model serves as a spontaneous pulmonary metastatic model [128]. A combination of alginate, marine collagen, and agarose possesses higher cytocompatibility, and the usage of cell-tracking chemicals has been kicked out due to its transparency [129]. Adapting alginates in regenerative medicines in the form of 3D culture supports and improves the function of the organ [130].

14.6.2 Impressions

Natural alginates are found to be one of the commonly used impression materials due to their easy manipulation, comfort to the users, and being relatively inexpensive for the dentists. Irreversible hydrocolloids (alginates) are the impression material widely used in making dental cast for diagnosis, treatment, and fabrication. Elastic recovery is one of the abilities of the alginates. The greater the elastic recovery, the more accurate the impression will be. According to Frey *et al.*, [131], three different alginate polymers (Identic, Jeltrate, and Kromopan) were tested for their mechanical property and tear energy using two different mixing procedures, which finally proved that both the mixing method and three polymers were found to be effective. According to Cook [132], both the permanent set and tear energy were improved by increasing the alginate content and reducing the filler level.

A study carried out by Cohen *et al.*, [133] explains the dimensional stability of three different alginates (Jeltrate, Hydrogum, and a new hydrophilic alginate) stored in three different conditions, where 160 casts were fabricated and measurements were taken buccolingually, mediodistally, and diagonally. Few alginates were found to possess higher compatibility with dental stones. According to Morrow *et al.*, [134], few alginates used in dental impression possess compatibility with dental stone (gypsum); the article also states that it is important to remember that for compatibility with gypsum, the alginate impression material must fulfill only one of the 13 requirements to achieve certification.

14.6.3 Cell-Based Microparticles

Controlled or sustained drug delivery has been used for several ailments in which a natural or synthetic polymer is combined with an active compound. This controlled delivery of drugs in living environment helps in various ways by prolonging the period of drugs' survival, protecting them from digestive enzymes, etc. Controlled drug release also eliminates the under- and overdosing of drugs. While several polymers have been used for the controlled release combinations, alginates were found to be an efficient candidate. The biocompatibility of alginates in living systems and also their efficient ability to entrap several ranges of active compounds make them useful as carriers for sustained release of biomolecules as well as cells.

Several studies have been conducted on examining the encapsulation of active compounds by alginates. Ciofani *et al.*, [135] reported the controlled release of Netrin-I, an axonal guidance protein in the development of neuronal cells. *In vivo* experiments on alginate microspheres with Netrin-I on embryonic neuronal cells showed that neuronal axons grow toward Netrin-I, indicating the efficient release of Netrin-I from the alginate microsphere, thus showing the efficiency of alginates as microspheres. Alginates have also been coupled along with other molecules such as polyethylene glycol, Poly-L-Ornithine (PLO), and Arginylglycylaspartic acid (RGD) for the immobilization of human foreskin fibroblasts, islets, and human mesenchymal stem cells, respectively [136–138]. Khanna *et al.*, [138] studied alginates' layered coating for controlled release theirin which alginates have been a core onto which angiogenic protein FGF-1 has been coated. After the coating, alginate has been layered above them, making them a layered microsphere. It has been found that the release of FGF-1 from the layered alginate microspheres prolonged for 30 days.

The properties of alginates have also been studied for efficient microsphere formation. Ramesh *et al.*, [139] showed that high modulus hydrogels of alginates demonstrated better proliferation of Wharton's jelly mesenchymal stromal cells (WJMSCs) and differentiation of osteogenic cells by upregulation of osteospecific genes. Similarly, a study by Mahou and Wandrey [140] showed that the permeability of poly ethylene glycol (PEG)-linked alginate gels was affected by the arm length of the PEG, and the swelling behavior was affected by the concentration of alginates. Therefore, by altering the concentration, more stable alginates can be prepared. Apart from their use as microspheres, microdroplets have also been prepared. Bigdeli *et al.*, [141] proposed the microdroplet preparation of *Rhodobacter sphaeroides* diluted in alginate polymer. The process was found to be an efficient single-step process for application and also detection. In order to study the efficiency of alginates in controlled drug disease, confocal Raman microscopy has been used by Kroneková *et al.*, [142]. It has been used for both qualitative and quantitative mapping of alginates throughout the microbead volume. The degrading efficiency of alginates in living system for the release of active compounds has been studied by Ashton *et al.* [143]

in which the alginate poly(lactic-co-glycolic acid) (PLGA) is loaded with the alginate lyases into alginate hydrogels. The alginates were found to be degraded enzymatically, and neural progenitor cells' (NPC) culture in these degrading alginates showed a better expansion rate than cells cultured in nondegradable alginates.

14.6.4 Alginate Oligosaccharides

Alginate oligosaccharides have found their application in the field of medicine for various ailments. Guluronate oligosaccharide (GOS-ED), an alginate oligosaccharide with G blocks prepared by enzymatic degradation, was found to have effective immunomodulatory effects such as inducing nitric oxide synthase expression and stimulating reactive oxygen species (ROS) and tumor necrosis factor α (TNFα) expression. The activity of this GOS-ED was found to be maximal when compared with no or minimal effect of alginate oligosaccharides (AOS) [144]. The inhibitory action of AdO (alginate-derived oligosaccharides) in neuroinflammatory BV2 microglial cell lines showed that AdO are active in inhibiting nitric oxide and prostaglandin E2, and promoting expression of inducing nitric oxide synthase, cyclooxygenase 2, and proinflammatory cytokines. They were also found to inhibit the expression of nuclear factor (NF-κB), Lipopolysaccharide (LPS)-activated overexpression, and toll-like receptor 4 in BV2 cells. These findings on neuroinflammatory attenuation of alginate oligosaccharides suggested that they can be used as therapeutic agents in treating neurodegenerative disorders [145].

Similar studies on alginate oligosaccharides on their medicinal application show that alginate oligosaccharides and guluronate-derived oligosaccharides can efficiently activate macrophages, thereby increasing immune response, inducing cytokine secretion and (TNF) secretion, and they also act as antitumor agents in osteosarcoma patients [146–148]. Exploration of alginase from biological sources like microorganisms has been studied for degradation of alginates. Alginase obtained from *Flavobacterium* sp. has been found to be used in the production of alginate oligosaccharides [149]. Similarly, an alginate-degrading enzyme capable of degrading alginate oligosaccharides except sodium alginate has been studied in *Alteromonas* sp. [150]. The use of these biologically derived alginases makes the process of alginate oligosaccharide production more biological and harmless.

Alginate oligosaccharides find their application in food industry and also in plant growth to be expanding. A study by Falkeborg *et al.*, [151] showed that alginate oligosaccharides inhibited lipid peroxidation and promoted radical scavenging activity toward 2,2'-Azino-bis(3-ethylbenzothiazoline-

6-sulfonic acid) (ABTS), hydroxyl, and superoxide radicals. They were also found to chelate very weakly with iron molecules. These make them useful as an effective antioxidant in food products for efficient preservation and protection from oxygen molecules and oxidative damage. Khan *et al.*, [152] showed that alginate oligosaccharide produced by gamma irradiation of sodium alginate promoted plant growth and increased the content of morphine, codeine, and opium per plant. Apart from their various applications, the detection of alginate oligosaccharides was also studied and it has been found that they can be easily detected by LC-MS/MS equipped with anion exchange column and negative-ion electron tandem mass spectrometry [153, 154].

14.6.5 Drug Targeting

Site-targeted drug delivery has become an effective method for various diseases, where the drugs are carried by a vector to the targeted sites. Alginates are preferable natural polymers used in the field of medicine due to their biocompatibility, improved dispersability, and stability under physiological pH. Graphene oxide-based drug delivery system, modified by natural peptides protamine sulfate and sodium alginate, was found to establish higher drug-releasing property on the target Michigan Cancer Foundation-7 (MCF-7) cancer cell lines [155]. According to Tawfik *et al.*, [156], anionic natural polymer acts as a functionalizing agent for up conversion nanoparticles (UCNPs) and shows higher stability and compatibility. UCNPs, when incorporated with doxorubicin (DOX), were found to inhibit the growth of Ubiquitous Keratin forming tumor cell line HeLa (KB)cancer cell line compared with free DOX.

The antiviral drug *Zidovudine* encapsulated inside amide-functionalized alginate nanoparticles (AZT-GAAD NPs) was found to be a promising candidate in antiviral drug delivery for HIV/AIDS therapy. *In vitro* cellular internalization studies indicate that the NPs were found to possess higher internalization efficiency in glioma cell lines [157]. Iron-linked alginate aerogel beads were found to release the loaded oral drug ibuprofen with higher acceleration in both phosphate-buffered saline (PBS) and hydrochloric acid (HCL) by incorporating the beads with ascorbic acid [158]. Nano lipid carriers (NLCs) loaded with alginates and the hydrophobic drug Amphotericin B were found to have a high swelling ratio, where the amount of drug delivery directly depends on the size of the alginate bead, which enables them to act as an effective oral drug delivery vector [159].

14.6.6 Nanoparticulate Systems

The application of alginates in medicinal field has also been experimented with nanoparticle formation. Katuwavila *et al.*, [160] showed that chitosan–alginate nanoparticles can efficiently deliver doxorubicin to MCF 7 cells, making them an efficient candidate when compared with chitosan–doxorubicin combinations. Similarly, chitosan–alginate nanoparticles were used for the controlled release of the drug nifedipin [161]. In addition to these, chitosan–alginate nanoparticle has also been studied for the delivery of nonviral DNA via oral route. Plasmid DNA expressing humanized secreted Gaussia Luciferase as a reporter gene encapsulated in chitosan–alginate nanoparticle was found to show effective transfection of human epithelial cell lines with distinct cell differentiation [162]. Alginate nanoparticles have been combined with several molecules such as polyvinyl alcohol, piperazine, cyclodextrin, chitosan, rhodamine 6G, etc., and are studied for their delivery of compounds such as vitamin C, molecules with superficial carboxylate groups such as epidermal growth factor, ketoprofen, insulin, and other drugs, respectively [163–165].

14.7 Conclusion

Being abundant in uronic acid obtained from ocean and also the flexibility of producing them in bacteria are the best attributes of alginate for its easy production. Despite its easy production, the properties of alginates discussed in this chapter prove that they are an efficient candidate to be used in various applications. Apart from food industry and medicinal industry, the use of alginates has to be explored in many other industries. Apart from this, the immunogenic response produced by alginates as mentioned in biocompatibility can be or may be reduced by further insights into them. To conclude, alginate is a best candidate for its application, which can be put forward to the betterment of life.

References

1. Paterson, I. and Anderson, E.A., The renaissance of natural products as drug candidates. *Science*, 310, 5747, 451–453, 2005.
2. Khodse, V.B., Fernandes, L., Bhosle, N.B., Sardessai, S., Carbohydrates, uronic acids and alkali extractable carbohydrates in contrasting marine and estuarine sediments: Distribution, size fractionation and partial chemical characterization. *Org. Geochem.*, 39, 265–283, 2008.

3. Nikolić, G.S. and Cakić, M.D., Analysis of bioactive polysaccharide-metal complexes by modern FTIR spectroscopy: Copper complexes, in: *Fourier Transforms—New Analytical Approaches and FTIR Strategies*, pp. 15–44, InTech., 2011.
4. Technical Evaluation Report, Compiled by OMRI for the USDA National Organic Program, 1–23, 2015.
5. Handa, N. and Tominaga, H.A., detailed analysis of carbohydrates in marine particulate matter. *Mar. Biol.*, 2, 228–235, 1969.
6. Laurienzo, P., Marine polysaccharides in pharmaceutical applications: An overview. *Mar. Drugs.*, 8, 2435–2465, 2010.
7. Coviello, T., Matricardi, P., Marianecci, C., Alhaique, F., Polysaccharide hydrogels for modified release formulations. *J. Control. Release*, 119, 5–24, 2007.
8. Siddhesh, N.P. and Edgar, K.J., Alginate derivatization: A review of chemistry, properties and applications. *Biomaterials*, 3729–3305, 2012.
9. Susan, L.H. and Kraan, S., Bioactive compounds in seaweed: Functional food applications and legislation. *J. Appl. Phycol.*, 543–597, 2011.
10. Draget, K.I., Smidsrød, O., Skjåk-Bræk, G., Alginates from algae, in: *Polysaccharides and Polyamides in the Food Industry, Properties, Production, and Patents*, A. Steinbüchel and S.K. Rhee (Eds.), pp. 1–30, Verlag, Wiley, 2005.
11. Rehm, B.H.A. and Valla, S., Bacterial alginates: Biosynthesis and applications. *Appl. Microbiol. Biotechnol.*, 48, 281–288, 1997.
12. Arasaki, S. and Arasaki, T., *Low calorie, high nutrition vegetables from the sea*, p. 60, Japan Publications, Inc., Tokyo, 1983.
13. Morrissey, J., Kraan, S., Guirym, M.D., A guide to commercially important sea weeds on the Irish coast. 1–66, 2007.
14. Langer, R. and Vacanti, J.P., Tissue engineering. *Science*, 260, 5110, 920–926, 1993.
15. Thomas, S., Alginate dressings in surgery and wound management—Part 1. *J. Wound Care*, 9, 2, 56–60, 2000.
16. Neill, K., Nelson, W., Kelly, M., Herr, B., Beautiful browns—A guide to the large seaweeds of New Zealand. *NIWA*, 1, 1–41, 2016.
17. Vauchel, P., Kaas, R., Arhaliass, A., Baron, R., Legrand, J., A new process for extracting alginates from *Laminaria digitata*: Reactive extrusion. *Food Bioprocess Technol.*, 1, 3, 297–300, 2008.
18. Fertah, M., Belfkira, A., Dahmane, E., Taourirte, M., Brouillette, F., Extraction and characterization of sodium alginate from Moroccan *Laminaria digitata* brown seaweed. *Arab J. Chem.*, 10, S3707–S3714, 2017.
19. Rhein-Knudsen, N., Ale, M.T., Ajalloueian, F., Meyer, A.S., Characterization of alginates from Ghanaian brown seaweeds: *Sargassum* spp. and *Padina* spp. *Food Hydrocolloids*, 2017, https://doi.org/10.1016/j.foodhyd.2017.05.016.
20. Latifi, A.M., Nejad, E.S., Babavalian, H., Comparison of extraction different methods of sodium alginate from brown alga *Sargassum* sp. localized in the southern of Iran. *J. Appl. Biotechnol. Reports*, 2, 2, 251–255, 2015.

21. Vijayaraghavan, G. and Shanthakumar, S., Efficacy of alginate extracted from marine brown algae (*Sargassum* sp.) as a coagulant for removal of direct blue2 dye from aqueous solution. *Global NEST J.*, 17, 4, 716–726, 2015.
22. Kleinübing, S.J., Gaia, F., Bertagnolli, C., Gurgel, M., Silva, C.D., Extraction of alginate biopolymer present in marine alga *Sargassum filipendula* and bio-adsorption of metallic ions. *Mater Res.*, 16, 2, 481–488, 2013.
23. Costa, C.S.D., Cardoso, S.L., Nishikawa, E., Vieira, M.G.A., Silva, M.G.C., Characterization of the residue from double alginate extraction from *Sargassum filipendula* seaweed. *Chem. Eng. Trans.*, 52, 133–138, 2016.
24. Indrani, D.J. and Budianto, E., A study of extraction and characterization of alginates obtained from brown macroalgae *Sargassum duplicatum* and *Sargassum crassifolium* from Indonesia. *Dental J.*, 46, 2, 65–70, 2013.
25. Mushollaeni, W., The physicochemical characteristics of sodium alginate from Indonesian brown seaweeds. *African J. Food Sci.*, 5, 6, 349–352, 2011.
26. Belattmania, Z., Tahiri, S., Bentiss, F., Sahibed-dine, A., Jama, C., Zrid, R., Reania, A., Sabour, B., The residue of alginate extraction from *Sargassum muticum* (brown seaweed) as a low-cost adsorbent for hexavalent chromium removal from aqueous solutions. *Desalin Water Treat*, 75, 107–114, 2017.
27. Torres, M.R., Sousa, A.P.A., Filho, E.A.T.S., Melo, D.F., Feitosa, J.P.A., DePaula, R.C.M., Lima, M.G.S., Extraction and physicochemical characterization of Sargassum vulgare alginate from Brazil. *Carbohydr. Res.*, 342, 2067–2074, 2007.
28. Gholamipoor, S., Nikpour-Ghanavati, Y., Oromiehie, A.R., Mohammadi, M., Extraction and characterization of alginate from *Sargassum angustifolium* collected from northern coasts of Persian Gulf. *International Symposium on Advances in Science and Technology, 7th SASTech*, pp. 1–5, 2013.
29. Torres, M.L., Cortizo, A.M., Oberti, T.G., Fernández, J.M., Characterization of commercial and algae (*Undaria pinnatifida*) extracted sodium alginate for future application in bone tissue engineering. 1–13, 2016.
30. Chee, S.Y., Wong, P.K., Wong, C.L., Extraction and characterisation of alginate from brown seaweeds (Fucales, Phaeophyceae) collected from Port Dickson, Peninsular Malaysia. *J. Appl. Phycol.*, 23, 191–196, 2011.
31. Viswanathan, S. and Nallamuthu, T., Extraction of sodium alginate from selected seaweeds and their physiochemical and biochemical properties. *IJIRSET*, 3, 4, 10998–11003, 2014.
32. Fenoradosoa, T.A., Ali, G., Delattre, C., Laroche, C., Petit, E., Wadouachi, A., Michaud, P., Extraction and characterization of an alginate from the brown seaweed *Sargassum turbinarioides* Grunow. *J. Appl. Phycol.*, 22, 131–137, 2010.
33. Jayasankar, R., On the yield and quality of sodium alginate from *Sargassum wightii* (Greville) by pre-treatment with chemicals. *Seaweed Res. Utiln.*, 16, 1&2, 63–66, 1993.
34. Eluvakkal, T., Shanthi, N., Mrugan, M., Arunkumar, K., Extraction of antibacterial substances, galactofucoidan and alginate successively from the Gulf

of Mannar brown sea weed *Sargassam wightii* Greville Ex. J. Aagarth. *IJNPR*, 5, 3, 249–257, 2014.
35. Hernandez- Carmona, G., Pelgrin, Y.F., Hernandez- Garibay, E., Conventional and alternative technologies in extraction of algal polysaccharides, https://doi.org/10.1533/9780857098689.3.475
36. Jork, A., Thurmer, F., Cramer, H., Zimmermann, G., Gessner, P., Hamel, K., Hofmann, G., Kuttler, B., Hahn, H.J., Josimovic-Alasevic, O., Fritsch, K.G., Zimmermann, U., Biocompatible alginate from freshly collected *Laminaria pallida* for implantation. *Appl. Microbiol. Biotechnol.*, 53, 2, 224–229, 2000.
37. Horton, J.C., Harding, S.E., Mitchell, J.R., Morton-Holmes, D.F., Thermodynamic non-ideality of dilute solutions of sodium alginate studied by sedimentation equilibrium ultracentrifugation. *Food Hydrocolloids*, 5, 112, 125–127, 1991.
38. Indergaard, M., Skj kBraek, G., Jensen, A., Studies on the influence of nutrients on the composition and structure of alginate in *Laminaria saccharina* (L.) Lamour. (Laminariales, Phaeophyceae). *Bot. Mar.*, 33, 277–288, 1990.
39. Venegas, M., Matsuiro, B., Edding, M.E., Alginate composition of *Lessonia trabeculata* (Phaeophyta: Laminariales) growing in exposed and sheltered habitats. *Bot. Mar.*, 36, 47–51, 1993.
40. Sánchez, M., Vásquez-Quitral, P., Castañeda, E., Vargas, A., Neira-Carrillo, A., Alginate from *Lessonia nigrescens* as template for $CaCO_3$ crystals obtained by electro-crystallization. *Proceedings of the World Congress on New Technologies (NewTech 2015) Barcelona, Spain*, p. 394, 2015.
41. Sabra, W. and Zeng, A.P., Microbial production of alginates: Physiology and process aspects, in: *Alginates: Biology and Applications, Microbiology Monographs*, vol. 13, B.H.A. Rehm (Ed.), pp. 153–173.
42. Linker, A. and Jones, R.S., A new polysaccharide resembling alginic acid isolated from. *Pseudomonads. J. Biol. Chem.*, 241, 3845–3851, 1966.
43. Gorin, P.A.J. and Spencer, J.F.T., Exocellular alginic acid from *Azotobacter vinelandii.. Can. J. Chem.*, 44, 993–998, 1966.
44. Govan, J.R.W., Fyfe, J.A.M., Jarman, T.R., Isolation of alginate-producing mutants of *Pseudomonas fluorescens, Pseudomonas putida* and *Pseudomonas mendocina. J. Gen. Microbiol.*, 125, 217–220, 1981.
45. Chen, W.P., Chen, J.Y., Chang, S.C., Su, C.L., Bacterial alginate produced by a mutant of *Azotobacter vinelandii. Appl. Environ. Microbiol.*, 49, 3, 543–546, 1985.
46. Fett, W.F., Osman, S.F., Fishman, M.L., SIEBLES, T.S., Alginate production by plant-pathogenic pseudomonads. *Appl. Environ. Microbiol.*, 52, 3, 466–473, 1986.
47. Osman, S.F., Fett, W.F., Fishman, M.L., Exopolysaccharides of the phytopathogen *Pseudomonas syringae* pv. *glycinea. J. Bacteriol.*, 166, 66–71, 1986.
48. Fialho, A.M., Zielinski, N.A., Fett, W.F., Chakrabarty, A.M., Berry, A., Distribution of alginate gene sequences in the *Pseudomonas* rRNA homology group I—*Azomonas–Azotobacter* lineage of superfamily B procaryotes. *Appl. Environ. Microbiol.*, 56, 436–443, 1990.

49. Gacesa, P., Bacterial alginate biosynthesis—Recent progress and future prospects. *Microbiology*, 144, 1133–1143, 1998.
50. Rehm, B.H.A., Alginates from bacteria, in: *Biopolymers*, E.J. Vandamme, S. De Baets, A. Steinbuchel (Eds.), pp. 179–21, Wiley VCH, New York, N.Y., 2002.
51. Vazquez, A.P., Kidambi, S.P., Chakrabarty, A.M., Bender, C.L., Characterization of the alginate biosynthetic gene cluster in *Pseudomonas syringae* pv. *syringae*. *J. Bacteriol.*, 179, 4464–4472, 1997.
52. Muhammadi and Ahmed, N., Genetics of bacterial alginate: Alginate genes distribution, organization and biosynthesis in bacteria. *Curr. Genom.*, 8, 191–202, 2007.
53. Cook, W.H. and Smith, D.B., Molecular weight and hydrodynamic properties of sodium alginate. *Can. J. Biochem. Phys.*, 32, 3, 227–239, 1953.
54. Moe, S.T., Draget, K.I., Skjak Braek, G., Smidsrod, O., Alginates, in: *Food Polysaccharides and Their Applications*, A.M. Stephen (Ed.), pp. 245–286, Marcel Dekker Inc., N.Y., 1995.
55. Tonnesen, H.H. and Karlsen, J., Alginate in drug delivery systems. *Drug Dev. Ind. Pharm.*, 28, 621–630, 2002.
56. Stokke, T.B., Kurt, D., Smidsrod, O., Yuguchi, Y., Urakawa, H., Kajiwara, K., Small-angle x-ray scattering and rheological characterization of alginate gels Ca-alginate gels. *Macromolecules*, 33, 1853–1863, 2000.
57. Masuelli, M.A. and Illanes, C.O., Review of the characterization of sodium alginate by intrinsic viscosity measurements. Comparative analysis between conventional and single point methods. *Int. J. BioMat. Sci. Eng.*, 1, 1, 1–11, 2014.
58. Wedlock, D.J., Fasihuddin, B.A., Phillips, G.O., Comparison of molecular weight determination of sodium alginate by sedimentation–diffusion and light scattering. *Int. J. Biol. Macromol.*, 8, 57–61, 1986.
59. Sperger, D.M., Fu, S., Block, L.H., Munson, E.J., Analysis of composition, molecular weight, and water content variations in sodium alginate using solid-state NMR spectroscopy. *J. Pharm. Sci.*, 100, 8, 3441–3452, 2011.
60. Zhang, L.Y., Yao, S.J., Guan, Y.X., Diffusion characteristics of solutes with low molecular weight in sodium alginate/cellulose sulfate-$CaCl_2$/poly (methylene-*co*-guanidine) capsules. *J. Chem. Eng. Data*, 48, 864–868, 2003.
61. Mancini, F., Montanari, L., Peressini, D., Fantozzi, P., Influence of alginate concentration and molecular weight on functional properties of mayonnaise. *Lebensm.-Wiss. Technol.*, 35, 517–525, 2002.
62. Haug, A. and Larsen, B., The solubility of alginate at low Ph. *Acta Chem. Scand.*, 17, 1653–1662, 1963.
63. Haug, A., Larsen, B., Smidsrod, O., A study of the constitution of alginic acid by partial acid hydrolysis. *Acta Chem. Scand.*, 20, 183–190, 1966.
64. Carré, M.C., Delestre, C., Hubert, P., Dellacherie, E., Covalent coupling of a short polymer on sodium alginate—Synthesis and characterization of the amphiphilic derivative. *Carbohydr. Polym.*, 16, 4, 367–379, 1991.

65. Babak, V.G., Skotnikova, E.A., Lukina, I.G., Pelletier, S., Hubert, P., Dellacherie, E., Hydrophobically associating alginate derivatives: Surface tension properties of their mixed aqueous solutions with oppositely charged surfactants. *J. Colloid Interface Sci.*, 225, 2, 505–510, 2000.
66. Smidsrød, O., Haug, A., Larsen, B., The influence of pH on the rate of hydrolysis of acidic polysaccharides. *Acta Chem. Scand.*, 20, 1026, 1966.
67. Liu, S., Kang, M., Hussain, I., Li, K., Yao, F., Fu, G., High mechanical strength and stability of alginate hydrogel induced by neodymium ions coordination. *Polym. Degrad. Stab.*, 2016, https://doi.org/10.1016/j.polymdegradstab.2016.07.022
68. Shafiq, U., Rahim, S., Saleem, A., Anwari, M., Effect of pouring time on the dimensional stability of alginate impression material. *PODJ*, 36, 3, 495–497, 2016.
69. Ralph, W.J., Gin, S.S.L., Cheadlet, D.A., Harcourts, J.K., The effects of disinfectants on the dimensional stability of alginate impression materials. *Aust. Dent. J.*, 35, 6, 514–517, 1990.
70. Rad, F.H., Ghaffari, T., Safavi, S.H., *In Vitro* evaluation of dimensional stability of alginate impressions after disinfection by spray and immersion methods. *J. Dent. Res. Dent. Clin. Dent. Prospect*, 4, 4, 130–135, 2010.
71. Walker, M.P., Burckhard, J., Mitts, D.A., Williams, K.B., Dimensional change over time of extended-storage alginate impression materials. *Angle Orthod.*, 80, 6, 1110–1115, 2010.
72. Smidsrød, O. and Haug, A., A light scattering study of alginate. *Acta Chem. Scand.*, 22, 797, 1968.
73. Kohn, R., Ion binding on polyuronates—Alginate and pectin. *Pure Appl. Chem.*, 42, 371, 1975.
74. Mancini, F. and McHugh, T.H., Fruit–alginate interactions in novel restructured products. *Nahrung*, 44, 3, 152–157, 2000.
75. Morris, E., Rees, D., Thom, D., Boyd, J., Chiroptical and stoichiometric evidence of a specific, primary dimerisation process in alginate gelation. *Carbohydr. Res.*, 66, 145–154, 1978.
76. Atkins, E.D.T., Mackie, W., Smolko, E.E., Crystalline structures of alginic acids. *Nature*, 225, 626, 1970.
77. Sletmoen, M., Draget, K.I., Stokke, B.T., Alginate oligoguluronates as a tool for tailoring properties of Ca-alginate gels. *Macromol. Symp.*, 291–292, 345–353, 2010.
78. Borgogna, M., Skjåk-Bræk, G., Paoletti, S., Donati, I., On the initial binding of alginate by calcium ions. The tilted egg- box hypothesis. *J. Phys. Chem.*, 117, 24, 7277–7282, 2013.
79. Fang, Y., Al-Assaf, S., Phillips, G.O., Nishinari, K., Funami, T., Williams, P.A., Li, L., Multiple steps and critical behaviors of the binding of calcium to alginate. *J. Phys. Chem. B*, 111, 2456–2462, 2007.
80. Rodrigues, J.R. and Lagoa, R., Copper ions binding in Cu-alginate gelation. *J. Carbohydr. Chem.*, 25, 2–3, 219–232, 2006.

81. Arıca, M., Arpa, C., Ergene, A., Bayramog lu, G., Genc., O., Ca-alginate as support for Pb(II) and Zn(II) biosorption with immobilized *Phanerochaete chrysosporium*. *Carbohydr. Polym.*, 52, 167–174, 2003.
82. Arıca, M., Bayramoglu, G., Yılmaz, M., Bektas, S., Genc, O., Biosorption of Hg^{2+}, Cd^{2+}, and Zn^{2+} by Ca-alginate and immobilized wood-rotting fungus *Funalia trogii*.. *J. Hazard. Mater.*, 109, 191–199, 2004.
83. Gotoh, T., Matsushima, K., Kikuchi, K.I., Adsorption of Cu and Mn on covalently cross-linked alginate gel beads. *Chemosphere*, 55, 57–64, 2004.
84. Ibanez, J. and Umetsu, Y., Uptake of trivalent chromium from aqueous solutions using protonated dry alginate beads. *Hydrometallurgy*, 72, 327–334, 2004.
85. Sano, I.M., Ogawa, S., Ueda, H., Kimura, Y., Satoh, N., Namiki, H., Effects of composition of iron-cross-linked alginate hydrogels for cultivation of human dermal fibroblasts. *Int. J. Biomat.*, Article ID: 820513, 1–8, 2012.
86. Draget, K.I., Moe, S.T., Skjak-Brek, G., Smidsrod, O., Alginate, in: *Food Polysaccharides and Their Application*, A.M. Stephen, G.O. Phillips, P.A. Williams (Eds.), Taylor & Francis Group, LLC, Boca Raton, 2006.
87. Smidsrod, O. and Draget, K.I., Alginate gelation technologies, in: *Food Colloids Proteins, Lipids and Polysaccharides*, E. Dickinson and B. Bergenstahl (Eds.), The Royal Society of Chemistry, Cambridge, 1997.
88. Qin, Y., Gel swelling properties of alginate fibers. *J. Appl. Polym. Sci.*, 91, 1641–1645, 2004.
89. Kokubo, T., Hanakawa, M., Kawashita, M., Minoda, M., Beppu, T., Miyamato, T., Nakamura, T., Apatite forming ability of alginate fibres treated with calcium hydroxide solution. *J. Mat. Sci. Mater. Med.*, 15, 1007–1012, 2004.
90. Hosoya, K., Ohtsuki, C., Kawai, T., Kamitakahara, M., Ogata, S., Miyazaki, T., Tanihara, M., *A Novel Covalently Cross-linked Gel of Alginate and Silane with the Ability to Form Bone-Like Apatite*, Wiley Interscience, 2004.
91. Soon-Shiong, P. and Heintz, R.E., Insulin independence in a type 1 diabetic patient after encapsulated islet transplantation. *Lancet*, 343, 950, 1994.
92. Otterlei, M., Ostgaard, K., Skjakbraek, G., Smidsrod, O., Soonshiong, P., Espevik, T., Induction of cytokine production from human monocytes stimulated with alginate. *J. Immunother.*, 10, 286–291, 1991.
93. Houghton, D., Wilcox, M.D., Peter, I., Biological activity of alginate and its effect on pancreatic lipase inhibition as a potential treatment for obesity, in: *Food Hydrocolloids*, vol. 49, A. Brownlee, C.J. Seal, J.P. Pearson (Eds.), pp. 18–24, 2015.
94. Goodman, C., Technical control of ice cream with sodium alginate. *Ice Cream Rev.*, 18, 42–48, 1935.
95. Kohajdová, Z., Karovičová, J., Schmidt, S., Significance of emulsifiers and hydrocolloids in bakery industry. *Acta Chim. Slovac.*, 2, 1, 46–61, 2009.
96. Cheong, K.W., Mirhosseini, H., Hamid, N.S.A., Osman, A., Basri, M., Tan, C.P., Effects of propylene glycol alginate and sucrose esters on the

physicochemical properties of modified starch-stabilized beverage emulsions. *Molecules*, 19, 6, 8691–8706, 2014.
97. Vilgis, T.A., Hydrocolloids between soft matter and taste: Culinary polymer physics. *Int. J. Gastronomy Food Sci.*, 1, 1, 46–53, 2012.
98. Krasaekoopt, W. and Watcharapoka, S., Effect of addition of inulin and galactooligosaccharide on the survival of microencapsulated probiotics in alginate beads coated with chitosan in simulated digestive system, yogurt and fruit juice. *LWT—Food Sci. Technol.*, 57, 761–766, 2014.
99. Zanjani, M.A.K., Tarzi, B.G., Sharifan, A., Mohammadi, N., Microencapsulation of probiotics by calcium alginate–gelatinized starch with chitosan coating and evaluation of survival in simulated human gastrointestinal condition. *Iran J. Pharm. Res.*, 13, 3, 843–852, 2014.
100. Rodríguez-Sánchez, J.A., Cuatzo-Lozano, M.I., Pérez-Loredo, M.G., Abarca-Sarro, D.I., Navarro, Y.G., Alginate encapsulation as a preservation method of pitaya fruit juice (*Stenocereus* spp.). *J. Food Sci. Eng.*, 7, 127–134, 2017.
101. Silva, C.M., Ribeiro, A.J., Figueiredo, I.V., Goncalves, A.R., Veiga, F., Alginate microspheres prepared by internal gelation: Development and effect on insulin stability. *Int. J. Pharm.*, 311, 1–10, 2006.
102. Ghavidel, R.A., Davoodi, M.G., AdibAs, A.F., Tanoori, T., Sheykholeslami, Z., Effect of selected edible coatings to extend shelf-life of fresh-cut apples. *Int. J. Agri. Crop Sci.*, 6, 16, 1171–1178, 2013.
103. Chiabrando, V. and Giacalone, G., Effect of chitosan and sodium alginate edible coatings on the postharvest quality of fresh-cut nectarines during storage. *Fruits*, 71, 2, 79–85, 2016.
104. Yu, X.L., Li, X.B., Xu, X.L., Zhou, G.H., Coating with sodium alginate and its effects on the functional properties and structure of frozen pork. *J. Muscle Foods*, 19, 4, 333–351, 2008.
105. Storebakken, T., Binders in fish feeds: I. effect of alginate and guar gum on growth, digestability, feed intake and passage through the gastrointestinal tract of rainbow trout. *Aquaculture*, 47, 1, 11–26, 1985.
106. Rychen, G., Aquilina, G., Azimonti, G., Bampidis, V., Bastos, M.L., Bories, G., Chesson, A., Cocconcelli, P.S., Flachowsky, G., Kolar, B., Kouba, M., Lopez-Alonso, M., Lopez Puente, S., Mantovani, A., Mayo, B., Ramos, F., Saarela, M., Villa, R.E., Wallace, R.J., Wester, P., Lundebye, A.K., Nebbia, C., Renshaw, D., Innocenti, M.L., Gropp, J., Scientific opinion on the safety and efficacy of sodium and potassium alginate for pets, other non food-producing animals and fish. *EFSA J.*, 15, 7, 4945, 1–13, 2017, https://doi.org/10.2903/j.efsa.2017.4945.
107. Joshy, K.S., Susan, M.A., Snigdha, S., Nandakumar, K., Pothen Laly, A., Sabu, T., Encapsulation of zidovudine in PF-68 coated alginate conjugate nanoparticles for anti-HIV drug delivery. *Int. J. Biol. Macromol.*, 107, 921–927, 2018.
108. Bera, H., Ippagunta, S.R., Kumar, S., Vangala, P., Core-shell alginate-ghatti gum modified montmorillonite composite matrices for stomach-specific flurbiprofen delivery. *Mater. Sci. Eng. C Mater. Biol. Appl.*, 76, 715–726, 2017.

109. Kumar, T.S.S., Madhumathi, K., Rajkamal, B., Zaheatha, S., Malar, A.R., Bai, S.A., Enhanced protein delivery by multi-ion containing eggshell derived apatitic-alginate composite nanocarriers. *Colloids Surf B Biointerfaces.*, 123, 542–548, 2014.
110. Orive, G., Ponce, S., Hernandez, R., Gascon, A., Igartua, M., Pedraz, J., Biocompatibility of microcapsules for cell immobilization elaborated with different type of alginates. *Biomaterials*, 23, 18, 3825–3231, 2002.
111. Augst, A.D., Kong, H.J., Mooney, D.J., Alginate hydrogels as biomaterials. *Macromol. Biosci.*, 6, 8, 623–633, 2006.
112. Lim, F. and Sun, A.M., Microencapsulated islets as bioartificial endocrine pancreas. *Science*, 210, 4472, 908–910, 1980.
113. Mosahebi, A., Wiberg, M., Terenghi, G., Addition of fibronectin to alginate matrix improves peripheral nerve regeneration in tissue-engineered conduits. *Tissue Eng.*, 9, 2, 209–218, 2003.
114. Doyle, J.W., Roth, T.P., Smith, R.M., Li, Y.Q., Dunn, R.M., Effect of calcium alginate on cellular wound healing processes modeled *in vitro. J. Biomed. Mater. Res.*, 32, 4, 561–568, 1996.
115. Tarun, K. and Gobi, N., Calcium alginate/PVA blended nanofibre matrix for wound dressing. *IJFTR*, 37, 127–132, 2012.
116. Adamson, A.W., *Physical Chemistry of Surfaces*, pp. 978–983, Wiley, 1990.
117. Qin, Y., Shi, B., Liu, J., Application of chitosan and alginate in treating waste water containing heavy metal ions. *Indian J. Chem. Technol.*, 13, 464–469, 2006.
118. Siripattanakul-Ratpukudi, S. and Togkliang, T., Municipal waste water treatment using barium alginate entrapped activated sludge: Adjustment of utilization conditions. *Int. J. Chem. Eng. App.*, 3, 5, 328–332, 2012.
119. Wu, Y., Qi, H., Shi, C., Ma, R., Liu, S., Huang, Z., Preparation and adsorption behaviours of sodium alginate-based adsorbent-immobilized β-cyclodextrin and grapheme oxide. *RSC Adv.*, 7, 31459–31557, 2017.
120. El-Tayieb, M.M., El-Shafei, M.M., Mahmoud, M.S., The role of alginate as polymeric material in treatment of tannery waste water. *IJST*, 2, 2, 218–224, 2013.
121. Shugaba, A., Wuyep, P.A., Nok, A.J., Ameh, D.A., Lori, J.A., Bioremediation of hexavalent chromium and tannic acid in synthetic tannery wastewater using free and calcium alginate-immobilized spores and mycelia of *Aspergillus niger* and *Aspergillus parasiticus.*. *Bioremed J.*, 14, 3, 142–149, 2010.
122. Wu, Z., Zhao, Y., Kaleem, I., Li, C., Preparation of calcium alginate microcapsuled microbial fertilizer coating *Klebsilla oxytoca* Rs-5 and its performance under salinity stress. *Eur. J. Soil Biol.*, 47, 152–159, 2011.
123. Kay, A.N., An alginate based fertilizer to reduce eutrophication of aquatic environments. *Trans. Kansas Acad. Sci.*, 107, 3/4, 155–164, 2004.
124. Frampton, J.P., Hynd, M.R., Shuler, M.L., Shain, W., Fabrication and optimization of alginate hydrogel constructs for use in 3D neural cell culture. *Biomed. Mater.*, 6, 015002, 1–18, 2011.

125. Chen, C.Y., Ke, C.J., Yen, K.C., Hsieh, H.C., Sun, S.J., Lin, F.H., 3D porous calcium alginate scaffold cell culture system improved Human Osteoblast cell clusters for cell therapy. *Theranostics.*, 5, 6, 643–655, 2015.
126. Xu, X.X., Liu, C., Liu, Y., Yang, L., Li, N., Guo, X., Sun, G., Ma, X., Enrichment of cancer stem cell-like cells by culture in alginate gel beads. *J. Biotechnol.*, 177, 1–12, 2014.
127. George, S.M. and Moon, H., Digital microfluidic three-dimensional cell culture and chemical screening platform using alginate hydrogels. *Biomicrofluidics*, 9, 2, 024116, 2015.
128. Akeda, K., Nishimura, A., Satonaka, H., Shintani, K., Kusuzaki, K., Matsumine, A., Kasail, Y., Masuda, K., Uchida, A., Three-dimensional alginate spheroid culture system of murine osteosarcoma. *Oncol. Rep.*, 22, 997–1003, 2009.
129. Shin, S., Ikram, M., Subhan, F., Kang, H.Y., Lim, Y., Lee, R., Jin, S., Jeong, Y.H., Kwak, J., Na, Y., Yoon, S., Alginate/marine collagen/agarose composite hydrogels as matrices for biomimetic 3D cell spheroid formation. *RSC Adv.*, 6, 46952–46965, 2016.
130. Andersen, T., Auk-Emblem, P., Dornish, M., 3D cell culture in alginate hydrogels. *Microarrays*, 4, 133–161, 2015.
131. Frey, G., Lu, H., Powers, J., Effect of mixing methods on mechanical properties of alginate impression materials. *J. Prosthodont.*, 14, 4, 221–225, 2005.
132. Cook, W., Alginate dental impression materials: Chemistry, structure, and properties. *J. Biomed. Mater. Res.*, 20, 1–24, 1986.
133. Cohen, B.I., Pagnillo, M., Deutsch, A.S., Musikant, B.L., Dimensional accuracy of three different alginate impression materials. *J. Prosthodont.*, 4, 3, 195–199, 1995.
134. Morrow, R.M., Brown, C.E., Stansbury, B.E., Delorimier, J.A., Powell, J.M., Rudd, K.D., Compatibility of alginate impression materials and dental stones, in: *Dental Technology*, D.H. Gehl and S.H. Payne (Eds.), pp. 556–566, 1971.
135. Ciofani, G., Raffa, V., Menciassi, A., Micera, S., Dario, P., A drug delivery system based on alginate microspheres: Mass-transport test and *in vitro* validation. *Biomed Microdevices*, 9, 395–403, 2007.
136. Mahou, R., Meier, R.P.H., Bühler, L.H., Wandrey, C., Alginate-poly(ethylene glycol) hybrid microspheres for primary cell microencapsulation. *Materials*, 7, 275–286, 2014.
137. Bidarra, S.J., Barrias, C.C., Barbosa, M.A., Soares, R., Granja, P.L., Immobilization of human mesenchymal stem cells within RGD-grafted alginate microspheres and assessment of their angiogenic potential. *Biomacromolecules*, 11, 1956–1964, 2010.
138. Khanna, O., Larson, J.C., Moya, M.L., Opara, E.C., Brey, E.M., Generation of alginate microspheres for biomedical applications. *J. Vis. Exp.*, 66, e3388, 2012.

139. Ramesh, A., Kanaf, M.M., Bhonde, R.R., Modulus-dependent characteristics of Wharton's jelly mesenchymal stem cells (WJMSC) encapsulated in hydrogel microspheres. *J. Biomater. Sci.*, 25, 17, 1946–1961, 2014.
140. Mahou, R. and Wandrey, C., Alginate-poly(ethylene glycol) hybrid microspheres with adjustable physical properties. *Macromolecules*, 43, 1371–1378, 2010.
141. Bigdeli, S., Dettloff, R.O., Frank, C.W., Davis, R.W., Crosby, L.D., A simple method for encapsulating single cells in alginate microspheres allows for direct pCR and whole genome amplification. *PLoS One*, 10, 2, e0117738, 2015.
142. Kroneková, Z., Pelach, M., Mazancová, P., Uhelská, L., Treľová, D., Rázga, F., Némethová, V., Szalai, S., Chorvát, D., McGarrigle, J.J., Omami, M., Isa, D., Ghani, S., Majková, E., Oberholzer, J., Raus, V., Siffalovic, P., Lacík, I., Structural changes in alginate based microspheres exposed to *in vivo* environment as revealed by confocal Raman microscopy. *Sci. Rep.*, 8, 1637, 2018.
143. Ashton, R.S., Banerjee, A., Punyani, S., Schaffer, D.V., Kan, R.S., Scaffolds based on degradable alginate hydrogels and poly(lactide-co-glycolide) microspheres for stem cell culture. *Biomaterials*, 28, 5518–5525, 2007.
144. Xu, X., Wu, X., Wang, Q., Cai, N., Zhang, H., Jiang, Z., Wan, M., Oda, T., Immunomodulatory Effects of alginate oligosaccharides on murine macrophage RAW264.7 cells and their structure–activity relationships. *J. Agric. Food Chem.*, 62, 3168–3176, 2014.
145. Zhou, R., Shi, X.Y., Cheng Bi, D., Fang, W.S., Wei, G.B., Xu, X., Alginate-derived oligosaccharide inhibits neuroinflammation and promotes microglial phagocytosis of β-amyloid. *Mar. Drugs*, 13, 5828–5846, 2015.
146. Fang, W., Cheng Bi, D., Zheng, R., Cai, N., Xu, H., Zhou, R., Lu, J., Wan, M., Xu, X., Identification and activation of TLR4-mediated signalling pathways by alginate-derived guluronate oligosaccharide in RAW264.7 macrophages. *Sci. Rep.*, 7, 1663, 2017.
147. Iwamoto, M., Kurachi, M., Nakashima, T., Kim, D., Yamaguchi, K., Oda, T., Iwamoto, Y., Muramatsu, T., Structure–activity relationship of alginate oligosaccharides in the induction of cytokine production from RAW264.7 cells. *FEBS Lett.*, 579, 4423–4429, 2005.
148. Chen, J., Hu, Y., Zhang, L., Wang, Y., Wang, S., Zhang, Y., Guo, H., Ji, D., Wang, Y., Alginate oligosaccharide DP5 exhibits antitumor effects in osteosarcoma patients following surgery. *Front. Pharmacol.*, 8, 623, 2017.
149. An, Q.D., Zhang, G.L., Wu, H.T., Zhang, Z.C., Zheng, G.S., Luan, L., Murata, Y., Li, X., Alginate-deriving oligosaccharide production by alginase from newly isolated *Flavobacterium* sp. LXA and its potential application in protection against pathogens. *J. Appl. Microbiol.*, 106, 161–170, 2009.
150. Sawabe, T., Sawada, C., Suzuki, E., Ezura, Y., Intracellular alginate-oligosaccharide degrading enzyme activity that is incapable of degrading intact sodium alginate from a marine bacterium *Alteromonas* sp. *Fish Sci.*, 64, 2, 320–324, 1998.

151. Falkeborg, M., Cheong, L., Gianfico, C., Sztukiel, K.M., Kristensen, K., Glasius, M., Xu, X., Guo, Z., Alginate oligosaccharides: Enzymatic preparation and antioxidant property evaluation. *Food Chem.*, 164, 185–194, 2014.
152. Khan, Z.H., Khan, M.M.A., Aftab, T., Idrees, M., Naeem, M., Influence of alginate oligosaccharides on growth, yield and alkaloid production of opium poppy (*Papaver somniferum* L.). *Front. Agric. China*, 5, 1, 122–127, 2011.
153. Nishikawa, T., Yokose, T., Yamamoto, Y., Yamaguchi, K., Oda, T., Detection and pharmacokinetics of alginate oligosaccharides in mouse plasma and urine after oral administration by a liquid chromatography/tandem mass spectrometry (LC-MS/MS) method. *Biosci. Biotechnol. Biochem.*, 72, 8, 2184–2190, 2008.
154. Zhang, Z., Yu, G., Zhao, X., Liu, H., Guan, H., Lawson, A.M., Chai, W., Sequence analysis of alginate-derived oligosaccharides by negative-ion electrospray tandem mass spectrometry. *J. Am. Soc. Mass Spectrom*, 17, 621–630, 2006.
155. Xie, M., Zhang, F., Liu, L., Zhang, Y., Li, Y., Li, H., Xie, J., Surface modification of graphene oxide nanosheets by protamine sulfate/sodium alginate for anti-cancer drug delivery application. *Appl. Surface Sci.*, 440, 853–860, 2018.
156. Tawfik, S.M., Sharipov, M., Huy, B.T., Gerelkhuu, Z., Biechele-Speziale, D., Lee, Y.I., Naturally modified nonionic alginate functionalized upconversion nanoparticles for the highly efficient targeted pH-responsive drug delivery and enhancement of NIR-imaging. *J. Industrial Eng. Chem.*, 2017, http://dx.doi.org/10.1016/j.jiec.2017.08.051.
157. Joshy, K.S., Susan, M.A., Snigdha, S., Nandakumar, K., Laly, A.P., Sabu, T., Encapsulation of zidovudine in PF-68 coated alginate conjugate nanoparticles for anti-HIV drug delivery. *Int. J. Biol. Macromol.*, 2017, http://dx.doi.org/10.1016/j.ijbiomac.2017.09.078.
158. Veresa, P., Sebőkb, D., Dékány, I., Gurikovc, P., Smirnova, I., Fábiána, I., Kalmár, J., A redox strategy to tailor the release properties of Fe(III)-alginate aerogels for oral drug delivery. *Carbohydr. Polym.*, 188, 159–167, 2018.
159. Senna, J.P., Barradas, T.N., Cardoso, S., Castiglione, T.C., Serpe, M.J., Silva, K.G.D.H., Mansur, C.R.E., Dual alginate-lipid nanocarriers as oral delivery systems for amphotericin B. *Colloids Surf B: Biointerf*, 2018, https://doi.org/10.1016/j.colsurfb.2018.03.015.
160. Katuwavila, N.P., Perera, A.D.L.C., Samarakoon, S.R., Soysa, P., Karunaratne, V., Amaratunga, G.A.J., Karunaratne, D.N., Chitosan-alginate nanoparticle system efficiently delivers doxorubicin to MCF-7 cells. *J. Nanomater.*, Article ID: 3178904, 1–12, 2016, http://dx.doi.org/10.1155/2016/3178904.
161. Li, P., Dai, Y.N., Zhang, J.P., Wang, A.Q., Wei, Q., Chitosan-alginate nanoparticles as a novel drug delivery system for nifedipine. *Int. J. Biomed. Sci.*, 4, 3, 221–227, 2008.

162. Goncalves, L.M.D., Cadete, A., Figueiredo, L., Calado, C.C.R., Almeida, A.J., Biodegradable nanoparticles of Alginate and chitosan as non-viral DNA oral delivery system. 1st Portuguese Meeting in Bioengineering. 1–4, 2011.
163. George, L., Bavya, M.C., Rohan, K.V., Srivastava, R., A Therapeutic polyelectrolyte–vitamin C nanoparticulate system in a polyvinyl alcohol–alginate hydrogel: An approach to treat skin and soft tissue infection caused by *Staphylococcus aureus.*. *Colloids Surf B: Biointerfaces*, 2017, http://dx.doi.org/10.1016/j.colsurfb.2017.09.030.
164. Román, J.V., Rodríguez-Rodríguez, J.A., Valle, E.M.M., Galán, M.A., Synthesis of a new nanoparticle system based on electrostatic alginate-piperazine interactions. *Polym. Adv. Technol.*, 27, 623–629, 2016.
165. Sudakar, C., Dixit, A., Regmi, R., Naik, R., Lawes, G., Naik, V.M., Vaishnava, P.P., Toti, U., Panyam, J., Fe_3O_4 incorporated AOT-alginate nanoparticles for drug delivery. *IEEE Trans. Magnetics*, 44, 11, 2800–2803, 2008.
166. Chen, Y.W., Kuo, C.W., Chuch, D.Y., Chen, P., Surface modified alginate microcapsules for 3D cell culture. *Surf. Sci.*, 2016, http://dx.doi.org/0.1016/j.susc.2015.10.009

Index

CPSIA information can be obtained
at www.ICGtesting.com
Printed in the USA
BVHW042234070319
542095BV00008B/25/P

9 781119 487913